Research in Mathematics Educ

Series Editors

Jinfa Cai, Newark, DE, USA

James A. Middleton, Tempe, AZ, USA

This series is designed to produce thematic volumes, allowing researchers to access numerous studies on a theme in a single, peer-reviewed source. Our intent for this series is to publish the latest research in the field in a timely fashion. This design is particularly geared toward highlighting the work of promising graduate students and junior faculty working in conjunction with senior scholars. The audience for this monograph series consists of those in the intersection between researchers and mathematics education leaders—people who need the highest quality research, methodological rigor, and potentially transformative implications ready at hand to help them make decisions regarding the improvement of teaching, learning, policy, and practice. With this vision, our mission of this book series is: (1) To support the sharing of critical research findings among members of the mathematics education community; (2) To support graduate students and junior faculty and induct them into the research community by pairing them with senior faculty in the production of the highest quality peer-reviewed research papers; and (3) To support the usefulness and widespread adoption of research-based innovation.

More information about this series at https://link.springer.com/bookseries/13030

Yan Ping Xin • Ron Tzur • Helen Thouless
Editors

Enabling Mathematics Learning of Struggling Students

 Springer

Editors
Yan Ping Xin
Department of Educational Studies
Purdue University
West Lafayette, IN, USA

Helen Thouless
Institute of Education
St Mary's University Twickenham
London, UK

Ron Tzur
School of Education and Human
Development
University of Colorado Denver
Denver, CO, USA

ISSN 2570-4729   ISSN 2570-4737 (electronic)
Research in Mathematics Education
ISBN 978-3-030-95218-1   ISBN 978-3-030-95216-7 (eBook)
https://doi.org/10.1007/978-3-030-95216-7

This Springer imprint is published by the registered company Springer Nature Switzerland AG
The registered company address is: Gewerbestrasse 11, 6330 Cham, Switzerland

Introduction

Yan Ping Xin, Ron Tzur, and Helen Thouless

It is not uncommon to hear these types of questions from teachers of students who are struggling in mathematics:

- How do I plan my instruction for students who come to me not ready?
- How do I build conceptual understanding of students with learning disabilities or difficulties?
- How do I help move my students from where they are to where they should be mathematically?
- What can teachers do (adapt) to foster students' growth?

Such questions, and many more, reflect teachers' concerns and caring for their students, not the least because of the dismal improvements in the learning and outcomes over time for this population. For example, during the past decade in the USA, national score gains were seen for higher performing students at the 75th and 90th percentiles at grades four and eight, while no significant gains were shown for lower-performing students. According to the Nation's Report Card (National Assessment of Educational Progress [NAEP], 2019), a great majority (85%) of American fourth graders with learning disabilities in mathematics performed at the below-proficiency level. At this grade level, while students without disabilities have increased their performance by three points since 2009, students with disabilities have dropped by seven points. The gap between students with and without disabilities seems wider over the past decade. Furthermore, the NAEP results showed that eighth grade students with learning disabilities in mathematics scored roughly at the level of typically achieving fourth graders. Many of these students are at significant risk of failing the secondary mathematics curriculum. These findings indicate a burning issue for all teachers and their educators, particularly because during the same period, expectations for all students have been elevated. For example, both the *Common Core State Standards for Mathematics* (Common Core State Standards Initiative, 2012) and the *English National Curriculum* (2013) emphasized conceptual understanding in problem solving as well as higher-order thinking and reasoning. Students with learning disabilities or difficulties in mathematics (LDM), for whom past instruction often focused primarily on computation and simple problem solving while following specific procedures, are now expected to *understand* and solve complex problems and develop solid *reasoning* skills in mathematics.

To help address this burning issue, this research-based book presents the outgrowth of a unique, ongoing collaboration among researchers in special education, psychology, and mathematics education from around the world. In particular, it reflects the decade-long work by members of the International Group for the Psychology of Mathematics Education (PME) and the North American Chapter of the PME Working Groups. Our main purpose in this book is to provide prospective and practicing teachers with research insights into the mathematical difficulties of students with LDM and with classroom practices that address these difficulties. By linking research and practice, we celebrate teachers as learners of their own students' mathematical thinking, thus contributing an alternative view of mathematical progression in which students are taught conceptually. The authors of chapters in this book are from Australia, Canada, the UK, and the USA. The commentary authors are renowned experts, including scholars from Belgium, the UK, and the USA.

One of the unique features of this book is the inclusion of a clear focus on understanding and using research insights, including theory about students' conceptual development, to guide a paradigm shift in existing teaching practices for students with diverse needs. Because we think highly of teachers and students, the heart of this shift is the inclusion of theory-driven instructional (intervention) practices that integrate heuristic and explicit instruction for inquiry in mathematics classrooms. Using theory and research-based learning trajectories recognizes teachers as capable professionals while being consistent with current curriculum standards and recommended teaching practices in mathematics education. A second key feature of this book is that it arises out of a longitudinal, genuine collaboration between mathematics education and special education researchers who teach students and teachers regularly. Accordingly, the book chapters have been written collaboratively by author teams composed of scholars from both fields. In addition, most chapters are driven by authentic questions gathered from teachers who work with students with diverse needs.

This book caters to teachers, as seen in the four central questions presented at the beginning of this *Introduction*. To address those questions, we share with you, the teacher, recent research findings and teaching methods that significantly promote mathematics learning and outcomes for students with LDM. These promising results reflect the aforementioned paradigm shift in how a teacher thinks about (theory) and carries out (practice) mathematics teaching – as compared to typical ways of explaining and studying students' difficulties. Specifically, this shift revolves around moving away from a focus on behavioral and procedural steps to a focus on conceptual learning and teaching marked by a "marriage" of heuristic and explicit strategies inspired by research-based practices from both fields.

The book consists of four Parts. Part I addresses equitable opportunities for mathematics learning. R. Ruttenberg-Rozen and B. Jacobs (Canada) open this Part with their chapter titled "Considerations of Equity for Learners Experiencing Mathematics Difficulties: At the Nexus Between Mathematics and Special

Education." They discuss what and how traditional practices for students with LDM promote deficit narratives and offer strategies and suggestions through connecting research in mathematics education and special education. In Chap. 2, K. Owens and S. Yates (Australia) delineate the characteristics of the learners with LDM pertinent to their physical, cognitive, social, and emotional well-being, while highlighting their strengths and the use of their capabilities for learning mathematics. In Chap. 3, "Discerning Learning as Conceptual Change: A Vital Reasoning Tool for Teachers," R. Tzur and J. Hunt (USA) embrace an asset instructional approach that centers on and starts from what students already know rather than what they do not know (i.e., from their deficit). Following these three chapters, A. Dowker (UK) provides a commentary on each of these chapters and discusses the impact of cultural practices on mathematics cognition.

Part II focuses on assessment and planning. Building on the framework for knowing and learning presented in Chap. 3, in Chap. 5, "Connecting Theory to Concept Building: Designing Instruction for Learning," J. Hunt and R. Tzur use an instructional case (the concept of ten as a unit) to illustrate how teachers can unpack "bridging, variation, and reinstating" to design their teaching. In Chap. 6, "What is Meaningful Assessment," D. Zhang, C. Maher, and L. Wilkinson (USA) focus on informative assessment for all students. In particular, they review alternative assessment methods that aim to identify what students know, monitor their learning trajectories, and evaluate their paths to deeper conceptual understanding. Following these two chapters in Part II, L. Ketterlin-Geller (USA) situates her commentary in a systems-level perspective framed within multi-tiered systems of support (MTSS) and highlights common themes surrounding designing and implementing assessments to support learning for students who are experiencing difficulties learning mathematics.

Part III focuses on teaching for equitable opportunities. This Part begins with Chap. 8, "Supporting Diverse Approaches to Meaningful Mathematics: From Obstacles to Opportunities," by C. Finesilver, L. Healy, and A. Bauer (UK). In this chapter, they explore tools, including manipulatives, calculation and communication devices, memory aids, and representational strategies, that may be particularly helpful in providing students with LDM access to mathematics instruction similarly to their normal achieving peers. Next, in Chap. 9, E. Smith and R. A. Smith (USA) provide some guidelines to help teachers effectively engage students with LDM, who are also multilingual learners, in mathematical discourses with particular attention to creating a learning environment that promotes the participation of these students. Then, in Chap. 10, "Equitable Co-teaching Practices in Mathematics," M. Stephan and L. Dieker (USA) illustrate a new vision for co-teaching mathematics in K-12 classrooms. Following these chapters, L. Dieker and A. Lannan (USA) situates her commentary chapter in the discussion of equity and access, while exploring how universal design for learning could provide an anchor for equity and access.

Part IV moves the focus to teaching specific mathematics content, which we elaborate a little more. It begins with Chap. 12, "Counting," in which H. Thouless, C. Hilton, and T. Webb (UK) focus on two of the early mathematical conceptualizations, counting and subitizing. Positioning counting as a prerequisite to other aspects of arithmetic (particularly the concept of number), the authors discuss some challenges students might face, including the language of counting, one-to-one correspondence, and formal symbolic notation(s) for counting. They also list some activities that would be useful for enhancing counting and subitizing skills of students with LDM.

Y. P. Xin and S. Kastberg (USA) open their Chap. 13, on additive reasoning and problem solving with two distinct paradigms in solving word problems: operational vs. relational. They emphasize the importance of students' understanding of mathematical relations presented in mathematical word problems as a crucial start for any solution attempt. They then illustrate how a computer-assisted intervention program, which integrates a constructivist view of learning and explicit teaching of mathematical model-based problem solving, can be used to enhance solving word problems by students with LDM.

In Chap. 14, R. Tzur and Y. P. Xin (USA) present a learning progression (trajectory) comprising six concepts (schemes) that were found to be crucial for student's construction of multiplicative reasoning and problem solving. They situate this presentation in the context of a computer tutor that shows promise in moving students onto solving multiplicative word problems using a unified mathematical model equation. In Chap. 15, R. Tzur and J. Hunt (USA) focus similarly on an 8-scheme learning trajectory in fractional reasoning for students with LDM. The early four schemes are based on the mental activity of iteration and the last four on recursive partitioning. Throughout the chapter, they illustrate these conceptual understandings with instructional tasks that help foster students' reflection on activities and abstraction of those concepts.

In Chap. 16, S. Kastberg, C. Hord, and H. Alyami (USA) use a series of cases with instructional tasks to illustrate ways in which teachers can support the development of ratios, proportions, and proportional reasoning while promoting abstraction through making connections among these concepts. They present a trajectory consisting of three benchmarks along with plenty of tasks to promote construction of those benchmarks by students with LDM. Their chapter is followed by a commentary chapter for the entire Part by L. Verschaffel and W. Van Dooren (Belgium), who both reflect on each of the Part's chapters and point out plausible alternative views.

Acknowledgments

In this introduction, we chose to use the term "learning disabilities/difficulties in mathematics (LDM)" rather than mathematics disabilities (MD), because typically schools do not have specific diagnosis for MD. More importantly, the term MD seems to indicate that the disability, or deficit, is fixed within the individual student. We intentionally chose the term LDM, rather than MD, to indicate that these students, while struggling in learning mathematics and possibly attaining low achievements in mathematics – are highly capable of learning if teaching adapts to their competencies. With that said, we respect the choice of term that each of the contributing authors used to describe students with LDM in their individual chapters.

An earlier product of members of this collaborative, special interest Working Group (of PME and PME-NA), is the "Cross-Disciplinary Thematic Special Series: Special Education and Mathematics Education" published in *Learning Disability Quarterly*. Most of the leading authors of each of the chapters in this book were members of that Working Group. The views/perspectives and programs/tools shared in this book are by no means all-inclusive or exhaustive. No doubt, the collaborative works among the Working Group members and the projects they have been undertaking were instrumental to the theoretical framework and strategies shared in this book. We take this opportunity to thank all contributing author teams for the effort and time they dedicated to share their unique perspectives. As noted above, co-authoring each chapter by scholars from mathematics education and special education was purposefully planned, so that ideas shared in each chapter reflect a collective, coordinated view from both fields.

We also express our sincere thanks to authors of the commentary chapters for their independent, insightful comments and critique. We strove to purposefully invite commentary authors who embrace independent views about the content covered in each Part of this book, with an intention to enrich the perspectives and strategies shared concerning teaching and learning of mathematics involving students with LDM.

We want to recognize contributions of the reviewers who provided initial and/or follow-up feedback to each of the chapters included in this book. Reviewers who reviewed at least one chapter or multiple chapters are listed below in alphabetical order by last name:

Carla Finesilver, Caroline Hilton, Casey Hord, Jessica Hunt, Signe Kastberg, Kay Owen, Robyn Ruttenberg-Rozen, Alex Smith, Erin Smith, Michele Stephan, Helen Thouless, Ron Tzur, and Yan Ping Xin.

Finally, we wish to thank Dr. Jinfa Cai for selecting this book as part of Springer's book series Research in Mathematics Education that he and James Middleton have been editing.

Contents

Part I

Chapter 1
Considerations of Equity for Learners Experiencing Mathematics Difficulties: At the Nexus Between Mathematics and Special Education

Robyn Ruttenberg-Rozen and Brenda Jacobs

Abstract In this chapter, we explore equity for learners experiencing difficulties in mathematics (MD) at the nexus of special education and mathematics education, with a focus on Gutierrez's four dimensions of equity. First, we explore narratives and positioning of learners experiencing difficulties: What and how are deficit narratives constructed around learners experiencing MDs? In the next section, we explore the consequences of those deficit narratives: How has the literature and school intervention typically responded with interventions for learners experiencing difficulties? And how are those interventions connected to deficit narratives about learners experiencing MDs? In the last sections, we offer a way forward: How can we provide mathematical opportunities to support learners experiencing MDs?

Keywords Mathematics difficulties · Special education · Equity · Anti-deficit narratives · Positioning · Inquiry · Mathematizing · Self-regulation

1.1 Introduction

The many visible and invisible roles of mathematics in society are far-reaching and entwined with much of our lived reality. When mathematics plays the role of savior, media, governments, and others tout mathematics as an integral and foundational component that will sustain countries (and people) in the digital economy. In the role of aesthetic for understanding, mathematics can also play a role of beauty.

R. Ruttenberg-Rozen (✉) · B. Jacobs
Ontario Tech University, Oshawa, ON, Canada
e-mail: Robyn.Ruttenberg-Rozen@ontariotechu.ca

© The Author(s), under exclusive license to Springer Nature Switzerland AG 2022   3
Y. P. Xin et al. (eds.), *Enabling Mathematics Learning of Struggling Students*,
Research in Mathematics Education,
https://doi.org/10.1007/978-3-030-95216-7_1

There is an aesthetic experience of playing with mathematics and discovering and creating mathematical patterns in the world around us (Sinclair, 2006), and many cultures and peoples have used mathematics to create beautiful products (Assis & Donovan, 2020). In the role of categorizer of others people with power use mathematics to describe and categorize other people (Swanson, 2017). People with power often do this in order to make claims and decisions about those inside and outside of the categories. Mathematics also plays a role as a tool of advancement or its opposite, impediment. Doing well in school mathematics creates pathways to societally deemed prestigious jobs, whereas lack of achievement in school mathematics can often prevent students from graduating high school, college, or university. Lack of achievement will prevent students from entering many professions (Ernest, 2002; Esmonde, 2009). Mathematics and especially school mathematics plays the role of gatekeeper. Mathematics, more than any other subject, "play(s) a special role in sorting out students and preparing them for and directing them to different social stations" (Ernest, 2007, p. 2). Success in mathematics becomes a mechanism for exclusion, funneling only the "intellectually worthy" toward certain opportunities. Certain learners, typically those with mathematics difficulties, are made to believe that mathematics is not for them (Allexsaht-Snider & Hart, 2001).

For all the reasons mentioned above, a belief that one cannot do mathematics can have far-reaching consequences for learners *and* society, because as a society we lose out on a diversity of ideas when we funnel only certain learners toward mathematics (Allen-Ramdial & Campbell, 2014).

Mathematics holds such power in society that those who possess the knowledge of mathematics acquire a cultural capital (Gutierrez, 2012). Thus, it is not surprising that the understanding of mathematics and achievement in mathematics would be considered a civil rights issue (Moses & Cobb, 2001) – it is important for all learners to have access to mathematics. Nevertheless, some learners, those experiencing difficulty, do not have access to mathematics which is likely to affect their future economic and academic prospects, and importantly, leaves them without access to the myriad societal and internal benefits, some already discussed above, of learning and interacting with mathematics. Children who experience mathematics difficulty (MD) include those with labeled learning disabilities, as well as those with sustained low mathematics achievement despite interventions and teaching (Lewis, 2014).

In this chapter, we explore equity for learners experiencing difficulties in mathematics (MD) at the nexus of special education and mathematics education. In the introduction, we have explored the importance of mathematics and the idea that there are systemic barriers that purposefully funnel away some learners (including those who experience MDs) from the learning and enjoyment of mathematics (Ernest, 2007). All learners should be given the opportunity to learn and enjoy mathematics. Consequently the systemic barriers and funneling is an issue of equity. At the root of the systemic barriers and funneling are deficit meta-narratives about who can and cannot learn mathematics, what type of mathematics people can learn, and in what ways people can learn mathematics. In the next section, we explore deficit narratives and positioning of learners experiencing difficulties.

There are significant inequitable consequences to learners as a result of the deficit narratives. In the section after, we explore the consequences of those narratives – how the literature and intervention practices limit learning for learners experiencing difficulties. Finally, we want to offer hope that things can be changed and we can move toward a more equitable situation for learners experiencing MDs. In the last sections, we offer a way forward. First, we situate our way forward in equity research of mathematics education and weave it together with research from special education. Finally, addressing the research in equity we offer three strategies and suggestions.

1.2 Narratives and Positioning

So how does it come to be that access is limited for a child experiencing difficulties?

In this section we examine deficit narratives (Adiredja, 2019) constructed around learners experiencing difficulties in mathematics, and how deficit narratives result in deficit positioning (Harré & van Langenhove, 1999) of learners experiencing MDs.

1.2.1 Story 1

Nine-year-old David was a student in a self-contained grade 4/5 split special education classroom and had a label of "slow learner." The label of slow learner was meant to convey low intelligence, certainly below average, and falling somewhere between 75 and 90 on a normal curve of intelligence (Williamson & Paul, 2012). Before the start of the school year the principal told the teacher not to worry about David, as he would not be able to do mathematics. Nevertheless, the teacher worked with David and created a program for him, teaching him mathematics. One day, in about February that year, David displayed a moment of deep abstract reasoning in multiplicative structures – he intuitively visually represented and explained the associative property of multiplication on an abacus. The teacher was excited at what she noticed and wanted to share her discovery with her principal. After school that same day, the teacher went to the principal's office to share the exciting noticing. In an excited voice, the teacher re-told David's story of displaying deep abstract reasoning of multiplicative structures. There was a pause after the story as the principal took in the events. The teacher waited in anticipation; surely the principal would "see" how amazing it was that David visually represented the associative property of multiplication. The principal finally answered: "what you saw was a fluke. David cannot really do math. David is unable to reason abstractly" (Ruttenberg-Rozen, 2019, p.213).

1.2.2 Story 2

Megan began with all 17 cubes in one pile in front of her. She moved one cube to the right and then one to the left continuously until she was left with one cube in her hand. She shifted the right pile to the side and said, "one pile," then Megan shifted the left pile to the side and said "two piles." Finally, she placed the one leftover cube in the middle of the piles and said "number one." I (Robyn) asked Megan how much was in each pile. Later, when I reviewed the video recording of our session, I could see that Megan counted all the cubes in one pile and began to count two of the cubes in the second pile. In my re-watching of the video Megan assuredly announced "seven and seven." However, in person, I could not hear Megan's answer and I asked her to repeat herself. In repetition, Megan's answer became tentative:

> Seven, (points to the right pile) and then I'm guessing seven here (points to the left pile), because seven… (pauses to count the left pile by twos). Yeah. (looks at me) I'm pretty sure. Yeah.

The fact that there were really eight cubes in each of Megan's piles is inconsequential here. It is common for all people, at times, to make small numerical errors especially between close numbers (Dehaene, 2011), in this case 7 and 8. Of interest was Megan's tentativeness to give an answer for the second pile only after I asked her to repeat herself. Megan's usage of symmetry in this case was strategic. Once she knew the number in one pile, there was no need to count the cubes in the second pile. Because of her motions of symmetry, making one-to-one correspondence between the two piles, the two piles were the same amount. Through asking Megan to repeat her answer, I gave her a cue to question her strategy and thus, her understanding. Notably, Megan began to defend that there were seven in the second pile: "because seven…" However, this did not last long, as still questioning herself, Megan paused to count and then looked at me for verification.

Both David and Megan experienced learning difficulties in mathematics, and both David and Megan had deficit meta-narratives (Adiredja, 2019) and positioning (Herbel-Eisenmann et al., 2015) that shaped and interfered with their development in mathematics. In David's case the meta-narrative was that he was labeled a slow-learner and slow-learners cannot understand or reason abstractly in mathematics. When the principal responded to David's story of reasoning abstractly, she was responding to David's deficit meta-narrative, and because of her acceptance of this narrative, the principal positioned David as "unable." The act of positioning is always a limiting action – it limits the variety of interpretations possible of any action (Harré & Slocum, 2003). Subsequently, once the principal positioned David as unable, the idea that David had some capabilities and could reason abstractly in mathematics was not even in her repertoire. The principal's only choice was to deny the possibility of David's ability. David's principal was influenced by a deficit meta-narrative and sadly, her subsequent deficit positioning of David is not unique. We find other cases of disbelief of ability of learners experiencing MDs in other research. For example, Houssart (2004) discusses elementary teachers attributing moments of insight by learners experiencing math difficulties to the "maths fairy."

The maths fairy was something outside the student granting momentary ability to the learner- as if the students were not capable of reasoning abstractly on their own.

In David's case the principal was positioning another person, David, according to a negative meta-narrative. However, narratives around positioning also influence the individual of whom the narrative is about: "Positioned in some given way, a person may be more or less tightly constrained as to what story line it is possible, proper, or even necessary to be living out" (Harré & Slocum, 2003, p. 107). Megan's actions were "tightly constrained" as a result of her own belief in her deficit meta-narrative and her positioning as someone who experiences difficulties in learning mathematics. Megan strategically used symmetry to help her keep track of the number of cubes in each pile, and when questioned about the number, she recounted her strategy. All this demonstrated important aspects of Megan's ability to reason mathematically. But like David, Megan was positioned as unable. Megan's positioning was not by an other – a teacher or principal like David's positioner; Megan is positioned by her own acceptance of her deficit narrative. Subsequently, Megan's assuredness was short-lived because her storyline is constrained by beliefs that she must be incorrect in mathematics, and that an authoritative other must be correct. Story 2 is an excerpt from a study where the first author (Robyn) worked with Megan for a month. This vignette is one small episode among many similar occurrences. Throughout the month, Megan often and repeatedly demonstrated mathematical reasoning only to negate her own reasoning in one way or another.

Megan and David both highlight how deficit narratives are constructed around learners who experience mathematical difficulties, and how those narratives serve to negate our and their own awareness of their mathematical reasoning abilities. Both David and Megan displayed abstract mathematical reasoning. Yet, in both cases that abstract reasoning was mislabeled and misacknowledged because of the constraints of the narratives around "mathematics difficulties." Importantly, those who are mislabeling and misacknowledging, whether done to an other or done to the self, may not even be aware that they are doing so. Significantly, these beliefs around deficit narratives are often implicit and may even contradict a person's explicit beliefs (Ruttenberg-Rozen et al., in press). It therefore becomes imperative for the holder of these beliefs to become aware (Mason, 2002) of their beliefs in order to circumvent them.

The negation of ability, as discussed in the introduction, can have far-reaching consequences for the learner and society. Research (e.g., Jamar & Pitts, 2005; Lambert, 2015; Louie, 2017; Shah, 2017) is replete with the steep consequences to learners' self-efficacy and achievement in mathematics when their abilities are dismissed. Significantly, teaching also changes as a result of educators holding onto these deficit narratives. In the next section, we explore the detrimental consequences to mathematics interventions in special education and "remediation"[1] that result from educators' belief in these deficit narratives.

---

[1] The term remediation itself is associated with deficit narratives. To re-mediate conveys a need by an other to fix deficits in a learner by mediating again.

1.3 Consequences of Deficit Narratives to Interventions

Because of the deficit (Gervasoni & Lindenskov, 2011) perspectives of learners experiencing MDs, interventions for them are often behaviorist at their root (Lambert, 2015). This means that there is a hyper-focus on procedural methods for computational procedures and rote fluency as interventions. Significantly, the computation and fluency that is stressed for these learners is only a small part of the complexity and beauty of mathematics. Mathematics is much broader than the four operations of addition, subtraction, multiplication, and division promoted in fluency interventions – mathematics consists of: integrals, angles, spirals, spatial dimensions, curves, proportionality, logic, conjectures, and generalizations, to name just a few. The hyper-focus on procedures and fluency of computation is often at the expense of conceptual understanding development. Ironically, for children experiencing MDs it may very well be conceptual understanding of mathematics that leads them to better fluency in procedures in mathematics (Landerl et al., 2004). The efficacy of typical interventions for children experiencing MDs, such as teaching rules to memorize steps, may receive mixed results (i.e., Zentall, 2007) because the child does not yet have a conceptual understanding of the underlying mathematics. It may also very well be that educators confuse teaching explicitly with teaching rules without understanding for rote procedural fluency. Explicit teaching, that is the modeling and explanation of steps in procedural fluency, has been an important teaching strategy in teaching those experiencing difficulties (i.e., Mills & Goos, 2011). However, explicit teaching does not preclude understanding; in fact, the modeling done in explicit teaching should include the modeling of thinking for conceptual understanding.

Added to the ignoring of a significant component of mathematics in favor of procedural fluency, many of the other interventions for learning mathematics proscribed for children experiencing MDs center around the peripheral supports (i.e., accommodations such as the way work is presented and discourse used by the teacher) to learning mathematics and not on the learning of mathematics itself (e.g., Houssart, 2004). Gervasoni and colleagues (2011), considering international input, summarized the state of affairs of mathematics interventions for students with special learning needs from an international perspective:

> …many mathematics programs and learning activities for students with 'special needs' attempt to teach mathematics using a conventional approach, but at a slower pace and with a more tunnelled view of a limited range of mathematics. In these cases, instructional innovations were based on deficit models of learners and focused mainly on designing tools to aid communication between the teacher and student that enabled students to access classroom mathematics programs and teaching (p. 308).

Indeed, through relying on deficit narratives and by placing a hyper-focus on fluency and supports peripheral to mathematics, interventions for children with MDs have become fixated on the child as the object of study, not the mathematics.

Educators cannot look to policy to rectify the current hyper-focus on fluency and the discrepancy between procedural and conceptual interventions in mathematics for children with special learning needs. Neoliberal educational policy documents and curricula, with their focus on skills, do not aid a move toward conceptual under-standing either. These documents are usually focused on a narrow view of achieve-ment as acquiring skills. Educators, in turn, interpret these policies by equating procedural fluency with mathematical understanding. As a result, interventions leading to robust understanding for children who are experiencing difficulty can be viewed as superfluous, especially when there are time constraints on teaching. However, in consideration of time constraints, practice of fluency does not have to be taught from a solely behaviorist method. Of note, there are other methods for developing fluency in computational procedures that extend beyond behaviorist methods (see Sect. 1.4 for discussions of some of these methods). Building rela-tional understandings (Fosnot & Dolk, 2001) in computation can also develop flu-ency while at the same time strengthening conceptual understanding (Landerl et al., 2004). Importantly, Fosnot (2010) has edited a book containing anecdotes of how building relational understandings has aided the growth of students experi-encing MDs.

Some special educators (e.g., Cawley, 2002) have called for a shift from only "doing mathematics" (i.e., procedural fluency) to also "knowing mathematics" (p. 3) (i.e., conceptual understanding). While fluency and procedural competency are important for learning mathematics, conceptual understanding is essential, espe-cially for access to higher order mathematical reasoning, and for appreciating aes-thetic components of mathematics. Mathematics education research, with its specific theoretical perspectives around education and the thinking and understand-ing of mathematics (Schoenfeld, 2000), including on the activity of doing mathe-matics, offers the possibility to explore "the dynamic "knowing" that portrays the growth of understanding" (Pirie, 1996, p. xv) of children with MDs. Importantly, it is the exploration of mathematical activity that may create more equitable spaces and counters deficit beliefs (Moschkovich, 2010) about mathematics learners.

In the first sections of this chapter, we have constructed our argument that deficit narratives resulting in negation of ability result in: (i) learner's constructing deficit beliefs about themselves and low achievement in school; (ii) behavioral interven-tions that do not focus on learning of mathematics, only the peripheral of mathemat-ics; and (iii) educators reifying the deficit narrative through preventing learners experiencing MDs from actually getting access to mathematics. Access to mathe-matics in all its diverse forms for learners experiencing MDs, then, is an imperative of equity. In the next section, we explore a theoretical framework for equity that is useful in framing these issues and a way forward.

1.4 Equity in Mathematics Education and Special Education Research

Although mathematics education research has addressed issues of equity for a quar-ter century already, and equity is a concern for every aspect of schooling, many inequities that were discussed 25 years ago sadly still exist in mathematics education today (Ellington & Prime, 2011). Wagner et al. (2012) assert that conversations about equity in mathematics education became more robust in recent years because of research using sociocultural theoretical frameworks. Sociocultural frameworks are important to equity research because these frameworks shift the considerations from personal access, and the individual, to include social considerations as well. As a consequence of these new considerations the view of the learner shifts from being an object to be studied, to being a part of situated, and complex interactions. In other words, equity, as viewed through a social constructivist lens is not a tangible possession, but relationships in and through people and their environment (Walshaw, 2011).

There have been a number of different explorations and definitions of equity in mathematics education research (e.g., Lubienski, 2002). Some of the scholars discussing equity take into account only a few of the aspects of equity (Bose & Remillard, 2011), or confuse equality with equity (Allexsaht-Snider & Hart, 2001). A more robust exploration, and one that is aligned with our current focus on the equity imperative for learners experiencing MDs can be found in the work of Rochelle Gutierrez. Gutierrez (2012, p. 18) gathered multiple aspects of equity and categorized them into four dimensions:

(i) access,
(ii) achievement,
(iii) identity, and
(iv) power.

Each of Gutierrez's dimensions is complex, interrelated, and relates to inherent systemic inequities within the school system and the teaching and learning of mathematics. All of these dimensions are integral to equity but equity does not exist when only one is present, as each dimension has discrepancies that are filled by another.

Of Gutierrez's (2012) four dimensions of equity, access is the first category. Access is related to "opportunity to learn," a rhetoric popular thirty years ago that assumes everyone is equivalent. It addresses the learning supports that create opportunities for learning in the classroom. Access is an important aspect of equity, and as such all learners should have opportunities to access mathematical learning. Yet the responsibility to provide opportunities for learners experiencing MDs does not fall on the system itself. Instead, inequitably, the responsibility to receive opportunities falls on the individual learner. The individual learner must demonstrate how they can achieve similar to other "normal" peers in their age group or grade level before they can gain access (Dudley-Marling & Burns, 2014). Mathematics learning is then not a right (Moses & Cobb, 2001) where learners can only gain access through perceived and compared ability, but a reward for school achievement. Importantly, access implies something that will benefit the learner (Dudley-Marling & Baker, 2012). Subsequently, what is called "access to mathematics" for learners experiencing MDs is not actually access. Mathematics teaching that involves only memorizing small disconnected pieces of information without understanding or placing it into the grand structure cannot benefit the learner and therefore cannot be access, and cannot be a genuine opportunity for learning.

Stemming from access, Gutierrez's second dimension is achievement. Achievement describes the outcomes from access. These achievement outcomes include not only school-based, both K-12 and post-secondary success in mathematics, but also post-academic, including access to mathematics careers. Achievement as equity originated as a discourse twenty years ago as a way to view equity in relation to standardized assessment. Achievement in schooling is highly intertwined with the structural constraints of schooling on the learner experiencing MDs. Difficulties are often created and then constructed by a system that seeks to categorize learners in terms of arbitrary referents of ability and normalcy (McDermott, 1993). For example, in Ontario, our home province in Canada, some school boards have created a "scope and sequence" for their elementary schools. For each grade, the Ontario Elementary Mathematics Curriculum (Ontario Ministry of Education, 2005) is broken down into a yearly schedule with prescribed days for coverage of topics. Because of the time constraints of the schedule, the result is that teachers feel a need to keep moving forward in the schedule even if their students are not ready to move forward mathematically or developmentally. The misalignment between students developing mathematically and the time constraints create gaps in understanding for learners. The system, having created difficulties, then rewards schools for creating these very difficulties through funding incentives that make it profitable for schools to diagnose difficulties (Stegemaan, 2016).

The third dimension of equity, identity, has arisen as an important aspect of achievement. School-based mathematics often implicitly vitiates the identities of those in the non-dominant culture by forcing them to conform to sometimes opposing perspectives of mathematics (Wagner & Borden, 2011). Thus, in order to achieve and participate in school-based and societal mathematics, students may begin to construct deficit views of themselves and their cultural backgrounds. These deficit views may then lead them to assume negative identities vis-a-vis mathematics. Like Megan, introduced above, learners experiencing MDs internalize the deficit meta-narratives that position them as unable. These meta-narratives are enacted through social interactions compelling learners to take on the identity of being "unable" (Heyd-Metzuyanim, 2013). Consequently, it is imperative that equity work also support learners experiencing MDs in their identity development as mathematics able learners and doers.

Finally, even if all the other dimensions of equity-access, achievement, and identity are in place, issues of power may still disrupt equity. Power may have an impact on equity concerning the personalized, social impact, and social justice components of mathematics – the spaces where students lose their agency: "voice in the classroom," "opportunities for students to use math as an analytical tool to critique society," "alternative notions of knowledge," and "rethinking the field of mathematics as a more humanistic enterprise" (Gutierrez, 2012, p. 20). There is no mathematical equity without mathematical agency.

To sum up the dimensions: a learner has to have access to the mathematics that allows for school and life achievement, societal participation, and societal contribution (Gutierrez, 2012, p. 158), while simultaneously gaining the opportunity to critique the inequities of the dominant mathematics and their situation at the foundations of their learning and achievement. In the final section, we use Gutierrez's four dimensions (2012) of equity to support our way forward.

1.5 A Way Forward

An imperative of equity dictates that there must be change for learners experiencing MDs. However, change cannot happen in a vacuum; it must incorporate access, achievement, identity, and power. To do so we must endeavor to change the meta-narratives around learners experiencing MDs. This,

> requires a focus on what these students can do, rather than lamentations of what they do not. That is, they argue that, instead of attempting to determine "normal" or "ideal" achievement and positioning those who deviate from supposed norms as problematic and in need of remediation, attention should be directed to how students' mathematical ideas develop differently and the pedagogical strategies appropriate to support these developmental trajectories (Healy & Ferreira Dos Santos, 2014, S125).

The meta-narrative of inability vs. ability around learners experiencing MDs is a false binary because it assumes that mathematics is one thing detached from all other knowings and abilities (Boylan & Povey, 2014). The first imperative of the way forward is to shift from a deficit discourse to an empowerment discourse. We can only accomplish this through identifying and becoming aware of our implicit biases (Ruttenberg-Rozen et al., in press) toward learners experiencing MDs and creating alternative narratives that supplant the deficit narratives.

The second imperative of a way forward is to provide many and rich opportunities for learners experiencing MDs to interact with real mathematics. The focus of mathematics interventions should be shifted toward "knowing mathematics." It is important that mathematics be the medium, object or tool, through which intervention takes place. Mathematics should not serve a secondary role in interventions. For access, achievement, identity, and power, the learner must interact and have access to real mathematics, not just computation. We cannot label someone as "unable" to do mathematics when they have not even had the opportunity to interact with mathematics. Mathematics on its own does not limit access. It is school mathematics through its testing, tracking, and rote teaching that limits access to the very few deemed worthy (Ernest, 2007).

The third and final imperative of a way forward is to provide inquiry activities that develop self-regulation in order to support rich opportunities to interact with real mathematics. We must change teaching practices to align with access, achievement, identity, and power. Many of the chapters in this book will introduce you to innovative and equitable practices that engage learners and are aligned with access, achievement, identity, and power considerations. We use this last section to explicate two promising ideas we have been exploring in our research. Learners experiencing MDs can be empowered through inquiry-type practices (Ruttenberg-Rozen, 2020) and activities that develop self-regulation (Jacobs, 2022).

In recent years, mathematics education has increasingly shifted to inquiry and problem-based learning, which are closely related teaching practices that encourage active learning and critical thinking through student led short and long term investigations guided by interesting questions or problems. Inquiries can emerge from either the children's or teachers' interests, encounters with materials in the classroom, or unexpected events (Jacobs, 2022). Problem-based learning is in effect a type of inquiry-based learning where the teacher presents a problem for the students to investigate. In open inquiries students formulate their own questions (Banchi & Bell, 2008). As inquiries unfold, students and teachers co-construct knowledge by sharing multiple perspectives.

During inquiries teachers observe and reflect on children's interests, listen to children's theories and ideas, watch how the children are engaging with the classroom materials, interact and think about what concepts the children are exploring, document the children's learning, and respond to the children in thoughtful ways through reciprocal actions (Fraser, 2012). Reciprocal actions occur when teachers ask children questions to provoke further thought, provide provocations that scaffold the children's learning, adapt the classroom environment to accommodate these interests, and take the children on outings to enhance their understanding (Stacey, 2015). Inquiry and problem-based learning also enable students to build authentic personal relationships, develop their communication skills, share their own ideas and theories, and nurture their reasoning and problem-solving capabilities.

Inquiry is an especially valuable teaching practice because it enables students to learn how to self-regulate (Jacobs, 2022), which supports a positive mathematical identity and can empower learners. There is a broad consensus in the educational research that how well students do in school depends on how well they can self-regulate (Blair & Diamond, 2008; Blair & Razza, 2007; Fitzpatrick & Pagani, 2013; Jacobs, 2022; McClelland et al., 2006; McClelland & Cameron, 2011; Ponitz et al., 2009; Rimm-Kaufman et al., 2009; Shanker, 2013; Shanker & Barker, 2016; Welsh et al., 2010). Self-regulation is a reflective learning process where children become aware of what it feels like to be overstressed, recognize when they need to up-regulate or down-regulate, and develop strategies to reduce their stress and restore their energy. This process empowers children to see themselves as self-regulated learners (Shanker, 2013). Self-regulated learners have the ability to reflect critically on their strengths and weaknesses, set goals, monitor their progress, use strategies to assist in problem-solving, and seek assistance when needed (Zimmerman, 2002). Learning to self-regulate in the early years lays the foundation for these higher metacognitive functions as children grow older.

Self-regulation supports a growth mindset as it fosters the belief that intellectual and academic abilities are not fixed but can be developed, combatting deficit narratives learners construct about themselves. Wang et al. (2019) found that when students experiencing mathematics difficulties were encouraged to apply their self-regulation strategies on fraction word problems it improved their performance. Butler et al. (2017) argue that students bring motivationally charged beliefs to contexts. When students believe they can improve their ability by using learning strategies effectively, they view errors as opportunities to learn and will persist when challenged. In contrast, students can feel anxious and disengaged from learning if they think they have little control over the outcomes.

1.6 Conclusion

Deficit narratives about learners experiencing mathematics difficulties are pervasive and destructive. But these deficit narratives can be replaced (Ruttenberg-Rozen et al., in press). It takes a lot of work to replace deficit narratives. Deficit narratives situate themselves deep into our psyche. The starting point to rooting out these limiting biases is awareness (Mason, 2002).

In this chapter, we have sought to bring awareness to the myriad ways that deficit narratives impede learners experiencing mathematics difficulties. It is inequitable to limit opportunity and access to the myriad of benefits of mathematics. Yet, that is the situation for many learners experiencing MDs. It behooves us to make changes to increase opportunities for all learners. It is not that these learners cannot "do," or "learn," or "think about," or whatever other action comes to mind, mathematics. It is that there are a variety of ways that learners learn. Societal meta-narratives have influenced us to discount learning, knowing, and ability that is not like the "norm," whatever that may be. We have to make changes and these changes have to be based on Gutierrez's conception of access, achievement, identity, and power, for the changes to be truly equitable. We have introduced 3 strategies as a way forward. Of course, there are other strategies that can be used – we have provided a starting point for equitable change.

References

Adiredja, A. P. (2019). Anti-deficit narratives: Engaging the politics of research on mathematical sense making. *Journal for Research in Mathematics Education, 50*(4), 401–435. https://doi.org/10.5951/jresematheduc.50.4.0401

Allen-Ramdial, S. A. A., & Campbell, A. G. (2014). Reimagining the pipeline: Advancing STEM diversity, persistence, and success. *BioScience, 64*(7), 612–618. https://doi.org/10.1093/biosci/biu076

Allexsaht-Snider, M., & Hart, L. E. (2001). "Mathematics for all": How do we get there? *Theory into Practice, 40*(2), 93–101. https://doi.org/10.1207/s15430421tip4002_3

Assis, M., & Donovan, M. (2020). Origami as a teaching tool for indigenous mathematics education. In D. Bailey, N. S. Borwein, R. P. Brent, R. S. Burachik, J. H. Osborn, B. Sims, & Q. J. Zhu (Eds.), *From analysis to visualization. JBCC 2017* (pp. 171–188). Springer. https://doi.org/10.1007/978-3-030-36568-4_12

Banchi, H., & Bell, R. (2008). The many levels of inquiry. *Science and Children, 46*, 26–29.

Blair, C., & Diamond, A. (2008). Biological processes in prevention and intervention: The promotion of self-regulation as a means of preventing school failure. *Development and Psychopathology, 20*, 899–911.

Blair, C., & Razza, R. P. (2007). Relating effortful control, executive function, and false belief understanding to emerging math and literacy ability in Kindergarten. *Child Development, 78*(2), 647–663. https://doi.org/10.1111/j.1467-8624.2007.01019.x

Bose, E., & Remillard, J. (2011). Looking for equity in policy recommendations for instructional quality. In B. Atweh, M. Graven, W. Secada, & P. Valero (Eds.), *Mapping equity and quality in mathematics education* (pp. 177–190). Springer. https://doi.org/10.1007/978-90-481-9803-0_13

Boylan, M., & Povey, H. (2014). Ability thinking. In H. Mendick & D. Leslie (Eds.), *Debates in mathematics education* (pp. 7–16). Routledge.

Butler, D., Schnellert, L., & Perry, N. (2017). *Developing self-regulating learners*. Pearson.

Cawley, J. F. (2002). Mathematics interventions and students with high-incidence disabilities. *Remedial and Special Education, 23*(1), 2–6. https://doi.org/10.1177/074193250202300101

Dehaene, S. (2011). *The number sense: How the mind creates mathematics* (2nd ed.). Oxford University Press.

Dudley-Marling, C., & Baker, D. (2012). The effects of market-based school reforms on students with disabilities. *Disability Studies Quarterly, 32*(2). https://doi.org/10.18061/dsq.v32i2.3187

Dudley-Marling, C., & Burns, M. B. (2014). Two perspectives on inclusion in the United States. *Global Education Review, 1*(1), 14–31.

Ellington, R., & Prime, G. (2011). Reconceptualizing quality and equity in the cultivation of minority scholars in mathematics education. In B. Atweh, M. Graven, W. Secada, & P. Valero (Eds.), *Mapping equity and quality in mathematics education* (pp. 423–435). Springer. https://doi.org/10.1007/978-90-481-9803-0_30

Ernest, P. (2002). Empowerment in mathematics education. *Philosophy of Mathematics Journal, 15*, 1–16.

Ernest, P. (2007). Why social justice. *Philosophy of Mathematics Education Journal, 21*, 1–5.

Esmonde, I. (2009). Ideas and identities: Supporting equity in cooperative mathematics learning. *Review of Educational Research, 79*(2), 1008–1043. https://doi.org/10.3102/0034654309332562

Fitzpatrick, C., & Pagani, L. (2013). Task-oriented Kindergarten behavior pays off in later childhood. *Journal of Developmental & Behavioral Pediatrics, 34*, 94–101. https://doi.org/10.1097/DBP.0b013e31827a3779

Fosnot, C. (Ed.). (2010). *Models of intervention in mathematics: Reweaving the tapestry*. NCTM.

Fosnot, C. T., & Dolk, M. (2001). *Young mathematicians at work, 2: Constructing multiplication and division*. Heinemann.

Fraser, S. (2012). *Authentic childhood: Experiencing Reggio Emilia in the classroom* (3rd ed.). Nelson Educational.

Gervasoni, A., & Lindenskov, L. (2011). Students with 'special rights' for mathematics education. In B. Atweh, M. Graven, W. Secada, & P. Valero (Eds.), *Mapping equity and quality in mathematics education* (pp. 307–323). Springer. https://doi.org/10.1007/978-90-481-9803-0_22

Gutierrez, R. (2012). Context matters: How should we conceptualize equity in mathematics education. In B. Herbl-Eisenmann, J. Choppin, D. Wagner, & D. Pimm (Eds.), *Equity in discourse for mathematics* (pp. 17–33). Springer. https://doi.org/10.1007/978-94-007-2813-4_2

Harré, R., & Slocum, N. (2003). Disputes as complex social events: On the uses of positioning theory. *Common Knowledge, 9*(1), 100–118. https://doi.org/10.1215/0961754X-9-1-100

Harré, R., & van Langenhove, L. (1999). The dynamics of social episodes. In R. Harré & L. van Langenhove (Eds.), *Positioning theory* (pp. 1–13). Blackwell.

Healy, L., & Ferreira dos Santos, H. (2014). Changing perspectives on inclusive mathematics education: Relationships between research and teacher education. *Education as Change, 18*(S1), S121–S136. https://doi.org/10.1080/16823206.2013.877847

Herbel-Eisenmann, B. A., Wagner, D., Johnson, K. R., Suh, H., & Figueras, H. (2015). Positioning in mathematics education: Revelations on an imported theory. *Educational Studies in Mathematics, 89*(2), 185–204. https://doi.org/10.1007/s10649-014-9588-5

Heyd-Metzuyanim, E. (2013). The co-construction of learning difficulties in mathematics— teacher–student interactions and their role in the development of a disabled mathematical identity. *Educational Studies in Mathematics, 83*(3), 341–368. https://doi.org/10.1007/s10649-012-9457-z

Houssart, J. (2004). *Low Attainers in primary mathematics: The whisperers and the Maths fairy.* Routledge. https://doi.org/10.4324/9780203464151

Jacobs, B. (2022). *Self-regulation and inquiry-based learning in the primary classroom.* Canadian Scholars Press.

Jamar, I., & Pitts, V. R. (2005). High expectations: A "how" of achieving equitable mathematics classrooms. *Negro Educational Review, 56*(2/3), 127–134.

Lambert, R. (2015). Constructing and resisting disability in mathematics classrooms: A case study exploring the impact of different pedagogies. *Educational Studies in Mathematics, 89*(1), 1–18. https://doi.org/10.1007/s10649-014-9587-6

Landerl, K., Bevan, A., & Butterworth, B. (2004). Developmental dyscalculia and basic numerical capacities: A study of 8–9-year-old students. *Cognition, 93*(2), 99–125. https://doi.org/10.1016/j.cognition.2003.11.004

Lewis, K. E. (2014). Difference not deficit: reconceptualizing mathematical learning disabilities. *Journal for Research in Mathematics Education, 45*(3), 351–396. https://doi.org/10.5951/jresematheduc.45.3.0351

Louie, N. L. (2017). The culture of exclusion in mathematics education and its persistence in equity-oriented teaching. *Journal for Research in Mathematics Education, 48*(5), 488–519.

Lubienski, S. T. (2002). Research, reform, and equity in US mathematics education. *Mathematical Thinking and Learning, 4*(2-3), 103–125.

Mason, J. (2002). *Researching your own practice: The discipline of noticing.* Routledge-Falmer. https://doi.org/10.4324/9780203471876

McClelland, M., Acock, A., & Morrison, F. (2006). The impact of Kindergarten learning-related skills on academic trajectories at the end of elementary school. *Early Childhood Research Quarterly, 21*, 471–490. https://doi.org/10.1016/j.ecresq.2006.09.003

McClelland, M., & Cameron, C. (2011). Self-regulation and academic achievement in elementary school children. *New Directions for Child and Adolescent Development, 133*, 29–44. https://doi.org/10.1002/cd.302

McDermott, R. (1993). The acquisition of a child by a learning disability. In S. Chaiklin & J. Lave (Eds.), *Understanding practice* (pp. 269–305). Cambridge University Press. https://doi.org/10.1017/CBO9780511625510.011

Mills, M., & Goos, M. (2011). Productive pedagogies in the mathematics classroom: Case studies of quality and equity. In B. Atweh, M. Graven, W. Secada, & P. Valero (Eds.), *Mapping equity and quality in mathematics education* (pp. 479–491). Springer. https://doi.org/10.1007/978-90-481-9803-0_34

Moschkovich, J. N. (2010). Language(s) and learning mathematics: Resources, challenges and issues for research. In J. N. Moschkovich (Ed.), *Language and mathematics education: Multiple perspectives and directions for research* (pp. 1–28). Information Age Publishing.

Moses, R., & Cobb, C. E. (2001). *Radical equations: Civil rights from Mississippi to the algebra project.* Beacon Press.

Ontario Ministry of Education. (2005). *The Ontario curriculum grades 1–8 mathematics.* Available at: http://www.edu.gov.on.ca/eng/curriculum/elementary/

Pirie, S. (1996). Foreword. In B. Davis (Ed.), *Teaching mathematics: Toward a sound alternative* (pp. xi–xvi). Garland. https://doi.org/10.4324/9780203054802

Ponitz, C. C., McClelland, M. M., Matthews, J. S., & Morrison, F. J. (2009). A structured observation of behavioral self-regulation and its contribution to Kindergarten outcomes. *Developmental Psychology, 45*, 605–619. https://doi.org/10.1037/a0015365

Rimm-Kaufman, S., Curby, T., Grimm, K., Nathanson, L., & Brock, L. (2009). The contribution of children's self-regulation and classroom quality to children's adaptive behaviors in the Kindergarten classroom. *Developmental Psychology, 45*, 958–972. https://doi.org/10.1037/a0015861

Ruttenberg-Rozen, R. (2019). Growth of mathematical understanding for learners who experience difficulties in mathematics. In J. Holm (Ed.), *Canadian mathematics education study group.* Nova Scotia.

Ruttenberg-Rozen, R. (2020). A wonder-full task leads to a wonder-full intervention. *Mathematics Teacher: Learning and Teaching PK-12, 113*(6), 474–479.

Ruttenberg-Rozen, R., Mahendirarajah, S., & Brady, B. (in press). Educating awarenesses in an online reflective practice course: Becoming aware of implicit biases leaps to judgement. In E. Mikulec & T. Ramalho (Eds.), *Best practices for teaching critical pedagogy online*. Springer.

Schoenfeld, A. H. (2000). Purposes and methods of research in mathematics education. *Notices of the AMS, 47*(6), 641–649. https://doi.org/10.1007/0-306-47231-7_22

Shah, N. (2017). Race, ideology, and academic ability: A relational analysis of racial narratives in mathematics. *Teachers College Record, 119*(7), 1–42.

Shanker, S. (2013). *Calm, alert, and learning: Classroom strategies for self-regulation*. Pearson.

Shanker, S., & Barker, T. (2016). *Self-Reg: How to help your child (and you) break the stress cycle and successfully engage with life*. Viking/Penguin Canada.

Sinclair, N. (2006). The aesthetic sensibilities of mathematicians. In N. Sinclair, D. Pimm, & W. Higginson (Eds.), *Mathematics and the aesthetic* (pp. 87–104). Springer. https://doi.org/10.1007/978-0-387-38145-9_5

Stacey, S. (2015). Pedagogical documentation in early childhood: Sharing children's learning and teachers' thinking. St. Paul, MN: Redleaf Press.

Stegemaan, K. C. (2016). Learning disabilities in Canada. *Learning Disabilities: A Contemporary Journal, 14*(1), 53–62.

Swanson, D. M. (2017). Mathematics education and the problem of political forgetting: In search of research methodologies for global crisis. *Journal of Urban Mathematics Education, 10*(1), 7–15.

Wagner, D., & Borden, L. L. (2011). Qualities of respectful positioning and the connection to quality mathematics. In B. Atweh, M. Graven, W. Secada, & P. Valero (Eds.), *Mapping equity and wuality in mathematics education* (pp. 379–392). Springer. https://doi.org/10.1007/978-90-481-9803-0_27

Wagner, D., Herbel-Eisenmann, B., & Choppin, J. (2012). Inherent connections between equity and discourse in mathematics classrooms. In B. Herbel-Eisenmann, J. Choppin, D. Wagner, & D. Pimm (Eds.), *Equity in discourse for mathematics* (pp. 1–13). Springer. https://doi.org/10.1007/978-94-007-2813-4_1

Walshaw, M. (2011). Identity as the cornerstone of quality and equitable mathematical experiences. In B. Atweh, M. Graven, W. Secada, & P. Valero (Eds.), *Mapping quality and equity in mathematics education* (pp. 91–104). Springer. https://doi.org/10.1007/978-90-481-9803-0_7

Wang, A., Fuchs, L., Fuchs, D., Gilbert, J., Krowka, S., & Abramson, R. (2019). Embedding self-regulation instruction within fractions intervention for third graders with mathematics difficulties. *Journal of Learning Disabilities, 52*, 337–348. https://doi.org/10.1177/0022219419851750

Welsh, J., Nix, R., Blair, C., Bierman, K., & Nelson, K. (2010). The development of cognitive skills and gains in academic school readiness for children from low-income families. *Journal of Educational Psychology, 102*, 43–53. https://doi.org/10.1037/a0016738

Williamson, J., & Paul, J. (2012). The "Slow Learner" as a mediated construct. *Canadian Journal of Disability Studies, 1*(3), 91–128. https://doi.org/10.15353/cjds.v1i3.59

Zentall, S. S. (2007). Math performance of students with ADHD: Cognitive and behavioral contributors and Interventions. In D. B. Berch & M. M. Mazzocco (Eds.), *Why is math so hard for some children. The nature and origins of mathematical learning difficulties and disabilities* (pp. 219–243). Paul H. Brookes Pub. Co..

Zimmerman, B. (2002). Becoming a self-regulated learner. *Theory into Practice, 41*, 64–70.

Chapter 2
Characteristics of the Learners

Kay Owens and Shirley Yates

Abstract The emphasis in this chapter is on individual students' physical, cognitive, social, and emotional characteristics and on the strengthening and use of their capabilities for learning mathematics. Strengths such as recognizing and using patterns or having a good sense of structure or size, combine with alternative perceptual paths and long-term memory to assist with mathematical inquiry and learning. In particular, these strengths help in making and having imagery for developing mathematical concepts. Students then notice characteristics of the concept through visual and verbal processes leading to establishing relationships between physical and mental objects, attention and intention, conceptual development, heuristics and approaches to inquiry, generalizing, and reasoning. The learner's attitudes, beliefs, emotions, confidence, willingness to try, and learned attitudes to themselves as mathematics learners impact on their mathematical identity.

Keywords Psychology of students learning · Inclusive education · Inquiry learning · Perception · Anxiety · Strengths of learners · Self-regulation · Responsiveness · Visuospatial reasoning · Working memory · Motivation · Executive function · Learned helplessness

K. Owens (✉)
Charles Sturt University, NSW, Dubbo, Australia
e-mail: kowens@csu.edu.au

S. Yates
Tatachilla Lutheran College, McLaren Vale, South Australia
e-mail: shirley.yates@tatachilla.sa.edu.au

2.1   Introduction

The meso-aspects of schooling, such as education funding and curricula, teacher and school attitudes and beliefs about students' capabilities as well as the immediate class, all impact on learning. Professional development, parent-school relationships, the stipulated curriculum, and available resources (for a specific learner and in terms of human support) also impact on the inclusive classroom with its specific sociocultural approach to teaching. Inclusive teachers who notice student variability and support students concentrate on each individual's mathematical thinking (Tan et al., 2019; Turner et al., 2012) and avoid a deficit approach that focuses on difficulties or labels. The teacher takes account of what the student brings to the classroom although it may not be observable and measurable.

Early research by Reisman and Kauffman (1980) confirmed that students with mathematical learning difficulties (MLD) acquire content involving verbal comprehension, perceptual organization, and numerical reasoning at varying rates. Students' searching strategies, retrieval of labels, or connecting of thoughts were found to impact the development of mathematical thinking. In turn, working and/or long-term memory affected these learning processes. Many students were identified as having poor spatial relations, communication difficulties, or tiring easily (Reisman, 1978). However, there were also social and emotional issues such as the need for acceptance by others and uncertainty of their learning environment. Table 2.1 provides a summary of their findings upon which the rest of this chapter will expand and build with recent research.

Table 2.1 Aspects of learning relevant to students with special needs (Reisman & Kauffman, 1980)

Cognitive issues	Physical abilities	Social and emotional issues
Rate and amount of learning compared with peers	Movement resulting from mental processes	*Social*
Speed of learning-specific content	Visual perception disorders	Rules of conduct, moral codes, values, customs
Ability to retain information	Poor visual discrimination	Modeling others' behavior, awareness of environmental cues, interacting with others
Need for repetition	Visual figure-ground distractibility	Diplomacy, understanding another's point of view, empathizing & enjoying others company including others' intentions in decisions
Verbal skills	Form constancy problems	
Ability to learn symbol systems & arbitrary associations	Visual-sequential memory deficits	
Size of vocabulary	Spatial relationship difficulties	
Ability to form concepts, relationships, generalizations	Auditory perception disorders	Accepting help, balance between autonomy and dependency
Ability to attend to salient situational aspects	Poor auditory discrimination	
Problem-solving strategies	Auditory background distractibility	*Emotional*
Decision making and judgments	Poor auditory analysis	Feeling afraid, anxious, frustrated, joyous, angry, surprised
Ability to draw inferences, conclusions and hypotheses	Sound-blending difficulty	Becoming overly upset, moody, sad, happy
General ability to abstract and cope with complexity	Auditory sequential memory problems	
	Rules of general language and mathematics	
	Language rules related to phonology, morphology, syntactics, and semantics	

Teachers can identify some learning difficulties but, to assist further, the strengths of the individual student need to be ascertained. Earlier identification methods for mathematics included Piagetian stage development theories related to conservation of number, length, area, or volume (Margetts & Woolfolk, 2019), but these are no longer considered necessary prerequisites for early arithmetic or measurement (Newton & Alexander, 2013). Conservation may develop during learning experiences but is less significant than being able to compare lengths perceptually or by counting units (Jirout & Newcombe, 2018). Stage development has been largely replaced by trajectories, or learning progressions, that are tied to particular concepts. Students with disabilities face the same issues as others on specific mathematical topics (Marcone & Atweh, 2015; Booker, 2011).

2.1.1 The Importance of Variability

Students with MLD vary in the degree and mix of specific characteristics they manifest, but also in how they learn (Reisman & Kauffman, 1980). Studying variance of learning assists teachers to recognize student capabilities and limitations (Vale et al., 2019). Variance also occurs with the particular topic together with possible developmental delays in strategy use (Geary et al., 1991). Further variance occurs with language and sociocultural factors (Hunter et al., 2018).

An example of this approach to noticing individual learning within inclusive classrooms is provided by Faragher and Clarke (2014), who conducted a year-long study of 15 classes that included a student with Down syndrome (DS). They found that with professional learning, teachers stopped focusing on individual student labels and began to realize the importance of including the student in class activities. Students with DS may learn well with visual approaches and/or blocks and symbols rather than oral word counting and number word reading. Their sense of number size is closely linked to these visuals and symbols. Teachers learned how to motivate and engage these students in specific mathematics by ensuring visuals or concrete materials were available and enabling them in the inquiry approach in ways that could be applicable to other students. Importantly, the curriculum was not modified and students were given challenging mathematical tasks (Faragher & Clarke, 2016; Faragher & Clarke, 2014).

Teachers can support students in the moment with specific questions and information (Faragher & Clarke, 2016; Lambert & Sugita, 2016). To assist students with sharing, they might first ask students to share with a small group or they might check in with specific students to find out how the student is thinking in order to support his/her explanation. Teachers need to know the mathematical content well to do this. Furthermore, suitable materials such as linking blocks to represent a group may assist strategy development and explanation. Thus, students with MLD are better engaged in standards-based inclusive classrooms with multimodal problem-solving approaches (Faragher & Clarke, 2016; Lambert & Sugita, 2016).

2.2   Overview of Learning

Three models of learning and problem solving are particularly relevant to students
with MLD. The first model (Osborne & Wittrock, 1983)[1] emphasizes the role of
perception and memory in learning. It indicates students will test their immediate
perceptions and conceptual development with long-term memory. Not only are perceptions variable, e.g., visual, verbal, or kinesthetic, but also long-term memory
varies since it may be retrieved more as a visual or a verbal memory. Difficulties
might occur with any one of these aspects or their connections and so having this
overview is particularly useful when thinking about a particular student (Fig. 2.1).

Osborne and Wittrock (1983) and Goldin (1987) emphasized visual and verbal
processing for learning, but Goldin showed the importance of affective characteristics in problem solving and indeed learning that is seen as active problem solving.

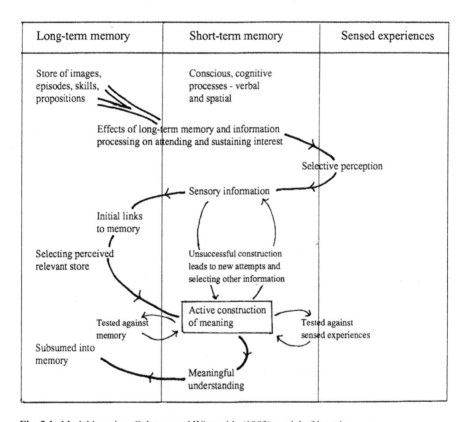

Fig. 2.1 Model based on Osborne and Wittrock's (1983) model of learning

---

[1] This model is influenced by information processing theories.

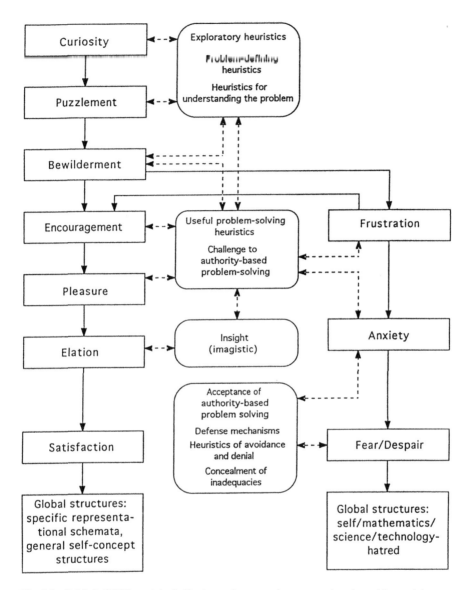

Fig. 2.2 Goldin's (2000) model of affective pathways and representations in problem solving

"Problem solvers use affect to make planning and heuristic choices" (Goldin, 2000, p. 214). The interactions of columns 1 and 2 in Fig. 2.2 indicate that students progress from the initial curiosity and puzzlement by exploring the situation. Having clues prepared to help students through a problem or inquiry encourages them through a mental block, especially if they are unfamiliar with a particular part of the problem. Clues might be a diagram, definition, enabling question, or hint about trying something. For an inclusive classroom working in groups, these clues might be in a set place where any group can access them when needed. Commonly heard in mathematics classrooms in Australia is the encouragement "Are your brains sweaty?" indicating students have been sustaining effort to solve the problem. For students with MLD who are accustomed to putting in extra effort to overcome their difficulties, effort in mathematics problem solving or inquiry is a strength. Students may have imagery insights and gradually develop pleasure and establish both mathematics conceptual knowledge and a positive self-concept. Utilizing expected learning trajectories to plan inquiry tasks (Johnson et al., 2017) and hints to avoid discouragement is important for positive attitudes and beliefs.

One approach for task design is to take account of Pirie and Kieren's (1994) model (Fig. 2.3), which shows that for learning, a primitive knowing occurs before the making of an image that is held and used for noticing properties. Encouraging students with experiences for developing imagery, noticing and articulating what they notice, comes before generalizing or setting up the method. Attention to appropriate features and assessing for relevance may require the student to return to an earlier stage (called folding back) before moving forward, reminiscent of the checking in Osborne and Wittrock's (1983) model (Fig. 2.1).

A summary model of learning (Fig. 2.4), which emphasizes the ecocultural context for mathematical learning, is based on the above models together with Owens

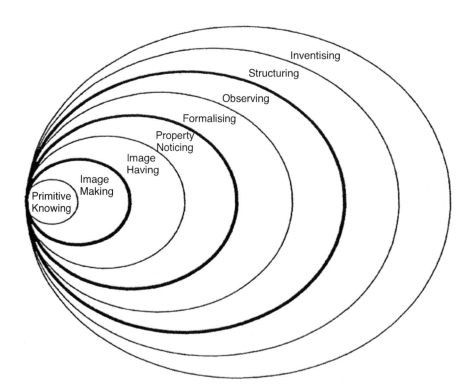

Fig. 2.3 Pirie and Kieren's (1994) model of learning

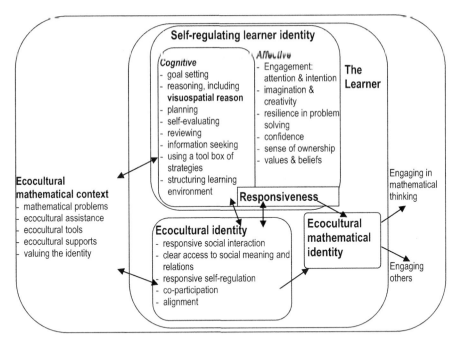

Fig. 2.4 Owens (2015) model of self-regulation and mathematical identity

(2015) ecocultural (cultural and ecological) research and that of Jonassen et al. (1999) on self-regulation. An ecocultural context that values all students developing a mathematical identity relevant to their circumstances will have a positive impact on their mathematical development and identity. The multifaceted description of self-regulation indicates that it is more than just controlling distractions that take the learner's attention away from the focus of learning (Lonigan et al., 2015).

Student responsiveness results from ecocultural identity and the cognitive and affective aspects of their learning. They develop a wide range of emotional and affective characteristics that impact on their learning. In particular, two aspects of engagement, attention and intention, influence responsiveness. Further, perception by which a students' surroundings (ecocultural environment) enter cognition is a significant aspect affecting the learning of students with MLD and often plays a part in their difficulties. However, perceptual modalities are malleable resulting in improved capabilities (Sinclair et al., 2016).

The aspects of self-regulation that are key to learning and establishing a mathematical learning identity are the ability to use strategies, make decisions and have ownership of mathematical thinking and learning despite the complexities that arise from context and relationship building. Further, students' responsiveness to the mathematical inquiry situation is to be valued. It is important to realize learning occurs through natural inquiry, as well as through guided learning in schools and other settings. Through these opportunities students can develop self-regulation, be empowered, and develop an identity related to mathematics learning.

Self-regulation is complex and includes executive control such as organizing capabilities as well as impulse control, working memory, and attention shifting (Joswick et al., 2019; Lonigan et al., 2015). Adding to this complexity, there may be developmental delays in either language or mathematical capabilities as well as other language issues such as cultural and language differences or speech and hearing issues.

These general understandings about learning are further explored by expanding on the following characteristics of learners, how these characteristics impact mathematics learning, how to manage the learning and to build on learners' hidden strengths:

- Self-regulation: Executive function and organizing capabilities
- Perceptual modalities and memory
- Incidental learning and intuition
- Verbal reasoning
- Responsiveness: Attention and intention
- Emotional and other affective characteristics
- Inquiry, ownership, self-regulation, empowerment, and identity

2.3   Self-Regulation: Executive Function and Organizing Capabilities

Self-regulation, managed through the executive function, involves heuristics that work toward students' goals. This function is a multi-faceted capability (Cirino & Willcutt, 2017). It includes controls on acting impulsively or shifting attention away from the learning space, working memory, and cognitive flexibility or deliberately shifting attention (Joswick et al., 2019). If a student shows unhelpful repetitive behavior then encouraging a slight shift in attention may harness persistence and encourage cognitive flexibility. Joswick et al. (2019) provide an example of a game to increase attention and flexibility while also reducing impulsive behavior. In this game, the player has to remember the position of cards as if by x-ray vision but actually by focusing on the distance the numbers are from the end of the line of numbers. Attention shifting and cognitive flexibility benefit students classified as hyperactive or lacking attention, when shifts in attention are noticed by the student and reflected on to see if they assist with the problem at hand. Generally, thinking flexibly is regarded as a positive for mathematical problem solving and thinking (De Corte et al. 2011).

Self-regulation and executive function involve planning, attention, and simultaneous and sequential processing. Difficulties with simultaneous or sequential processing are not necessarily associated with a particular kind of disability (Kroesbergen et al., 2003). However, it is important to be aware of these issues and take account of them when planning and teaching individual students. Students who process visual inputs simultaneously may shift attention readily. However, those with visual impairments who process sequentially or use kinesthetic learning tools may scan or connect more slowly to existing schema in long-term memory. Further learning structures are needed to assist their attention and flexibility.

Executive function also relates to adaptive motivational patterns that promote the establishment, maintenance, and attainment of personally challenging and valued achievement goals (Dweck, 1986). Student learning is focused on seeking challenges and persisting in the face of obstacles. By contrast, a maladaptive motivational pattern is seen in those who fail to establish reasonable and valued goals, display resistance toward the achievement of such goals, or set goals that easily sit within their reach. This is also described as a "helplessness" pattern with low persistence and challenge avoidance, showing negative cognitions when faced with adversity or challenge (Dweck, 1986, p. 1040). During mathematics lessons, such students behave as if they believe they are powerless to influence the outcomes of their learning (Yates, 2009). However, teachers may encourage these students to set small achievable challenges and increase self-regulation. Interestingly, Joswick et al. (2019) concluded that participating in mathematical inquiry will develop executive function.

2.4 Perceptual Modalities and Memory

Students vary in their capacities to perceive. While sight, hearing, touch, and movement senses are important in learning mathematics, how this information is received in the brain varies. Perception and visuospatial capabilities[2] are complex and include scanning, panning, zooming, rotating, reflecting and translating images, inspecting and classifying patterns, folding complex figures, combining figures, and seeing parts of figures (Kosslyn & Pomerantz, 1977; Pylyshyn, 1994; Tartre, 1990). In addition, perceptual speed, serial integration, closure speed, visual memory, and kinesthetic integration are involved (Owens, 2015). The term "visuospatial" recognizes that mental imagery may develop from various perceptual sources (e.g., physical actions, touch, sight, and sound) and that visualizing has a spatial component (arrangement and meaning in 2D and 3D space). Visuospatial reasoning incorporates "the ability to represent, transform, generate, communicate, document, and reflect on visual information" (Hershkowitz, 1990, p. 75). It involves developing and relating concepts to physical embodiments and computer-generated imagery through which students develop certain conceptualizations (Bauersfeld, 1991). Teachers need to be aware that visuospatial capabilities may not be readily described by standard testing (Newcombe, 2017), but an awareness of these aspects will assist teachers to notice particular children's difficulties and so assist them.

---

[2] Research on spatial abilities has expanded so the term visuospatial capabilities represents the research better.

2.4.1   Visuospatial Capabilities and Number Learning

Many training studies have shown visuospatial capabilities are malleable through experiences involving touch, movement, sight, and recognition of the position of the body (Sinclair, et al., 2016). Further, early intervention and recognition of a perceptual difficulty may lead to strengthening learning through alternative pathways. For example, a very able student with good sight but poor visual perceptual connectivity in the brain found it difficult to learn new material by reading. He participated in physical balancing activities and learned well from aural perceptual connectivity. With his family and later computer-aided facilities to read textbooks and notes, he completed his schooling, university, and doctorate and became a world-renowned professor in his business management field (name withheld for privacy reasons, personal communications, 1983, 1993, 2020).

When there was an emphasis on learning to count to understand number, studies showed students may rote learn basic counting and arithmetic but develop understanding later (Reisman & Kaufman, 1980). More recently, there have been alternatives to counting to learn about number size and arithmetic. Sense of size seems to be a natural ability (Yen et al., 2017), which is often associated with visualizing and an approximate number system (Geary, 2013). Physically comparing sizes assists this sense. Young students may numerically order or position numbers on a visual line to understand number size (Sella et al., 2017). They may then appreciate "greater than" and "smaller than" without counting and relying on counting order. Difficulties with remembering counting order may create anxiety in students. Thus alternative ways are needed for developing associations in early arithmetic such as learning doubles or chunking a group such as five within the group or making associations with the end number of the group such as 8 being two number before 10.

One aspect some students use is the ability to subitize numbers, that is, recognizing the number of a group without counting, just by a quick look. Students can innately distinguish 2 from 1 and 3 but other small numbers are quickly recognized and indeed quite large numbers may be subitized if configured in familiar patterns (Miravete et al., 2017; Moscoso et al., 2020). Their visuospatial capability encourages recognizing, say 5 and 1, and associating it with 6 (see Chap. 12 for more details) although spacing between objects or dots may be a difficulty (Macdonald, 2015).

Visual representations can provide a holistic view of a mathematical concept. When a particular model is consistently used with the length(s) varying then a holistic approach cements the concept. For example, knowing multiplication deals with multiple equal groups permits equal concrete rods, tapes, virtual rectangles, or jumps along the number line to be repeated to represent the multiplication. Using the empty number line or the tape or rectangular model for arithmetic word problems (see Xin, 2012; also Chaps. 13 and 14 in this book for this conceptual model) provides a consistent visual representation by which the arithmetic operations may be established. Interestingly, Soltanlou et al. (2015) found multiplication task performance was correlated with verbal working memory (WM) in third graders but with visuospatial WM in grade four. Visuospatial processes develop from pictorial static images to ones that are dynamic and associated with patterns, playing an increasingly important role in enhancing mathematical proficiency (Owens, 2015).

Visuospatial WM processes are important for the representation and manipulation of quantity information. Visuospatial WM is part of the executive function and helps scaffold the early stages of learning by providing support for building new semantic representations (Soltanlou et al., 2015). The central executive component is also required at subsequent stages for more complex problem-solving procedures, including the active maintenance of intermediate results and rule-based problem solving (Menon, 2016). The phonological loop may be essential in remembering word order when counting but the central executive and attention filters are more important in calculations that require some updating in the WM (e.g., in exchanging 1 ten for 10 ones) (Iuculano et al., 2011). Visualizing also assists students to focus on key facts and avoid the expectation that rote learning will result in recall because memory can be a major issue if too many steps are involved (Bird, 2013). Students will fold back to their visual imagery (see Pirie & Kieren's model in Fig. 2.3) to focus on the needed fact rather than fail in rote learned recall. Visual methods such as the empty number line mentioned above, and area models for fractions and multiplication presented in later chapters, are important holistic tools once manipulatives have been used to support meaning making.

Students with hearing impairments who use sign language and blind students who have heightened spatial awareness may have a strength in visualizing differences in shapes and size of spaces (van Dijk et al., 2013). In another study, deaf children were able to explain their mathematical reasoning in solving problems through drawings and diagrams (Nunes & Moreno, 2002). Faragher and Clarke (2014) found children with Down syndrome could achieve size differences visually but were unable to recall counting word order. Many students with MLD were able to visualize the mathematics involved and draw the mathematics or make a model (Lafay et al., 2016).

Some students have low visuospatial capability that is manifest when they are required to respond in a written form. Venneri et al. (2003) found students with poor visuospatial capability could carry out a number of calculations orally at the same level as those with average visuospatial capability but had difficulty with written subtraction involving the equal addend method in which 10 is added to the ones column at the top and to the tens column at the bottom (referred to as "borrowing and paying back"). As an alternative approach to this 3-digit vertical algorithm for number subtraction, the use of the empty number line or box method (see above and Chap. 13) could be more fruitful despite involving visuospatial capability because it promotes better connections to physical representations and meaning. Hence, the utilization of a multi-perceptual approach to teaching is important to ensure students with differing capabilities can participate. Using only symbols and numbers was found in a longitudinal study on magnitude processing to be an issue rather than WM for children with persistent MLD (Vanbinst et al., 2014).

If there is very limited visual sensory receptiveness, then another input may be required, such as touch assisting people with limited vision (e.g., to read Braille).

An alternative perceptual route may be improved under careful training and motivating circumstances. There may be mathematical occasions when this is particularly relevant. Healy and Fernandes (2011) argued that there is a need to take account of immediate cultural circumstances that arise from limited vision:

> gestures are illustrative of imagined re-enactions of previously experienced activities and that they emerge in instructional situations as embodied abstractions, serving a central role in the sense-making practices associated with the appropriation of mathematical meanings. (p. 157)

Providing physical experiences of mathematical concepts for students with limited vision enables concepts to be perceived and mentally reasoned with (visuo)spatially. The activity can mediate mental change through the use of a manipulative tool.

> The inclusion of the tool in activity hence alters the course both of the activity and of all the mental processes that enter into the instrumental act. In this way, tools do not only facilitate mental processes, they transform, they re-organise and they shape them. (p. 158)

One difference from the holistic perception possible through sight is that touch perception is sequential with a gradual analysis from the parts to the whole. However, the tool empowers a different yet effective cognitive practice. Coordination within the embodiment using touch, tools, and gestures is one form of making mathematical meaning along with speech and symbols (Radford, 2009).

2.4.2   Visuospatial Reasoning, Fractions, and Proportional Reasoning

Visuospatial reasoning, attending, and overall sense of size are important for fraction learning. Doing and making fractions by folding paper establishes the visuospatial relationship of the part to the whole so the kinesthetic perception strengthens the visual perception. However, Fuchs et al. (2017) have shown it is important to concentrate on the magnitude of the fractions for students to learn effectively about them. A group of at-risk students were physically and visually noting the size of different fractions by comparing areas and positioning them on a number line. Over the 5 years of the study, this group surprisingly outperformed the regular class in improvements on number magnitude, addition and subtraction of fractions, and conceptual understanding of fractions in terms of size and part-whole meaning. Not only is it helpful for students moving on to ratio and rate problems to have a sense of the magnitude of a fraction represented on the number line but also for them to understand that equivalent fractions are multiple ways of expressing the same fraction (see Chap. 15 for more details).

Another study of the value of visualizing for ratios or scaling was carried out by Möhring et al. (2015). In a map task, 4- and 5-year-olds (N = 50) were asked to point to the same position shown on a map in a larger referent space on a touch screen. The sizes of the maps were varied systematically, such that some trials required scaling and others did not. In a proportional reasoning task, children were presented with different relative amounts of juice and water and asked to estimate each mixture on a rating scale. Again, some trials required scaling, but others could be solved by directly mapping the proportional components onto the rating scale. Children's errors in locating targets in the map task requiring scaling were closely related to their performance in the proportional reasoning task even after controlling for age and verbal intelligence. These results shed light on the mechanisms involved in the close connection between spatial and mathematical thinking early in life (Möhring et al., 2015).

2.4.3 Visuospatial Reasoning and Holistic Learning

Schema and holistic learning are important for grounding conceptual development. For this reason schema-based instruction provides a teaching strategy that encourages students to "recognize underlying problem structures, represent problems using visual-schematic diagrams, plan and solve problems, check the reasonableness of answers", think about what they are doing and thinking (often called metacognition), and make connections with other knowledge (Im & Jitendra, 2020, p. 5). The researchers' schema-based instruction helped students organize their thinking, decide if they were doing a part-part or part-whole proportional reasoning question, represent the problem visually and record initial numbers so there was an opportunity to devise a plan to solve the problem.

Visual representations were less likely to benefit students with lower mathematical ability or younger students (Booth & Koedinger, 2012). For this reason, it is important to supplement with language and active inquiry so the initial steps of finding out about the concept can be initiated and the general abstraction or sorting out of commonalities and differences may then be grasped by students. At the same time, students will have some ownership and self-direction of the work involved. The bigger picture or schema is not lost and the problem for inquiry keeps them engaged (Owens, 2015, p. 257).

2.5 Incidental Learning and Intuition

One issue for students who have limited hearing or sight is that some incidental learning may not occur (Norris & Norwich, 2005), as the field of perception may be limited. Incidental learning was considered by van Hiele (1986) as the key to understanding children's intuitive learning. Teacher awareness of this feature is important so students can begin the important primitive knowing mentioned in the Pirie and Kieren (1994) model (Fig. 2.3). Similarly, a learned difficulty may arise for students who have developed their own conceptual understanding based on limited experiences. They may have missed some schooling or been physically unable to participate at the same rate as other students. Procedural teaching can reduce mathematics learning for students with these limitations. By contrast, conceptual teaching and asking students the right question about what they understand about the problem situation and why they carry out a procedure will focus their attention on relationships and concepts rather than rote-learned actions (Booth, 2011). As Siemon et al. (2015) point out:

> A useful approach is to ask two questions: What have you tried? What are you trying to find out? These questions allow students to summarise what they have done, often clarifying their approach to themselves, and also to look to what they need to find out. These questions also allow the teacher time to understand the approach the learner is taking. (p. 156)

When these questions are asked any missed incidental learning or mislearning may be noticed. Intuitive beginnings can be an advantage in getting started but they may also make some later mathematics more difficult. For example, a problematic intuition occurs for children who see an equals sign as related to something happening on the left side creating the right side so a statement like $10 = 7 + 3$ fails to have meaning (Kieran, 1981). For this reason, students need to fold back to an earlier level of image making such as the balance for equality (Pirie & Kieran, 1994) to develop their understanding of the equality of this statement. Then the visual symbol for equality is not limited to getting an answer to an operation.

Intuitive understanding of repeated addition for multiplication does not apply easily to multiplying by a decimal fraction or a number greater than one with a decimal in which the unit digit face value is small compared with the decimal fraction (e.g., 1.25 is harder than 3.25) (Fischbein et al., 1985). For division, decimals become more problematic in the construct of division as equal parts removed or equal sharing. When the decimal is the operator, the lack of intuitive understanding is worse. The teacher needs to bridge from multiplication and division with first concrete and then visuospatial representations with the rectangular area model and extend this for the fraction and decimal fraction to aid intuition or in Pirie and Kieren's model (Fig. 2.3) making a visual image upon which to return if needed in order to move forward. Visuals such as shown in Fig. 2.5 for 1.25 x 3 help.

Perceptual difficulties occur with numerical transcoding, that is reading written decimals, and with conceptual transcoding, that is connections between representations of percentages, decimal fractions, and vulgar fractions (Berch, 2016). Additional time is needed to make these connections. Thus, the intuitive proportional reasoning sense that is said to exist is not sufficient to assist with these perceptual and conceptual transcodings. Spatial awareness of proportions and proportional reasoning (see Fig. 2.5) builds on intuitive understanding of proportions and seems to be a fruitful way forward for students for whom the symbolic does not contain much meaning (Matthews & Hubbard, 2017).

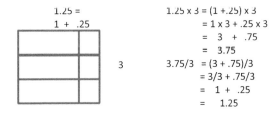

Fig. 2.5 Visualizing multiplication and division of decimals

$$1.25 =$$
$$1 + .25$$

$$3$$

$$1.25 \times 3 = (1 + .25) \times 3$$
$$= 1 \times 3 + .25 \times 3$$
$$= 3 + .75$$
$$= 3.75$$
$$3.75/3 = (3 + .75)/3$$
$$= 3/3 + .75/3$$
$$= 1 + .25$$
$$= 1.25$$

Butterworth (2018) continues to show that innate numerosity that acts like an accumulator (linked to active[3] spatial memory) varies from person to person. Low numerosity levels may be hereditary or cultural and not linked to general cognitive capabilities. Culture provides the language to support numerosity for identifying numerical size and calculations. Early recognition and intervention is needed to support students with active spatial numeracy for establishing relative size. However, this point raises the issue of verbal reasoning and cultural and language differences.

2.6 Verbal Reasoning

2.6.1 Awareness of Cultural and Language Differences

Students with cultural and language differences may also require teachers to adjust their strategies. Most of these students, even if able, may feel undervalued unless the teacher and peers are aware of the importance of learning and adjusting to new cultural and linguistic requirements of schooling and mathematical ways of learning (Atweh et al., 2011). Tiredness is common, as the student is often processing so much more than the rest of the class, having to establish meaning in two languages. Parental expectations and assistance vary, but may include extra tuition, additional learning similar to their own schooling, and different ways of working mathematically (Planas & Civil, 2013). The teacher needs to understand the student better and parents should be invited to assist in planning appropriate learning and sharing of cultural mathematics, realizing that this will benefit the whole class.

Murray (n.d.) provided a number of approaches for assisting students with language differences. For example, she encouraged teachers and students to understand technical mathematical words by reading, saying, and writing them, to understand comparative words like "more than" or "less than", and to play games in which pairs of students take turns with one reading words or definitions and the other choosing a picture or drawing, or one describing the picture and the other giving the word to go with the description.

---

[3] Activities like jumping and moving objects a number of spaces, arranging objects into patterns, or estimating and walking a longer or shorter length assists this memory.

2.6.2 Complexity and Variability in Verbal Reasoning

In a comprehensive study analyzing a range of measures in a large cohort of children, Peterson et al. (2017) confirmed that although numerosity difficulties can be independent of verbal capabilities, verbal comprehension accounts for more difficulties in language and mathematics than attentiveness. A large study, with data collected over 24 years (Barbaresi et al., 2005), showed that difficulties with mathematics (between 7% and 14% diagnosed each year) correlated with reading difficulties (between 43% and 64% of this MLD cohort), usually with left-right visual issues. Other factors included slower than usual developmental cognitive growth, low sight, aural perceptual capabilities, or movement difficulties.

Students with Down syndrome (DS) present a similar range of reading difficulties as other students (Næss et al., 2011). Næss et al.'s meta-analysis of 15 articles showed children with DS have low capabilities in all language domains except receptive vocabulary. This may be due to the generally low cognitive skill required for nonverbal responses, such as pointing to a picture, whereas measures of grammar and expressive vocabulary require multiple levels of processing, such as defining and pronouncing words, and understanding situations and pictures. Grammatical skills are cognitively more complex including sentence memory and listening skills. Children with DS have lower verbal short-term memory compared with those of the same non-verbal mental age. Hearing loss occurs in about two-thirds of children with DS (Roizen, 2002). It is therefore important to recognize how these issues may impact on mathematics learning. For example, understanding word problems or remembering a string of multiplication facts if there are no associated relationships and patterns makes learning difficult. In addition, words in mathematics that may sound the same, like "side" and "size," need to be carefully articulated and the meaning explained in a way that is supported by non-verbal physical or visual explanations.

Some students vary with their ways of understanding words. For example, some students with an autism spectrum disorder (ASD) are more likely to use words literally rather than metaphorically. For this reason, some of the metaphorical and technical language of mathematics such as slope, tangent, or area may be misunderstood. General receptive and expressive communication may also be difficult. However, a meta-analysis of 18 studies showed the average mathematical capabilities for high functioning students with ASD to be just below standardized expected results. This was considered clinically manageable[4] (Chiang & Lin, 2007). The proportion of students with ASD who were below average in arithmetic (around 23%) was far less than for reading difficulties (around 60%) (Mayes & Calhoun, 2006).

Teachers have found specific language difficulties can be managed with strategies such as using socially appropriate, pragmatic language for communication; clear, stepped tasks or responses; consistent expectations; and avoiding students being overwhelmed by words, noise, other stimuli, or change. Thus the processing

---

[4] There may be some students who are outstanding mathematically.

of sensory information needs to be considered. Tiredness may necessitate further breaks, interests may need to be established and other speech difficulties may also need addressing (Stokes et al., 2017). Reading and understanding teacher and peer gestures is also relevant whether as social communication or in mathematical explanations. Visuals (gestures, objects, and pictures) for activity and concepts may assist and need to be associated adequately with meaning. There may be gender differences in the occurrence of some characteristics, such as more boys than girls having to learn socially appropriate pragmatic communication.

Another language area is that of expressive communication, which may also vary. For some this is physical or it may be the lack of cortical executive control. For others, augmentative or assistive technology may provide the means for expressing thoughts and words but some education for associating the output of the technology with its purpose may be required. For explaining early mathematical thinking, concrete materials may be moved or arranged appropriately to provide expressive communication, and software may replace handwritten mathematical symbols. Some of the outputs of technology are also learning tools. For example, having a calculator add the same number repeatedly[5], the screen produces the numbers for group (or skip) counting, that is multiples of the number. This may be an effective learning tool. Similarly, work with a program like *GeoGebra* assists students to see tables and graphs develop together and to associate these with algebraic expressions to carry out inquiries with technological assistance.

2.7 Responsiveness: Attention and Intention

As students use various heuristics for inquiry, their responsiveness depends on their attention and intention (Owens & Clements, 1998). Intention is particularly influenced by the ecocultural identity and affective characteristics of the learner. Attention results from past experiences and knowledge but also the immediate context and perceptions (see Fig. 2.1 and Fig. 2.4). In Pirie and Kieren's (1994) model (Fig. 2.3), and in common with many current mathematics discussions on inquiry, the importance of noticing certain aspects of a mathematical situation is highlighted (see, for example, Owens, 2020). Chief among these is noticing patterns, especially when establishing efficient arithmetic operations, a geometric shape category, algebraic relationships, and proportional reasoning.

Focus of attention is often considered an issue for students with MLD. Furthermore, attention may be influenced by the surroundings of the image, object, or language being learned. Complexity, colors, directionality, multiple, or limited sensory inputs can be distractors. However, the role of attention and how it can be influenced is worth considering. There are some in-the-moment approaches, such as carefully planned encouragers such as hints, useful in-the-moment information, and

---

[5] For example, +3 = = = provides the multiples of 3.

mirroring the students' words or actions, to improve immediate attention. Noticing is encouraged by attending to expectations. For example, if students know the number of sides often provides the clue for naming a shape, then they will attend to and notice the number of sides. If students know a good procedure for adding numbers is to count on from the bigger number then they will look for and notice the bigger number. These are executive controls or heuristics.

Attending to relevant features can be influenced by intentions, affective characteristics, and the past and immediate contexts including teachers, peers, physical objects, routines, and expectations. Contexts can be culturally established, so it is important for teachers to ensure mathematics is made as culturally relevant as possible. The culture and climate of the classroom is also significant. Teachers' modeling of positive affects and facilitative attitudes, beliefs, expectations, and attributions can encourage students to enjoy mathematics and be confident in the learning situation (Yates, 2009). A desire to take ownership of the mathematics and to try out something new in an inquiry will help students to attend. Hence, a student who is less attentive needs to have a strong intention to do the mathematics, which can happen if the mathematics is relevant.

Attention is also assisted if the immediate recall of memory related to the inputs can be readily drawn upon. These might be a word, object, or diagram, preferably chosen by the student or otherwise very familiar to them. For example, a student who readily remembered even numbers (2, 4, 6, etc.) and loved cars had difficulties with multiplication concepts. However, he could count the four wheels on cars rhythmically (2, **4**, 6, **8**, 10, **12**) in order to count in fours. His prior counting interest and social interest were used for the multiplication concept and his mathematics bloomed. Some students are particularly good at remembering sequences or seeing patterns, for example, students with ASD, and these strengths should be utilized by the teacher.

Attentional control is important in keeping information relevant to the problem in mind, no matter how basic, while processing other information (Goldin, 1987; Owens, 2015). Attentional control impacts during early number learning on both "numeral magnitude mapping and early explicit number system knowledge" (Geary, 2013, p. 24). The already existing approximate number system (Pirie and Kieren's image making, Fig. 2.3) influences thinking about magnitude and involves the recognition of differences in the size of numbers. The larger the difference, the more likely students with MLD will recognize the different sized numbers. Symbols and words begin to have meaning when associated with numbers in an approximate number system. By comparison, the explicit number system incorporates knowing, for example, the number one more than or one less than a given number. Imagining patterns will draw attention to size differences for close numbers as illustrated in Fig. 2.6.

Mathematical processing can also be encouraged through triggers. These might be a word or acronym, such as letter mnemonic strategies like SOLVE "**S**tudy the problem, **O**rganize the facts, **L**ine up a plan, **V**erify your plan with action, **E**xamine

Fig. 2.6 Patterns for comparing numbers with little difference in quantity

your results" (Freeman-Green, O'Brien, Wood & Hitt, 2015). Similar steps may be used regularly, like the inquiry approach of "tuning in, finding out, sorting out, going further, act and reflect" (Murdoch, 2019), or a flexible diagram like the conceptual model method for word problems (see Chap. 13).

2.8 Emotional and Other Affective Characteristics

Emotional issues may arise for students with MLD. They may internalize and externalize problems (Prior et al., 1999). In a six-year longitudinal study, 29 students who continued to have severe issues with number and arithmetic were matched for gender and intelligence test scores (minus the arithmetic subtest) with students who had mild issues with arithmetic (Auerbach et al., 2008). All students showed some difficulties with attention and externalizing behaviors but these difficulties were less evident in those whose number and arithmetic knowledge improved slightly. This was a comparative study in which gender, language difficulties, and attention-deficit hyperactivity disorder symptoms were taken into account. Compared with the norms given in the testing manual, at the start of the study males had more internalizing, externalizing, and attention difficulties whereas females showed difficulties only with attention. In terms of the model of mathematical thinking identity, it is argued that a persistent severe difficulty with arithmetic is likely to indicate a low sense of identity that is portrayed in externalizing behaviors rather than in using mathematics productively.

MLD may be a result of emotional states such as anxiety, frustration, and disinterest that may have developed in association with another cognitive, physical, or social difficulty (Smith et al., 2012). Anxiety commonly occurs when students need to complete an activity under time constraints, lack procedural knowledge, or find problem-solving onerous. If teachers encourage students to go from a basic strategy like the conceptual model method (Chap. 13), it increases understanding and success but it might take them longer to complete a question than students who associate a word problem immediately with the required mathematical operation and quickly recall the appropriate number fact. Students who take longer to process sensory information, recall conceptual or procedural knowledge, or compare themselves to others can become anxious and/or frustrated. Anxieties can then mask students' capability and reduce the opportunity to learn (Chinn, 2009). The teacher needs to recognize the anxiety and its cause by knowing the mathematics, how students might learn it, and providing what might be the best approach for the particular student to learn the topic and overcome the anxiety. Timed activities and unnecessary repetitive activities like rote learning may both cause anxiety.

Negative affective responses to learning resulting from consistent failure to get a correct answer or feeling like a failure compared with others may result in avoidance behaviors or intentions that are unrelated to working mathematically, such as expressions of helplessness (McLeod, 1992). The student may ask a question before the teacher finishes stating the problem or explanation, expect help at every step, or show frustration, refusal, sadness, disruption, or disinterest (Yates, 2009).

However, avoidance and anxiety behaviors may also be associated with the teacher dismissing students' different approaches to doing the problem. The teacher needs to problem solve to decide if the strategy choice is mathematically valid (Hopkins & de Villiers, 2016) or can be linked before dismissing the approach (Allsopp, 2018). Students, too, can be given an opportunity to explain, for other children to see how it relates to their method or to self-evaluate the efficacy of their approach in the light of all students' approaches (Hopkins & de Villiers, 2016). In fact, disaffection may be giving a voice to the student (Hartas, 2011). Until some of these affective triggers can be broken, it is difficult to assess other capabilities needed for learning. For example, by attributing the reason for limited achievement to effort rather than to lack of ability or luck (Yates, 2000), students may overcome their sense of helplessness that has been learned. Goldin (2000) suggests these negative affects can build up over time but one "aha" moment can move students from negative responses to positive responses (Fig. 2.2). Teachers can assist students to view failure less catastrophically and cope productively with frustration and discouragement (Brophy, 1998).

Adults who are mathematically anxious are poorer than their non-anxious peers at counting objects, deciding which of two numbers represents a larger quantity, and mentally rotating 3D objects (Chang et al., 2017). People who lack knowledge in a particular domain are often easily swayed by negative self-assessment so children who start formal schooling with limited or different mathematical foundations may be especially predisposed to pick up on social cues (e.g., their parents' or teacher's behavior) that highlight their mathematical attempts in negative terms (Gentile & Monaco, 1988). Students with language and cultural differences may also develop less than positive attitudes toward their mathematical thinking just because they struggle with word meanings.

2.9 Inquiry, Ownership, Self-Regulation, Empowerment, and Identity

One characteristic of students with MLD that is often overlooked is their identity. A positive mathematical identity can develop through students' responsiveness to the inquiry or problem-solving situation (see Fig. 2.4). As they self-regulate their learning through their visuospatial reasoning and heuristics, together with verbal understanding and affective aspects such as a sense of self-efficacy (Pajares & Miller, 1994) and self worth (Covington, 1992), they became responsive and develop a positive mathematical identity. The culture and climate of the mathematics classroom to which a student belongs and is valued assists to establish this identity. Through their empowerment, students can do mathematics and share with others. Highly structured drill and practice programs with small incremental and seemingly disconnected procedures reduce the opportunities to develop ownership of the mathematics, self-regulation, and mathematical identity.

Empowerment to do mathematics by students with special needs was exemplified by the students with Down syndrome who were expected to think mathematically with the rest of the class and the teachers noted that these students were indeed thinking mathematically. The teachers were focusing on developing each student's mathematics. The classroom milieu and planning for diversity and inclusion were evident in resources such as the enabling questions (or hints) available for all students in the classroom (see earlier comments). The teaching approach diminished labeling and maladaptive behavior (Faragher & Clarke, 2014, 2016).

Hanich (2011) explained repeated failure may result in maladaptive avoidance behavior or impulsive responding rather than controlling impulsive and unrelated responses. Teachers have a critical role in identifying learned helplessness as it is manifest in students' responses to situations of actual or perceived failure in the classroom (Yates, 2009) and in overcoming conditioned helplessness in mathematics (Wieschenberg, 1994).

Student responsiveness may also be assisted if they have a voice in planning for their learning. This may be subtly expressed by affective responses of interest and "having a go" as students become engaged. The student and teacher might problem pose and problem solve together for meaningful and often practical purposes (such as life after school). However, the teacher needs to be prepared to adjust the curriculum for the class (Leat & Reid, 2012). For example, in one school the curriculum was adjusted so that music and school musicals involved behavioral learning and mathematical actions providing a sense of belonging in the learning community and mathematical learning (Rusinek, 2008). In terms of diversity of cultural background, Civil and colleagues (e.g., Planas & Civil, 2013) have emphasized the funds of knowledge that students bring to the classroom from their home background, their ecocultural background. When teachers capitalized on these funds of knowledge, difference in terms of language was valued and positioning of the students in the class changed and there were improvements in mathematical learning. This applies equally to all students who are thinking differently. Looking at the affordances that difference brings to the classroom, rather than as deficits, is empowering when the different ways of learning are used as resources for learning (Planas & Civil, 2013). The studies referred to in this section indicate that autonomy, peer acceptance, and supports such as resources and group work in inclusive classrooms should not be underestimated in students' learning mathematics (Moriña Diez, 2010).

2.10 Conclusion

Learning begins with perceptions and testing newly constructed thoughts in working memory against long-term memory and perceptions (Osborne & Wittrock, 1983, Figure 2.1). Students vary in all these aspects of learning mathematics. Areas involved in learning include senses, perception, accessing schema, working memory, language, and functions such as attending, planning, and self-regulation. Learning begins with a preliminary knowing, developing into an image with properties that need noticing for the mathematics to be formalized in students' minds (Pirie & Kieren, 1994). The key to noticing is seeing patterns and relationships between examples and constructs of a topic, between representations and language or symbols. Teachers' awareness of these facets of learning, while working with a diversity of individuals, encourages the planning and in-the-moment adaptation of the mathematical inquiry.

Students' cultural and language backgrounds, together with the classroom climate, have important roles in mathematical learning for all, and especially for students with MLD. An inquiry approach, together with other learned approaches to supporting oneself and the group, enhances the opportunity for students with special needs to think mathematically and be integrated into inclusive classrooms. The affective aspects of self-regulation along with heuristics for problem solving, no matter how basic, are to be considered by teachers to ensure students' sense of being mathematical thinkers and belonging to a community of mathematical thinkers (Atweh et al., 2011; Owens, 2015).

In the following chapters, specific supports for particular areas of mathematics are presented, drawing on a general understanding of the learning principles outlined in this chapter. In many cases, holistic and visuospatial approaches are recommended. Connecting information and valuing mathematical inquiry for all students is important. A sense of being able to work mathematically and belonging to the mathematics classroom, strengthened by teachers' planning and approach to the classroom learning environment, becomes critical.

Often, teachers need to work with other professionals, parents, and augmentative resources in order to notice strengths and ways of learning for individual students with MLD. Building on those strengths, the following chapters provide the mathematical understandings and research to assist teachers to plan students' learning of specific topics.

References

Allsopp, D. H. (2018). *Teaching mathematics meaningfully: Solutions for reaching struggling learners* (2nd ed.). Brookes Publishing Co.

Atweh, B., Graven, M., Secada, W., & Valero, P. (2011). *Mapping equity and quality in mathematics education* (1st ed.). Springer.

Auerbach, J. G., Gross-Tsur, V., Manor, O., & Shalev, R. S. (2008). Emotional and behavioral characteristics over a six-year period in youths with persistent and nonpersistent dyscalculia. *Journal of Learning Disabilities, 41*(3), 263–273. https://doi.org/10.1177/0022219408315637

Barbaresi, W. J., Katusic, S. K., Colligan, R. C., Weaver, A. L., & Jacobsen, S. J. (2005). Math learning disorder: Incidence in a population-based birth cohort, 1976–82, Rochester, Minn, *Ambulatory Pediatrics, 5*(5), 281-289. https://doi.org/10.1367/A04-209R.1

Bauersfeld, H. (1991). The structuring of the structures: Development and function of mathematizing as a social practice. In L. Steffe (Ed.), *Constructivism and education* (pp. 1–26). Lawrence Erlbaum.

Berch, D. B. (2016). Why learning common fractions is uncommonly difficult: Unique challenges faced by students with mathematical disabilities. *Journal of Learning Disabilities, 50*(6), 651–654. https://doi.org/10.1177/0022219416659446

Bird, R. (2013). *The dyscalculia toolkit*. Sage.

Booker, G. (2011). *Building numeracy: Moving from diagnosis to intervention*. Oxford University Press.

Booth, J. (2011). Why can't students get the concept of math? *Perspectives on Language and Literacy, 37*(2), 31–35.

Booth, J., & Koedinger, K. (2012). Are drawings always helpful tools? Developmental and individual differences in the effect of presentation format on student problem solving. *British Journal of Educational Psychology, 82*, 492–512.

Brophy, J. (1998). *Motivating students to learn*. McGraw-Hill.

Butterworth, B. (2018). The implications for education of an innate numerosity-processing mechanism. *Philosophical Transactions of the Royal Society of London. Series B, Biological Sciences, 373*(1740). https://doi.org/10.1098/rstb.2017.0118

Chang, H., Sprute, L., Maloney, E. A., Beilock, S. L., & Berman, M. G. (2017). Simple arithmetic: Not so simple for highly math anxious individuals. *Social Cognitive and Affective Neuroscience, 12*(12), 1940–1949. https://doi.org/10.1093/scan/nsx121

Chiang, H.-M., & Lin, Y.-H. (2007). Mathematical ability of students with Asperger syndrome and high-functioning autism: A review of literature. *Autism, 11*(6), 547–556. https://doi.org/10.1177/1362361307083259

Chinn, A. (2009). Mathematics anxiety in secondary students in England. *Dyslexia, 15*(1), 61–68.

Cirino, P., & Willcutt, E. (2017). An introduction to the special issue: Contributions of executive function to academic skills. *Journal of Learning Disabilities, 50*(4), 355–358. https://doi.org/10.1177/0022219415617166

Covington, M. V. (1992). *Making the grade: A self-worth perspective on motivation and school reform*. Cambridge University Press.

De Corte, E., Mason, L., Depaepe, F., & Verschaffel, L. (2011). Self-regulation of mathematical knowledge and skills. In B. Zimmerman & D. Schunk (Eds.), *Handbook of self-regulation of learning and performance* (pp. 155–172). Routledge.

Dweck, C. (1986). Motivational processes affecting learning. *American Psychologist, 41*(10), 1040. https://doi.org/10.1037/0003-066X.41.10.1040

Faragher, R., & Clarke, B. (Eds.) (2014). Educating learners with Down Syndrome. Research, theory, and practice with children and adolescents. Routledge.

Faragher, R., & Clarke, B. (2016). Teacher identified professional learning needs to effectively include a child with Down Syndrome in primary mathematics. *Journal of Policy and Practice in Intellectual Disabilities, 13*(2), 132–141. https://doi.org/10.1111/jppi.12159

Fischbein, E., Deri, M., Nello, M. S., & Marino, M. S. (1985). The role of implicit models in solving verbal problems in multiplication and division. *Journal for Research in Mathematics Education, 16*(1), 3–17. https://doi.org/10.2307/748969

Freeman-Green, S., O'Brien, C., Wood, C., & Hitt, S. (2015). Effects of the SOLVE strategy on the mathematical problem solving skills of secondary students with learning disabilities. *Learning Disabilities Research and Practice, 30*(2), 76–90.

Fuchs, L., Malone, A., Schumacher, R., Namkung, J., & Wang, A. (2017). Fraction intervention for students with mathematics difficulties: Lessons learned from five randomized controlled trials. *Journal of Learning Disabilities, 50*(6), 631–639. https://doi.org/10.1177/0022219416677249

Geary, D. (2013). Early foundations for mathematics learning and their relations to learning disabilities. *Current Directions in Psychological Science, 22*(1), 23–27. https://doi.org/10.1177/0963721412469398

Geary, D., Brown, S., & Samaranayake, V. (1991). Cognitive addition: A short longitudinal study of strategy choice and speed-of-processing differences in normal and mathematically disabled children. *Developmental Psychology, 27*(5), 787–797.

Gentile, J. R., & Monaco, N. M. (1988). A learned helplessness analysis of perceived failure in mathematics. *Focus on Learning Problems in Mathematics, 10*(1), 15–28.

Goldin, G. (1987). (a) Levels of language in mathematical problem solving; (b) Cognitive representational systems for mathematical problem solving. In C. Janvier (Ed.), *Problems of representation in the teaching and learning of mathematics* (pp. 59-65, 125–145). Lawrence Erlbaum.

Goldin, G. (2000). Affective pathways and representation in mathematical problem solving. *Mathematical Thinking and Learning: An International Journal, 2*(3), 209–219.

Hanich, L. (2011). Motivating students who struggle with mathematics: An application of psychological principles. *Perspectives on Language and Literacy, 37*(2), 41–45.

Hartas, D. (2011). Young people's participation: Is disaffection another way of having a voice? *Educational Psychology in Practice, 27*(2), 103–115.

Healy, L., & Fernandes, S. (2011). The role of gestures in the mathematical practices of those who do not see with their eyes. *Educational Studies in Mathematics, 77*(2), 157–174. https://doi.org/10.1007/s10649-010-9290-1

Hershkowitz, R. (1990). Psychological aspects of learning geometry. In P. Nesher & J. Kilpatrick (Eds.), *Mathematics and cognition: A research synthesis by the International Group for the Psychology of Mathematics Education* (pp. 70–95). Cambridge University Press.

Hopkins, S., & de Villiers, C. (2016). Improving the efficiency of problem-solving practice for children with retrieval difficulties. In B. White, M. Chinnappan, & S. Trenholm, (Eds.). *Opening up mathematics education research: Proceedings of the 39th annual conference of Mathematics Education Research Group of Australasia, MERGA39*. Adelaide, SA, Australia: MERGA. https://merga.net.au/Public/Public/Publications/Annual_Conference_Proceedings/2016_MERGA_Conference_Proceedings.aspx.

Hunter, R., Civil, M., Herbel-Eisenmann, B., Planas, N., & Wagner, D. (2018). *Mathematical discourse that breaks barriers and creates space for marginalized learners*. SensePublishers.

Im, S.-H., & Jitendra, A. (2020). Analysis of proportional reasoning and misconceptions among students with mathematical learning disabilities. *Journal of Mathematical Behavior, 57*, 1–20. https://doi.org/10.1016/j.jmathb.2019.100753

Iuculano, T., Moro, R., & Butterworth, B. (2011). Updating working memory and arithmetical attainment in school. *Learning and Individual Differences, 21*, 655–661.

Jirout, J., & Newcombe, N. S. (2018). How much as compared to what: Relative magnitude as a key idea in mathematics cognition. In *Visualising mathematics* (pp. 3–24). Springer.

Johnson, H. L., Coles, A., & Clarke, D. (2017). Mathematical tasks and the student: Navigating "tensions of intentions" between designers, teachers, and students. *ZDM, 49*(6), 813–822. https://doi.org/10.1007/s11858-017-0894-0

Jonassen, D., Peck, K., & Wilson, B. (1999). *Learning with technology: A constructivist perspective*. Prentice Hall.

Joswick, C., Clements, D., Sarama, J., Banse, H., & Day-Hess, C. (2019). Double impact: Mathematics and executive function. *Learning and Teaching Mathematics, 25*(7), 416–426. https://doi.org/10.5951/teacchilmath.25.7.0416

Kieran, C. (1981). Concepts associated with the equality symbol. *Educational Studies in Mathematics, 12*, 317–326.

Kosslyn, S., & Pomerantz, J. (1977). Imagery, propositions and the form of internal representations. *Cognitive Psychology, 7*, 341–370.

Kroesbergen, E. H., Van Luit, J. E. H., & Naglieri, J. A. (2003). Mathematical learning difficulties and PASS cognitive processes. *Journal of Learning Disabilities, 36*(6), 574–582. https://doi.org/10.1177/00222194030360060801

Lafay, A., St-Pierre, M.-C., & Macoir, J. (2016). The mental number line in dyscalculia: Impaired number sense or access from symbolic numbers? *Journal of Learning Disabilities, 50*(6), 672–683. https://doi.org/10.1177/0022219416640783

Lambert, R., & Sugita, T. (2016). Increasing engagement of students with learning disabilities in mathematical problems-solving and discussion. *Education Faculty Research and Articles Chapman University Digital Commons*, *11–2016*.

Leat, D., & Reid, A. (2012). Exploring the role of student researchers in the process of curriculum development. *Curriculum Journal, 23*(2), 189–205.

Lonigan, C. J., Allan, D. M., Goodrich, J. M., Farrington, A. L., & Phillips, B. M. (2015). Inhibitory control of Spanish-speaking language-minority preschool children: Measurement and association with language, literacy, and math skills. *Journal of Learning Disabilities, 50*(4), 373–385. https://doi.org/10.1177/0022219415618498

Macdonald, B. (2015). Ben's perception of space and subitizing activity: A constructivist teaching experiment. *Mathematics Education Research Journal, 27*(4), 563–584. https://doi.org/10.1007/s13394-015-0152-0

Marcone, R., & Atweh, B. (2015). A meta-research question about the lack of research in mathematics education concerning students with physical disability. In S. S. Mukhopadhyay & B. Greer (Eds.), *Proceedings of the Eighth International Mathematics Education and Society Conference* (pp. 551–558). Portland State University.

Margetts, K., & Woolfolk, A. (2019). *Educational psychology* (5th ed.). Pearson Education.

Matthews, P. G., & Hubbard, E. M. (2017). Making space for spatial proportions. *Journal of Learning Disabilities, 50*(6), 644–647. https://doi.org/10.1177/0022219416679133

Mayes, S., & Calhoun, S. (2006). Frequency of reading, math, and writing disabilities in children with clinical disorders. *Learning and Individual Differences, 16*, 145–157.

McLeod, D. (1992). Research on affect in mathematics education: A reconceptualization. In D. A. Grouws (Ed.), *Handbook of research on mathematics teaching and learning* (pp. 575–596). NCTM and Macmillan.

Menon, V. (2016). Working memory in children's math learning. *Current Opinion in Behavioral Sciences, 10*, 125–132.

Miravete, B., Tricot, A., Kalyuga, S., & Amadieu, F. (2017). Configured-groups hypothesis: Fast comparison of exact large quantities without counting. *Cognitive Processing, 18*, 447–459.

Möhring, W., Newcombe, N., & Frick, A. (2015). The relation between spatial thinking and proportional reasoning in preschoolers. *Journal of Experimental Child Psychology, 132*, 213–220.

Moriña Diez, A. (2010). School memories of young people with disabilities: An analysis of barriers and aids to inclusion. *Disability and Society, 25*(2), 163–175.

Moscoso, E., Castaldi, E., Burr, D., Arrighi, R., & Anobile, G. (2020). Grouping strategies in number extend the subitising range. *Scientific Reports, 10*, 14979. https://doi.org/10.1038/s41598-020-71871-5

Murdoch, K. (2019). *A model for designing a learning of inquiry*. Retrieved from https://static1.squarespace.com/static/55c7efeae4b0f5d2463be2d1/t/5dcb82551bdcf03f365b0a6f/1573618265386/A+MODEL+FOR+DESIGNING+A+JOURNEY+OF+INQUIRY.pdf.

Murray, M. (n.d.) A very good literacy focus on mathematics (series). Mathematical Publications.

Næss, K.-A. B., Lyster, S.-A. H., Hulme, C., & Melby-Lervåg, M. (2011). Language and verbal short-term memory skills in children with Down Syndrome: A meta-analytic review. *Research in Developmental Disabilities, 32*(6), 2225-2234. doi:https://doi.org/https://doi.org/10.1016/j.ridd.2011.05.014

Newcombe, N. (2017). Harnessing spatial thinking to support STEM learning. *OECD Working Papers, 161*. https://doi.org/10.1787/7d5cae6-en

Newton, K. J., & Alexander, P. A. (2013). Early mathematics learning in perspective: Eras and forces of change. In L. D. English & J. T. Mulligan (Eds.), *Reconceptualizing early mathematics learning* (pp. 5–28). Springer.

Norris, A., & Norwich, B. (2005). *Special teaching for special children? Pedagogies for inclusion.* Open University.

Nunes, T., & Moreno, C. (2002). An intervention program to promote deaf pupils' achievement in numeracy. *Journal of Deaf Studies and Deaf Education, 7*, 120–133.

Osborne, R. J., & Wittrock, M. C. (1983). Learning science: A generative process. *Science Education, 67*(4), 489–508. https://doi.org/10.1002/sce.3730670406

Owens, K. (2015). *Visuospatial reasoning: An ecocultural perspective for space, geometry and measurement education.* Springer.

Owens, K. (2020). Noticing and visuospatial reasoning. *Australian Primary Mathematics Classroom, 25*(1), 12–15.

Owens, K., & Clements, M. (1998). Representations used in spatial problem solving in the classroom. *Journal of Mathematical Behavior, 17*(2), 197–218.

Pajares, F., & Miller, M. (1994). Role of self-efficacy and self-concept beliefs in mathematical problem solving: A path analysis. *Journal of Educational Psychology, 86*(2), 193–203. https://doi.org/10.1037/0022-0663.86.2.193

Peterson, R., Boada, R., McGrath, L., Willcutt, E., Olson, R., & Pennington, B. (2017). Cognitive prediction of reading, math, and attention: Shared and unique influences. *Journal of Learning Disabilities, 50*, 408–421.

Pirie, S., & Kieren, T. (1994). Growth in mathematical understanding: How can we characterise it and how can we represent it? *Educational Studies in Mathematics, 26*(2), 165–190. https://doi.org/10.1007/BF01273662

Planas, N., & Civil, M. (2013). Language-as-resource and language-as-political: Tensions in the bilingual mathematics classroom. *Mathematics Education Research Journal, 25*(3), 361–378. https://doi.org/10.1007/s13394-013-0075-6

Prior, M., Smart, D., Sanson, A., & Oberlaid, F. (1999). Relationships between learning difficulties and psychological problems in preadolescent children from a longitudinal sample. *Journal of the American Academy of Child Adolescent Psychiatry, 38*, 429–436.

Pylyshyn, Z. (1994). Some primitive mechanisms of spatial attention. *Cognition, 50*(1–3), 363–384. https://doi.org/10.1016/0010-0277(94)90036-1

Radford, L. (2009). Why do gestures matter? Sensuous cognition and the palpability of mathematical meanings. *Educational Studies in Mathematics, 70*(2), 111–126.

Reisman, F. (1978). *A guide to the diagnostic teaching of arithmetic* (2nd ed.). Charles E. Merrill.

Reisman, F., & Kauffman, S. (1980). *Teaching mathematics to children with special needs.* Charles E. Merrill.

Roizen, N. (2002). Down Syndrome. In M. L. Batshaw (Ed.), *Children with disabilities* (pp. 361–376). Brookes Publishing.

Rusinek, G. (2008). Disaffected learners and school musical culture: An opportunity for inclusion. *Research Studies in Music Education, 30*(1), 9–23.

Sella, F., Ilaria, B., Lucangeli, D., & Zorzi, M. (2017). Preschool children use space, rather than counting, to infer the numerical magnitude of digits: Evidence for a spatial mapping principle. *Cognition, 156*, 56–67.

Siemon, D., Beswick, K., Brady, K., Clark, J., Faragher, R., & Warren, E. (2015). Understanding diversity. In *Teaching mathematics: Foundations to middle years* (2nd ed.). Oxford.

Sinclair, N., Bartolini Bussi, M. G., de Villiers, M., Jones, K., Kortenkamp, U., Leung, A., & Owens, K. (2016). Recent research on geometry education: An ICME-13 survey team report. *ZDM*, 1–29. https://doi.org/10.1007/s11858-016-0796-6

Smith, T. E. C., Polloway, E. A., Patton, J. R., & Dowdy, C. A. (2012). *Teaching students with special needs in inclusive settings* (6th ed.). Pearson.

Soltanlou, M., Pixner, S., & Nuerk, H.-C. (2015). Contribution of working memory in multiplication fact network in children may shift from verbal to visuo-spatial: A longitudinal investigation. *Frontiers in Psychology, 6*, 1062. https://doi.org/10.3389/fpsyg.2015.01062

Stokes, M. A., Thomson, M., Macmillan, C. M., Pecora, L., Dymond, S. R., & Donaldson, E. (2017). Principals' and teachers' reports of successful teaching strategies with children with high-functioning autism spectrum disorder. *Canadian Journal of School Psychology, 32*(3-4), 192–208. https://doi.org/10.1177/0829573516672969

Tan, P., Lambert, R., Padilla, A., & Wieman, R. (2019). A disability studies in mathematics education review of intellectual disabilities: Directions for future inquiry and practice. *Journal of Mathematical Behavior, 54*. https://doi.org/10.1016/j.jmathb.2018.09.001

Tartre, L. (1990). Spatial skills, gender, and mathematics. In E. L. Fennema (Ed.), *Mathematics and gender*. Teachers College Press.

Turner, E., Drake, C., McDuffie, A., Aguirre, J., Bartell, T., & Foote, M. (2012). Promoting equity in mathematics teacher preparation: A framework for advancing teacher learning of children's multiple mathematics knowledge bases. *Journal of Mathematics Teacher Education, 15*(1), 67–82. https://doi.org/10.1007/s10857-011-9196-6

Vale, C., Widjaja, W., Doig, B., & Groves, S. (2019). Anticipating students' reasoning and planning prompts in structured problem-solving lessons. *Mathematics Education Research Journal, 31*, 1–25.

van Dijk, R., Kappers, A. M. L., & Postma, A. (2013). Haptic spatial configuration learning in deaf and hearing individuals. *PloS One, 8*(4), e61336–e61336. https://doi.org/10.1371/journal.pone.0061336

van Hiele, P. (1986). *Structures and insights*. Academic Press.

Vanbinst, K., Ghesquière, P., & De Smedt, B. (2014). Arithmetic strategy development and its domain-specific and domain-general cognitive correlates: A longitudinal study in children with persistent mathematical learning difficulties. *Research in Developmental Disabilities, 35*(11), 3001–3013. https://doi.org/10.1016/j.ridd.2014.06.023

Venneri, A., Cornoldi, C., & Garuti, M. (2003). Arithmetic difficulties in children with visuospatial learning disability (VLD). *Child Neuropsychology, 9*(3), 175–183. https://doi.org/10.1076/chin.9.3.175.16454

Wieschenberg, A. A. (1994). Overcoming conditioned helplessness in mathematics. *College Teaching, 42*(2), 51–54.

Xin, Y. P. (2012). *Conceptual model-based problem solving: Teach students with learning difficulties to solve math problems*. SensePublishers.

Yates, S. M. (2000). Student optimism, pessimism, motivation and achievement in mathematics: A longitudinal study. In N. T. Nakahara & M. Koyama (Eds.), *Proceedings of the 24th conference of the International Group for the Psychology of Mathematics Education (PME24) (Vol. 4)* (pp. 297–304). PME.

Yates, S. M. (2009). Teacher identification of student learned helplessness in mathematics. *Mathematics Education Research Journal, 21*(3), 86–106.

Yen, M.-H., Han, C.-C., Yu, P.-C., Yang, T.-H., Didino, D., Butterworth, B., & Yen, N.-S. (2017). The influence of memory updating and number sense on junior high school math attainment. *Learning and Individual Differences, 54*, 30–40. https://doi.org/10.1016/j.lindif.2017.01.012

Chapter 3
Discerning Learning as Conceptual Change: A Vital Reasoning Tool for Teachers

Ron Tzur and Jessica H. Hunt

Abstract Understanding what conceptual learning means and how it may take place can greatly improve teaching. We consider such understanding vital to teachers' pedagogical reasoning. In this chapter, we specify learning as a process of change in learners' concepts, including two types of reflection and two corresponding stages. This process entails an asset approach, as it guides instructional interventions that center on and start from concepts students do have rather than what students do not know (deficit). We provide instructional tenants to drive teachers' reasoning while planning and/or implementing tasks, activities, and questions to effectively advance their students' conceptual learning. Illustrating the rather abstract theoretical lens with a research-based case, we organize this chapter by three essential questions: "What does it mean for a student to have a mathematical concept?", "What mechanisms drive conceptual advancement (change)?", and "What tenants for teaching support such conceptual advancement as part of mathematical proficiency?"

Keywords Constructivist theory · Conceptual change · Activity · Reflection · Teaching conceptually · Tasks

We wrote this chapter in response to recurring questions of teachers with whom we work: What does conceptual learning mean and why is it important/relevant for me and for my students? To answer these questions while working with students with learning disabilities (LD) in mathematics, we have been using a constructivist (von Glasersfeld, 1995) theoretical lens. This lens, which discerns learning as a process of conceptual change, greatly empowered our own work while engendering

R. Tzur (✉)
University of Colorado Denver, Denver, CO, USA
e-mail: ron.tzur@ucdenver.edu

J. H. Hunt (✉)
North Carolina State University, Raleigh, NC, USA

momentous student growth (Hunt & Silva, 2020; Hunt et al., 2016a, 2019b; Xin et al., 2019, 2020). We recognize that, initially, making sense of and learning to use this lens as a tool to guide a teacher's pedagogical reasoning may prove challenging. Gradually, however, we believe it would empower your work, too. Our belief is rooted in what we understand to be a leading motivation for any teacher, namely, to become a professional who makes a difference in their students' lives. At the center of our profession is the aspiration to promote students' learning. Understanding what learning means and how it could be promoted is thus a necessary tool for being professional (think – a carpenter's understanding of wood and design principles). Because the notions that constitute this theoretical lens are rather abstract, we begin with an example of conceptual change we promoted in the concept of unit fractions (1/n) by a student with LD (Hunt et al., 2016a, b).

Why using a theory of conceptual learning to guide mathematics teaching for students with learning disabilities (LD) and mathematics difficulties (MD)? A key reason would be that it helps to both explain, and diminish, the persistent gap between students with LD (or MD) and their grade-level counterparts. For example, in the United States, the National Assessment of Educational Progress (NAEP, 2019) aptly includes many items to measure conceptual understanding and/or problem solving. Results in mathematics showed that students with LD and MD were strikingly outperformed by their grade-level peers: 214 vs. 245 average scale-scores at fourth grade; 247 vs. 287 at eighth grade (respectively). Similar results were found for the Number and Operations Category: 214 vs. 248 average scale-scores at fourth grade; 245 vs. 284 at eighth grade (respectively). That is, at eight grade students with LD and MD scored like their typically achieving peers at fourth grade. In the past 10 years, this four-year performance gap persisted for fourth graders and increased for eighth graders.

We point out the issue not to perpetuate a focus on performance gaps but to challenge and redirect it. Specifically, we argue that the issue is not students' inability to learn or develop mathematical concepts but their *opportunity* to do so. A growing body of empirical evidence across mathematics education and special education research suggests the disparities are caused by *opportunity gaps* stemming from policy that position instructional interventions that, by-and-large, work to remediate the learner (Gersten et al., 2009; Woodward & Tzur, 2017). Widely accepted practices for students with LD or MD consist of interventions in which a teacher explicitly models mathematical methods to students who are expected to follow those methods. While perhaps effective in increasing performance, such direct instruction has its primary focus on remediation measured in terms of student responses to instructional interventions (Fuchs et al., 2010). To us, the key here is that such approaches do not characterize *learning* in conceptual terms. Rather, the focus is on improvement in students' responses to interventions (RTI), that is, as improvements in problem-solving performance due to a teacher's direct modeling of core contents.

We contend that students with LD and MD have the capacity for conceptual advances in mathematics and thus design teaching to accomplish this goal. Our chapter provides a lens on such an advance by focusing on students' access to their own reasoning from which to build intended, new-to-them concepts (Hunt & Silva,

2020; National Mathematics Advisory Panel, 2008; Tzur, 2019c). Research studies in mathematics education have shown that children's conceptions of mathematics are qualitatively distinct from those of adults (Steffe & Olive, 2010), often commence in non-symbolic ways that are context-dependent (Pirie & Kieren, 1994; Tzur, 2000), and can progress in ways educators may not expect in advance of instruction (Hunt et al., 2019a; Hunt & Silva, 2020; Olive & Vomvoridi, 2006; Tzur, 2004). As a result, the challenge facing us as teachers is not how to lead students to solve tasks in ways deemed appropriate from the adult's frame of reference. Rather, the challenge is to articulate how, *from a student's conceptual frame of reference* (different from the teacher's), their solutions make sense and can serve as a basis for learning. Our contention about what constitutes worthy mathematics for students with LD or MD is consistent with Lockhart's (2009) emphasis on reasoning:

> By concentrating on *what*, and leaving out *why*, mathematics is reduced to an empty shell… If you deny students the opportunity to… pose their own problems, make their own conjectures and discoveries, to be wrong, to be creatively frustrated, to have an inspiration, and to cobble together their own explanations and proofs— you deny them mathematics itself. (p. 5)

Next, using the example of Lia's (pseudonym) conceptualization of unit fractions, we illustrate this conception-based, asset approach (Hunt, et al., 2016a; Tzur & Hunt, 2015).

3.1 Illustrating Learning as Conceptual Change

We present a case of a student with LD with whom we worked while using the theoretical lens we put forward. Our purpose is twofold – to provide an image of what such students are capable of when taught conceptually and to serve as a springboard to the abstract constructs of the theory.

Lia's Conceptualization of Unit Fractions To situate Lia's conceptual growth, we note that unit fractions (1/n) are considered a foundational concept for all fraction knowledge and beyond (Hackenberg et al., 2016; Steffe & Olive, 2010). Research about *all* children's difficulties to conceive of such units as a quantity, let alone understand the inverse relationship among them (e.g., 1/5 > 1/7 precisely because 7 > 5), repeatedly pointed out to teaching of fractions as parts-of-wholes as a key factor (Siegler et al., 2011). Indeed, NAEP items regularly pertained to this concept. For example, in 2013, one NAEP item asked: "Kim, Les, Mario, and Nina each had a string of 10 feet long. Kim cut hers into fifths, Les into fourths, Mario into sixths, and Nina into thirds. After the cuts were made, who had the longest pieces of string?" In 1996, another item asked simply: "How many fourths make a whole?" Results showed correct responses were given, respectively, by only 59% and 50%. Responses to those single NAEP items are not disaggregated by students' demographics. Nevertheless, we can assume the overall gap between students with LD and their counterparts entails a significantly larger proportion of the former would incorrectly answer such items.

We began working with Lia, a fifth grader, after she had received two and one-half years of direct instruction, including work on ordering unit fractions (see Hunt et al., 2016a, b). Initially, she appeared quite timid and reluctant to engage with us or in the mathematical tasks that we offered for her to think within. Lia exhibited what appeared to us as long-learned "coping" strategies of avoidance. She used a soft voice that could hardly be detected; she looked for hints in our language and/or actions about how to solve a task; and she initiated questions to elicit our "help" in the form of teacher demonstration and approval of her "question-like" responses.

Lia's initial understandings of the mathematics, and particularly of unit fractions, were equally precarious. She often referenced visuals she remembered from her classroom to justify magnitudes of unit fractions. For example, when shown a slip of paper folded into two parts, Lia easily explained and justified each part was one-half ("I have two equal parts"). Yet, when *one of those halves* was folded into halves again (Fig. 3.1), Lia very softly inquired, "one-third?" When we asked why one-third, she explained that [in class] she was shown a visual of fraction bars that "… goes, one-half, then a third, then a fourth. … (Tzur & Hunt, 2015, p. 149)." Lia also had difficulties comparing unit fractions. For example, when asked which fraction is larger, one-fourth or one-ninth, Lia said one-ninth was larger because nine is a larger number than four (Hunt et al., 2016a, b). We conjecture such reasoning was underlying incorrect responses of the majority of those 50% of students who incorrectly solved the 1996 NAEP item, "How many fourths make a whole?"

We thus set out to foster Lia's construction of unit fractions not just/mainly as "a part of a whole," but as a multiplicative relation between two units (Tzur, 2007, 2019a). For example, we would promote her reasoning that, in the strip-folding problem above, the right-most piece is 1/4 of the entire strip because the whole is 4 times as much as that unit fraction, called one-fourth. Accordingly, Lia would need to conceptualize 1/9 as smaller than 1/4 because 1/9 fits nine times within the whole whereas 1/4 fits only four times.

To this end, adhering to our asset approach, we first assessed Lia's available concepts. We found she had constructed a concept of number as a composite unit, which can serve as a basis for understanding unit fractions as multiplicative relation (Steffe & Olive, 2010; Tzur, 2019a). The rationale is that such a concept of number is rooted in her mental activity of iterating the unit of 1 to compose larger units (e.g., she anticipated that iterating 1 four times would produce the number 4). We conjectured she could thus use iteration of a unit to produce a whole, all along knowing each of those units has the same size (Tzur, 2007, 2019b).

We thus engaged Lia in the "French fry" task, a playful context in which a child is asked to equally share a continuous, linear whole among a few people (Tzur & Hunt, 2015). The series of gradually more challenging tasks began with an easy one: Share one whole fry – a yellow paper strip – among two people and show the

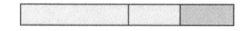

Fig. 3.1 Lia's strip folded in half, then the right half folded in half again

size of one person's share. Many children initially fold the paper strip to show the size, while others mark the fry into two parts and want to cut the fry apart as an alternate means of direct comparison (Hunt et al., 2016b). The initial task, which we conjectured to be rather simple for Lia, thus served to first bring forward her current, linked concepts of "one-half" and two.

In the follow-up tasks, we raised the level of challenge. Moving on to sharing the whole French fry among three people (and then four), we challenged Lia to figure out the "size of one person's share" without folding or cutting the paper. We introduced these constraints to *create Lia's need* to use her available activity of iterating a unit (Tzur, 2000, 2019a) as a means to accomplish her goal while advancing her concept of unit fractions. When using iteration within the equal sharing situation, we hypothesized that a student's strategy would be to (a) estimate the size of one person's share, (b) iterate that piece the number of times needed for people who share the entire French fry, (c) compare her iterated whole to the given one to be shared (yellow strips in Fig. 3.2), and (d) either terminate the work (if successfully equal) or adjust the length of the estimated, single piece and continue from the first step (Simon et al., 2004). For Lia, using her fingers to make the length of one person's share and moving her fingers across the length of the fry a number of times equal to the number of sharers confirmed that this was her goal-driven activity (see Steffe, 2001). We thus introduced to her a similar method in which an auxiliary piece of paper is used to mark the iterations of one person's share.

To orient Lia's reflection on the relationship she can notice between her goal-driven activity and its effects, we used two critical questions (see Hunt et al., 2016a, b). We asked the first, twofold question when she indicated noticing the iterated whole was either longer (dark blue) or shorter (light blue) than the original (yellow French fry) whole: "Would you make the next estimate shorter or longer than the previous attempt(s) and why?" This question focuses the child's attention onto the *direction of change* needed. The second, twofold question, usually suitable after some work on the first question, asked her to consider, "By how much the next estimate should be shorter or longer then the next estimate and why?" This second question focuses the child's attention onto the *amount of change* needed. For example, in the second attempt shown in Fig. 3.2, she correctly estimated her next

Fig. 3.2 Building a concept of unit fractions through iterating (the yellow strip is the whole to be shared)

estimate (light blue) should be shorter than the first estimate (dark blue), explaining she "was eyeballing it to be that much shorter." In the third attempt in Fig. 3.2, we present a new, improved way of estimating the amount of change. (Note: It took Lia several more attempts to arrive at and explain this solution.) She realized that partitioning the gap between the whole French fry and her iterated whole into three segments (i.e., the number of sharers) and adding just one of those small pieces to the previous estimate would lead to a correct solution. Such a realization is a major conceptual leap, shown to be a challenge not just for most children (with or without LD) but also for some teachers (Harrington et al., in press).

Using those two questions repeatedly for sharing the fry among a different number of people led Lia to reorganize her concept of unit fraction, from "part-of-whole" to a multiplicative relation (Tzur, 2019a). Specifically, her recurring adjustments to the length of one person's share led her to a novel anticipation. She realized, and abstracted, that for any given whole there would be only one piece (size) that fits precisely n times in the whole (e.g., the green piece in Fig. 3.2). Said differently, when she concluded a whole has been successfully shared – it implied *for her* the production of unique-size segment that forms a 1-to-n relation to that whole. When we inferred she had established this anticipation for herself, with reason, we introduced her to the fraction language used in the context of her iteration activity (Tzur, 2000; Tzur & Hunt, 2015). For example, the share of one person in five is called one-fifth, and symbolized 1/5, because it fits (iterated) five times within the whole.

An expansion of those two questions we then used challenged Lia to anticipate, ahead of taking any action, the direction (and later also the amount) of change needed when moving to a new task involving a different number of sharers. For example, once she would successfully complete sharing the French fry among 3 people, we said the next task would be to share a new paper strip (same-size fry as in the previous task) among four people. Before letting her start, we asked if she would make her first estimate shorter or longer than what she used for three people and why. Lia, using a computer software called JavaBars (Biddlecomb et al., 2013), then successfully solved tasks for four, five, and six people. These tasks, and the two reflection-orienting questions that accompanied each of her attempts, led her to realize the need to shorten one person's share in each of those new tasks. We thus engaged her in estimating the size of larger numbers of sharers, while changing tasks not only upward (e.g., 6 people to 11 people) but also downward (e.g., 11 people to 8 people). All in all, those tasks led Lia to arrive at the other intended anticipation (abstraction): A unit fraction, as a unique relation, must be smaller for a larger number of sharers, which she could then link with the denominator – and vice versa. The following transcript shows Lia's work after less than 10 teaching sessions (~ 30–40 minutes each), in a context (pizza) different than the French fry in which we taught her. Based on conceptualizing unit fractions as multiplicative relations with inverse relation among their sizes, Lia successfully compared unit fractions both in and out of context (see Hunt et al., 2016a, b, p. 202).

Teacher. I ate one-twelfth [of the pizza]. Here's you. You ate one-fourteenth [of the same-size pizza]. Who ate more [of the pizza])?

Lia: (Immediately, confidently) You.

Teacher: Why did I eat more?

Lia: Because the twelfths are bigger than the fourteenths; because fourteenths have to [be] shared among more people than twelfths. It only takes 12 to [make] the whole, and [here] 14 to [make the whole], so 14 needs more [pieces]. It's smaller.

(*In a later task, no context, numerical only, larger denominators*)

Teacher: OK. We've got one-thirty second [1/32] and one-thirty-eighth [1/38]. Which one would be larger and why?

Lia (correctly): The one-thirty second [1/32].

Teacher: How come?

Lia: Because if you have more people sharing, that means *you've got to* give smaller pieces to each. The size will get smaller because you have to bust [the whole] up among more people. The size [of one part] among 32 would be bigger. [Italics added by authors to emphasize the logical necessity expressed by Lia.]

To summarize Lia's learning as conceptual change, in a few lessons involving work on tasks, always accompanied by questions to explain her work (before and/or after taking action), she abstracted the need to adjust her next estimate relative to the number of iterations she would use to produce a whole. In this way, Lia inverted (reorganized) her prior concept of whole-number magnitude to that of unit fraction magnitude (Tzur, 2019a). That is, in whole numbers more iterations of the unit of 1 entailed for her the production of a larger number, whereas for unit fractions more iterations, which the larger number in the denominator then symbolized for her, must produce smaller parts/shares. This new concept served as the basis for her understanding why, for *any* two whole numbers (n and m), if $n > m$, $1/m$ must necessarily be larger than $1/n$ (e.g., $1/32 > 1/38$ precisely because $38 > 32$). Lia also began to understand – conceptually – that the magnitude of any unit fraction such that a remake of the whole comprises n iterations of one share entails a unique, multiplicative relation of size $1/n$ (i.e., the whole is n times as much as each share). We postulate that one or both of these conceptualizations were not available to those 50% and 41% of students who, respectively, incorrectly responded to the 1996 and 2013 NAEP items presented above. With this example of Lia's learning as a conceptual change of available (whole number) into new (unit fraction) concepts, we turn to the theoretical lens that explains such a process.

3.2   Articulating Learning as Conceptual Change

In three sub-sections, we explain conceptual learning using a constructivist theoretical lens that elaborates on Piaget's (1985) core notions of assimilation and reflective abstraction. First, we depict what "having a concept" means. We note that, like many constructivist researchers, in our articles we typically talk about "having a scheme." In this book, to avoid jargon that may sidetrack clarity and understanding, we use the term concept as a synonym. Next, we articulate a cognitive process, consisting of two types of reflections, which Simon, Tzur, Heinz, and Kinzel (2004) postulated as a mechanism by which the mental system transforms (reorganizes) available concepts into a new (to the learner) concept. Finally, we present Tzur and Simon's (2004) distinction between two stages through which a new concept emerges from previously available concepts. We could not emphasize enough the critical role that a teacher's appreciation of this stage distinction would serve in nurturing, or alas hindering, students' construction of a new concept at the empowering, transfer-enabling quality seen in Lia's example (Tzur, 2019a).

It is important for a teacher who attempts to understand and use this theoretical lens to clearly distinguish it from widespread notions of learning as change in performance (Fuchs & Deshler, 2007; Fuchs et al., 2010), often measured through change in students' responses to problems (e.g., answers, solution strategies). Rather, we focus on inferring and explaining what cognitive (invisible) processes could plausibly underlie particular performance and/or changes in it. For example, how would one explain Lia's correct solutions to any problem of ordering unit fractions, in real-world contexts or calculations? We explain those solutions as enabled by her concept of unit fractions as a (multiplicative) relation – between the number of iterations it takes to fit a piece within a given whole and the size of that piece relative to pieces that were iterated more/less times.

Critically, a teacher adhering to this theoretical lens would have to let go of pedagogical notions such as "giving concepts to students" or "modeling to help students see" what the teacher sees. Simply put (sadly?), our theoretical lens implies *concepts cannot be given or shown*. Rather, facilitated by competent, theoretically savvy teachers, learners must construct concepts that could become a mental apparatus with which to "see" the mathematics like their teachers "see" it. That is, we advance a stance that all learners, including students with LD, must construct concepts *for* themselves – though absolutely not *by* themselves.

Concept – An Abstracted, Goal-Activity-Effect Relationship We define a concept, the basic mental unit attributed to a learner, as a single structure comprised of three interrelated parts (von Glassersfeld, 1995): (i) a recognition template ("situation") that triggers a goal and intention to act, (ii) an activity the goal brings forth and regulates to determine stoppage, and (c) a result the mental system anticipates to ensue from that activity. As we shall explain, such an anticipation may precede the action or be retroactive in the sense of identifying why an effect one noticed should have followed the goal-directed activity (Piaget, 1985). Tzur and Simon (2004) picked the term result for the former, while introducing the term "effect" to explicitly distinguish an outcome of an activity if the mental system noticed and linked it with the activity only in retrospect.

In our example of Lia's work on unit fractions, like with many other students (LD or non-LD), we have regularly seen a child making an incorrect adjustment of the direction of one person's share while moving to a different number of sharers. If, for example, the child successfully completed equally sharing the fry among three people, the child would think the first (incorrect) estimate to share among four people needs to be longer. The child's explanation would often indicate their anticipation of the result of an activity of iterating 1 in whole numbers (e.g., "4 > 3, so I need to make the piece for one-in-four longer than what I got for one-in-three"). For a teacher, the crucial point in this explanation is to always remind oneself that an incorrect answer by a child reflects an underlying concept that, for the child, makes perfect sense. As our example indicates, a teacher striving to explain that sense for the child, erroneously as it may appear to the teacher, would not focus on remediating the error. Rather, the teacher would focus on understanding the child's anticipation of why a particular effect is expected to ensue from the goal-directed activity.

Just as important, a teacher would also attempt to identify (assess) the child's goal-activity-effect anticipation that underlies correct responses. In our example of Lia, she had eventually constructed a strong anticipation that the more a unit fraction is iterated to fit within a given whole the smaller that unit must be. It was this realization that our French fry activity (Tzur & Hunt, 2015) was designed to nurture in Lia as a basis for her abstraction of the inverse relationship between the number of iterations and the relative (inverse) size of the unit fraction. Having helped her also connect the number of iterations to the denominator (Tzur, 2000) "cemented" the concept underlying Lia's powerful transfer of reasoning shown in the protocol above (e.g., explaining why she anticipated 1/32 of a pizza would necessarily be larger than 1/38 of a pizza). Using the goal-activity-effect structure of a concept, we inferred her concept of unit fractions to be as follows: "A unit fraction ($1/n$) is the anticipated result of iterating a unit n times to fit within (hence, equally share) a given whole. The whole is n times as much as the unit fraction $1/n$. More iterations imply units are smaller, and thus if $n > m$ then necessarily $1/n < 1/m$." We now turn to address a question a teacher would naturally ask here: How could such learning – a change in anticipation of the effect that would follow a goal-directed activity – be explained?

Explaining Learning as Conceptual Change: Two types of reflection The three-fold notion of concept (situation/goal-activity-effect) explained above entails a view of learning as a conceptual process by which the mental system reorganizes prior concepts (anticipations) into a new concept (Tzur, 2019a). Elaborating on Piaget's theory, Simon et al. (2004) postulated such a reorganization comes about through a mental mechanism termed *reflection on activity-effect relationships*. This mechanism consists of two related-yet-distinct processes, Reflection Type-I and Reflection Type-II. Importantly, these authors emphasized the term reflection should not be taken to mean a conscious process (e.g., "metacognition"), as both types quite often occur outside the learner's awareness. Understanding each reflection type, and how it may be promoted, can greatly help teachers design opportunities for conceptual learning in the sense of change in anticipation, which we further articulate in Chap. 5.

Reflection Type-I is an automatic process attributed to the functioning of any mental system, because it underlies the start and stoppage of brain activities (Tzur, 2011). In essence, the mental system constantly monitors the extent to which an activity is progressing toward (or away from) the goal it was brought forth to accomplish and – crucially – the result it anticipates to ensue from the activity. The mental system terminates the activity when it detects (notices) an actual effect of the activity. Reflection Type-I then may take place, defined as *a comparison between the anticipated result and the actual effect*, leading to judging them as either matching or different. For example, Lia's concept of whole numbers included the anticipation that $4 > 3$ due to iterating the unit of 1 more times. *Her* goal of equally sharing a whole French fry among four people thus included an implied (incorrect) anticipation: one person's share should be longer than sharing among three people. Already after she iterated that estimate for a second time, and clearly after iterating it for the third and fourth times, her goal of creating an iterated whole equal to the given French fry drove *her* noticing the actual effect (iterated whole) was much longer than the anticipated result (being equal to the French fry). In turn, detecting this lack of match led to concluding *her* goal was not yet accomplished. For some children, the next attempt might be to (incorrectly) make the next estimate even longer. As teachers, we will let them give it a try. Eventually, and in Lia's case after just one incorrect attempt, she realized the correct adjustment for the next estimate would need to be shorter than the share of one-in-three people - a start of the conceptual change.

In our explanation of Reflection Type-I, we constantly italicized the references to Lia (*her, she*) to stress a key point: ensuring that the *learner's available concepts* provide the "ingredients" needed for comparing between the anticipated and the actual effect. A child's available concept (here, number as composite unit) may not yet be established to include a result *she* would anticipate and thus use as a benchmark in *her* comparison to the actual effect (here, the anticipated iteration of one person's share *n* times). Thus, in our example of the French fry task we would expect the child not to "see" the futility of her attempt to make the share of one-in-four people longer than for one-in-three, in spite of a teacher's relentless attempts to explicitly model this "painfully obvious" fact. The notion of Reflection Type-I equips the teacher with a more effective alternative, namely, doubt and thus focus on figuring out if the child's available concepts have available an anticipated result the teacher expects to be used in the comparison. In this example, when children with whom we work consistently use the wrong direction of adjusting one person's share we (a) first go back to assess their concept of number and, if it is not yet established, (b) focus on nurturing it as a conceptual prerequisite prior to teaching unit fractions.

Reflection Type-II is a process assumed to be a capacity of the mental system, that is, a comparison the brain can, but may not, carry out (Simon et al., 2004; Tzur, 2011). In essence, Reflection Type-II involves the mental system's *comparison among some instances* in which it linked an activity to its effect(s). For example, in each variation of the French fry task, say, from 3 people to 4 people, and then to 5 people, Lia's mental system could have recorded *her* choice to make one person's share shorter than for the previous (smaller) number of people. As we explained above, a child's correct choice of the direction of change might have happened after a few incorrect attempts to make that estimate longer and re-noticing the actual effect took them away from the anticipated result. Such a comparison involved Lia's noticing, against the background of things that did change between instances (number of people, size of one person's share), that some aspects of her activity and its actual effects did not change. In our example, *she could* notice an invariant: to share the fry among a larger number of people entailed making one person's share shorter than for a smaller number of people. Specifically, she could notice that one share for five people would be smaller not just than for four people but also for three people. Further work on variation of tasks in which she may go, say, from six people to 11 people, and then back to 8 people, could yield *her extension* of this logical, invariant anticipation called "inverse relation": more people entail smaller unit fractions and fewer people entail larger unit fractions.

We have again italicized the references to Lia as well as to the tentative nature (*could, may*) of Reflection Type-II to stress two key points. First, she needed to have those mental records of instances available in order to compare the anticipation involved in each instance (e.g., "to go from 3 to 4, I needed to make it shorter; to go from 4 to 5 I needed to make it shorter," etc.). That is, mental records of past instances in which an anticipation has been invoked serve as the input of Reflection Type-II. Typically, a learner is better off creating those instances through Reflection Type-I on their own actions, although some learners may be able to coordinate such instances by following others' actions and their ensued effects. Second, the tentative language we use conveys that Reflection Type-II is non-automatic for most brains (Tzur, 2011). For some learners it can develop as a propensity. Our experience working with LD and non-LD students, as well as with adults (e.g., teachers), indicated most of them would not carry out Reflection Type-II on their own but would greatly benefit from being oriented to do so. For example, we repeatedly asked Lia questions such as, "In what way has your work while moving from 3 to 4 people, or from 4 to 5, or from 5 to 6, was similar? Different?" (see Hunt & Tzur, 2017).

We cannot emphasize enough that such interactions differ markedly from a teacher's relentless attempts to "show" (model explicitly/directly) to the student a pattern that, for the teacher, is painfully obvious (e.g., "showing" that making the share of one-in-four larger than one-in-three did not work so we make it smaller). For us, to say "a person sees a pattern" means they have constructed the invariant anticipation of what they consider to necessarily be the effect in *all instances* of their activity – and why. This explanation of how an anticipation of activity-effect relationship is cemented as an invariant underlies transfer learning like the one demonstrated in Lia's case. Simply put, such an invariant, justified anticipation is the hallmark of "having a concept." It leads to the distinction of a stage at which the anticipation is constructed provisionally.

Explaining Learning as Conceptual Change: A Two-Stage Process When teaching a new concept, most teachers recognize a rather puzzling phenomenon. At first, students appear to learn the concept and correctly solve/explain problems. At some later time – a few days, a few hours, even a few minutes (e.g., within a single lesson) – the students regress to performance as if not having learned, or forgotten, the concept. Then, once somehow the activity underlying the new concept is being prompted for, the students resume proper work indicative of having learned the concept. For example, one time, Lia learned and began ordering unit fractions correctly (e.g., explaining why 1/5 of a pizza is larger than 1/8 of a same-size pizza). In our next session, she went back to explaining that 1/8 is larger than 1/5 because 8 is larger than 5. When asked, "What if you shared a fry among five people and then moved to sharing it among 8 people?" Lia soon corrected herself: "Oh, no; it's the other way around; because I need to repeat it more times so each of the 8 pieces is smaller." Tzur & Simon (2004) called this all-too-familiar sequence of correct reasoning ➔ incorrect (less advanced) reasoning ➔ prompt-dependent return to correct reasoning, "*the next day phenomenon.*" To explain it, they distinguished two stages in the learning of a new concept.

Here, before explaining those two stages, we appeal to our reader's sharpest attention to its imperative implication. To this end, let us consider Lia at two different points in her learning the concept of inverse relations among unit fractions (if $m > n$ then $1/n > 1/m$). An "earlier Lia" needed prompting not only to begin learning this new concept but also to resume reasoning after she already appeared to have learned it. A "later Lia" could independently and spontaneously not only use the concept to correctly solve problems in the context in which teaching took place (e.g., French fry) but also transfer it to novel situations/contexts and to context-free tasks. All teachers find the next-day phenomenon rather frustrating as they yearn to promote independence and concept transfer in their students. What we aim to pinpoint for teachers is a critical implication of the prompt-dependent, provisional stage in learning a new concept. Simply put, moving on to teach a concept for which a prompt-dependent concept is prerequisite is most likely to leave such students behind (Tzur, 2019a). For example, trying to teach an "earlier Lia" concepts of non-unit fractions (e.g., 2/5, 7/5) would most likely become not only unsuccessful but also an unbearable frustration for teacher and students alike.

Participatory stage is the term Tzur and Simon (2004) used in reference to the earlier, provisional stage of having a new concept, to emphasize the learner's "participation" in an activity that leads to anticipating its effects. As noted above, this stage is characterized by the sequence of reasoning, with a newly learned concept, that includes correct ➔ then "regressed" ➔ then returned to the new (by prompting). While people often think about "prompt" as something provided by another person, we emphasize that prompting, or being prompted, quite often occurs internally. A highly familiar manifestation of internal (mental) prompting is the "Oops" experience (Tzur, in press). In Lia's example, after she completed sharing a French fry among four people, we gave her a new paper strip and asked to equally share it among five people. Before starting, we asked her if the estimate for one person's share would be shorter or longer than for four people. Her initial, incorrect response ("longer") was followed by actions of producing that estimate and starting to iterate it. After the third iteration, she stopped: "Oops; I *should have* made it shorter, because I need to 'squeeze' more pieces." Two important aspects of an "oops" experience like the one Lia had can help distinguish it as indicating a participatory stage. First, the prompting leading to a change in her anticipation of the direction of change (e.g., from 4 to 5 people she needed to make it shorter) *arose out of her own activity*. Second, in spite of already having reasoned correctly about the shift from three to four people, just a few minutes later she "regressed" to her previous antici-pation (e.g., 5 is larger than 4, so one person's share needs to be longer).

Research had shown that learners at the participatory stage may carry out the entire activity before realizing their error; others (like Lia) may notice it during their activity; and still others would notice it only if prompted by someone else (see Simon et al., 2016, 2018b). Tzur & Lambert (2011) provided further analysis of such gradations within the participatory stage, explaining why internal prompting indicates a higher-level toward the new concept than being prompted by others. They also explained that more general prompts indicate a higher-level of the partici-patory stage than specific prompts (e.g., asking generally "Do you recall working on the French fry task?" vs. specifically, "How did you change your estimate when moving on from three to four people?"). Those subtle differences in levels can serve the teacher, say, in determining a proper time to introduce mathematical terms and symbols after a learner has constructed at least a higher-level of the participatory stage (Tzur, 2000). For example, Lia's self-prompting ("oops") experience, leading to a correct adjustment of an estimate for one person's share in five people, led us to introduce the term "one-fifth" and the symbol "1/5" in the context of her activity of equal-sharing by unit iteration.

Anticipatory stage is the term Tzur and Simon (2004) used in reference to the later, advanced stage of having a new concept. This term emphasizes the establish-ment of an invariant Goal ➔ Activity ➔ Effect(s). This stage is characterized by spontaneously, independently, and properly to the task/situation bringing forth and explaining the logical necessity of these anticipated links. The example of Lia's solution to the pizza problem (comparing 1/32 vs. 1/38) illustrates the anticipatory stage in constructing a concept of unit fractions. As soon as she heard the task, *she* brought forth the reasoning, which arose for her in a different context (French fry). She knew why the fractional number with a larger denominator must necessarily be a smaller unit fraction. Studies of other students with LD had shown they eventually constructed similar reasoning at the anticipatory stage (Hunt et al., forthcoming). In fact, after moving on to learning concept of non-unit fractions, many of them could further transfer their concept of unit fractions to tasks requiring to re-invert the inverse relation among unit fractions. For example, to solve the task, "Which frac-tion is larger, 12/13 or 13/14?" they would reason, "13/14; because it is just 1/14 away from One and the other is 1/13 away from One." Such reasoning differs mark-edly from the prevalent, incorrect reasoning that compares numerators and denomi-nators separately (e.g., 13 > 12 and 14 > 13, which incorrectly could lead to saying that, for example, 6/12 > 5/8). We note that, like with the participatory stage, researchers and teachers have noticed gradations in developing the anticipatory stage, but articles about those gradations are yet to be published.

To summarize this theoretical section, we answer a question that teachers often ask: How are the two types of reflection and two stages in constructing a new con-cept linked? Based on the Goal ➔ Activity ➔ Effect structure of a concept, and the changes it undergoes through each type of reflection, Tzur (2011, 2019c) postulated that Reflection Type-I is necessary for constructing a new concept at the participa-tory stage and Reflection Type-II for the anticipatory stage (a postulation still awaiting empirical corroboration). In Chap. 5, we will further specify implications of this theoretical lens, including the link between stages and reflection types, for planning, implementing, and assessing mathematics teaching. To culminate this chapter, we now turn to three tenants of teaching students with LD for mathematical proficiency that draw on the theoretical lens presented here.

3.3 Teaching for Mathematical Proficiency: Three Core Tenants

The story of Lia's learning the concept of unit fractions resembles many students with LD and MD with whom we have worked over the past two decades (Hord et al., 2016; Hunt & Empson, 2015; Xin et al., 2020). This story reflects how a theoretical lens can serve as a vital tool for teachers' reasoning, and a solid foundation for what it means for these students to know mathematics proficiently. Kilpatrick et al. (2001) explained mathematical proficiency as encompassing five interrelated strands – *conceptual understanding, procedural fluency, strategic competence, adaptive reasoning*, and *productive disposition* as shown in Fig. 3.3. Stressing the central (first) place of conceptual understanding, we briefly depict each strand in relation to Lia's story.

Conceptual understanding is defined as "comprehension of mathematics concepts, operations, and relations" (Kilpatrick et al., 2001, p. 5). As Lia's story showed, her conceptual understanding when we started the work could be characterized as pre-fractional reasoning (e.g., thinking that one-of-three unequal parts is one-third). Yet, her conceptual understanding grew tremendously, through a participatory to an anticipatory stage that enabled transfer of her knowledge to novel situations (e.g., 1/32 > 1/38 with clear rationale; see also Tzur, in press).

The changes evident in Lia's reasoning clearly connect to procedural fluency. Kilpatrick et al. (2001) defined *procedural fluency* as "skill in carrying out procedures flexibly, accurately, efficiently, and appropriately" (p. 5). When we began working with Lia, procedures were mostly what she drew upon, yet her use of and facility with procedures were limited and quite often mathematically incorrect. For example, Lia instantiated a classroom visual to compare unit fractions – a sort of routinization of a "rule" that she attempted to apply to the situation. The example of her ability to independently, spontaneously, correctly, and rather quickly order unit fractions support our contention that procedural fluency can be further understood as routine expertise and adaptive expertise.

Routine expertise is when students master procedures to become highly efficient and accurate. Adaptive expertise intertwines with conceptual understanding in that it allows the student to invent and connect new solutions to problems and even new procedures for solving problems. We view Lia's progress as moving from a limited version of procedural, routine expertise to the beginning of adaptive expertise. For example, she could connect her actions in equal sharing contexts (French fry) to

Fig. 3.3 NRC (2001) five interconnected strands of mathematics proficiency

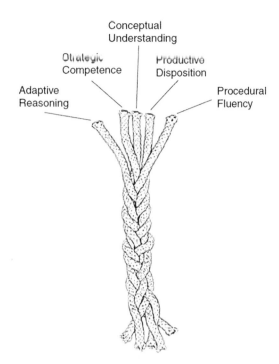

Conceptual
Understanding

Strategic
Competence

Productive
Disposition

Adaptive
Reasoning

Procedural
Fluency

situations in other contexts (pizza) and/or without context. All along, she would base her explanations and justifications on anticipations constructed through her work in the commencing context. Her progression highlights the interconnected nature of the newly built concept(s) with the advances in procedural fluency.

Kilpatrick et al. (2001) defined *productive disposition* as "habitual inclination to see mathematics as sensible, useful, and worthwhile, coupled with a belief in diligence and one's own efficacy" (p. 5). The trust students might place in their own mathematical potential has lasting implications both in terms of mathematics proficiency and in their propensity to choose math-related careers. We note that, at the start of our work with Lia, we did not measure her productive dispositions about mathematics directly. Yet, the timid and reluctant nature of her interactions, and the pensiveness regarding her own reasoning was telling – they were similar to many other students with whom we worked (see, for example, Tzur et al., 2009). Quite quickly, we have seen how those avoidance behaviors and uncertainty related to one's own mathematical thinking gave way to excitement in working on the tasks, to pride and confidence in their own answers, to expressed eagerness to engage in solving challenging tasks on their own, and to audible explanations of their work. We believe the theory-based teaching activities that led to their conceptual growth is a key factor in also explaining the productive dispositions we have witnessed.

Finally, Kilpatrick et al. (2001) defined *adaptive reasoning* as the "capacity for logical thought, reflection, explanation, and justification" (p. 5). The mechanism of reflection on activity-effect relationship that we have presented is both consistent with and contributes to further explicating this definition. We have seen how, when students with LD and MD have opportunities to build mathematics as reorganization in ideas they already have, they develop an identity of "motivated mathematical thinkers and doers." Being engaged in reflection and justification enables these students to see their journey in mathematics as one navigated through reasoning and sense making. Our theoretical lens clarifies that construction of adaptive reasoning is not the same thing as students "discovering" the mathematics on their own; indeed, the role of the teacher in this process is critical (see Chap. 5). Yet, we contend that students need to believe that the understanding they will gain starts within them, connects to what they already (intuitively/implicitly) know, and can grow over time through reflective and communicative processes guided by their teachers (Woodward & Tzur, 2017).

Short of articulating a pedagogical approach that draws on the theoretical lens we presented in this chapter as a vital tool for teachers' reasoning, we culminate this chapter with three core tenants that underlie such an approach. That is, we lay a basis for answering the question, "What core tenants of *student conceptual learning* were in place when planning instruction so that students like Lia are supported to access and advance (reorganize) their available knowledge?"

Core tenant #1: Bring forward students' prior knowledge The asset approach of learning as a change in concepts a *student already has available* entails a teacher much first bring forward those concepts. Furthermore, to promote learning that includes change in mental activity, students need opportunities to be active and interact within their environment (Piaget, 1985; von Glasersfeld, 1995). Within a school setting, the environment comprises their classroom, small groups, tasks, and tools/materials that a teacher uses to engage them in solving mathematical problems. Yet, too often, despite appropriate tools and tasks, teachers of students with LD and MD diminish "activity" to merely following the teacher modeling procedures, demonstrating concepts with manipulatives, or thinking aloud through a desired solution process (Woodward & Tzur, 2017). While well meaning, such instructional moves often perpetuate or lead to mathematically incorrect concepts on the part of the student (e.g., Hunt et al., 2019a).

A highly effective instructional method for engaging students in actively solving tasks as a means to bring forward what they do know, termed *bridging*, was explicated in research on mathematics teaching in China (Huang et al., 2015; Jin & Tzur, 2011a). Markedly different from the common use of "warm-up problems," bridging tasks are characterized by two features. First, they engage students in *independently* solving problems that their available concepts clearly enable. While warm-up problems often have that feature, it is the second feature that distinguishes bridging tasks. Teachers design bridging tasks in such a way that, as students engage in successfully solving them *on their own*, the goals, activities, and/or effects the students bring forward would be conceptually relevant to and supportive of those reorganized through the two types of reflection (first to participatory, then to anticipatory stage). That is, only tasks that students can both solve on their own (no help is needed) and bring forward concepts to be reorganized are considered bridging tasks (see also Simon et al., 2018b; Steffe, 1990; Tzur, 2018).

As we noted in Lia's example, the initial task we used as bridging engaged her in equally sharing a French fry between two people. She could easily solve it on her own by folding the paper strip in half, while bringing forward her notion of a whole and the possibility to create equal segments within it. In a follow-up bridging task, we added two constraints (no folding, no ruler) as a way to bring forward her available concept of number as a unit composed through iterating the unit of 1 so-many-times (Steffe & Olive, 2010). To start using iteration, her available concepts included the activity of "eyeballing" one person's share, of iterating it, and of comparing the resulting (iterated whole) to the given whole (fry). In this activity, her available concepts also included taking for granted each and every piece produced through iteration is equal to all other pieces. What Lia's available concepts at the start did not include, but we designed the bridging tasks to be relevant to and supportive of, is the direction and amount of adjustment to one person's share – reasoning whether to make it shorter or longer than her previous estimate(s). Importantly, her available concepts included knowing order of whole numbers, which was used to bring forth a new realization. Key here is that her use of the mental activities of iterating one person's share + comparing the iterated whole she produced to the given whole would require adjustment. This adjustment is needed not because anyone directed her to make it – but because *her* available concepts included a goal (sharing the given fry) that she could notice was yet to be accomplished. In this way, iteration + comparison (available activities) could lead to adjustments that could later engender the intended concept of inverse relation between the whole number of sharers and the size of each person's share (Simon et al., 2004).

We note that all teachers with whom we work find the idea of bridging tasks invigorating. When attempting to create their own bridging tasks, they also quickly realize it is a serious challenge. The good news is that researchers who have been using the theoretical lens presented in this chapter developed bridging tasks that teachers have found to greatly serve their students (the French fry is but one example of those). We encourage you, the teacher, to take advantage of such tried-out bridging tasks. Those tasks could help you learn to create your own bridging tasks, so they fit within what your students with LD or MD already know and the mathematics you thus intend for them to learn.

As a transition to the second core tenant, we remind you that tasks, by themselves, neither think for students nor "show" them the concept (Simon & Tzur, 2004). Likewise, materials used in tasks, such as manipulatives, drawings, or models, do not think for students. Tasks and materials are very important in that they support students' thinking while they are being active within an environment. Yet, we explicitly distinguish between tasks and materials on one hand and the thinking that students generate while working to solve them on the other hand (Tzur et al., 2013). We contend that using tasks to *bring forward and then promote student thinking* should be the backbone of mathematics instruction (Simon, 1995; Simon et al., 2018a). In the example of Lia, we engaged her in a bridging task of equally sharing a fry among three people without folding the paper strip or using a ruler. This task challenged her thinking while bringing forward the mental activities she (a) had available through her concepts of whole numbers and (b) were to undergo reorganization into a concept of unit fractions. We now turn to the second and third core tenants implied by the theoretical lens, which further elaborate on tasks that could help teachers engender in students intended conceptual changes (Sullivan et al., 2015; Zaslavsky, 2008).

Core tenant #2: Engage students in tasks that engender goal-directed activities relevant for the intended conceptual change

For Lia, a conception of one-half, of "three", and other small whole numbers was her prior knowledge and she used this knowledge to access ideas about unit fractions. Our job at that point was to promote the need for Lia to reorganize her available concepts, using as input for her activities what, at the outset, made sense to her. In other words, we needed situations within the learning environment in which Lia could both use her current knowledge yet find it problematic (Piaget, 1985). Within such situations, she would work to (a) interpret (assimilate) the task using her available knowledge and (b) adapt her knowledge to resolve the problematic aspects as she faces them (Steffe, 1990; von Glasersfeld, 1995).

Challenging a child to find ways to overcome problematic aspects of their work can promote *goal-directed activity*. For example, the French fry task integrated a challenge to equally share a given whole without using a ruler or folding the paper with questions to orient her attention onto the adjustment she was making – first before she would make them and then after she realized her goal was not yet accomplished. Similarly, we challenged her by asking if the share of one-person-in-four should be shorter or longer than the one she successfully produced for one-person-in-three. Simply put, to engender learning a teacher uses tasks that can make the child curious and interested in solving by using what they already know (Simon & Tzur, 2004; Sullivan et al., 2015). The student's current concepts set her goal and the activity she will ultimately carry out on some object (e.g., one person's share of a French fry as a manifestation of what would become a unit fraction) – leading to effects she would notice and reflect on (see Core Tenant #3).

As shown in Fig. 3.4, a teacher designs tasks – bridging and expanding – as an interface between a student's *available and evolving concepts* and the teacher's intentions for students' conceptual advance (Simon et al., 2018b; Simon & Tzur, 2004). Consider Lia's initial ideas about unit fractions – specifically, her thinking about the meaning of one-third. For Lia, our folding task (into half, then one-half into halves) brought forward her concept of fractions as parts of wholes, which reflects common teaching practices of fractions for all children around the world (Ministry of Education of the People's Republic of China, 2011; National Governors Association Center for Best Practices, 2010) and for students with LD or MD in particular (Fuchs et al., 2016). Lia folded the paper in a peculiar way that shows three parts (albeit not equal parts). Lia knew of a visual model in her school that, after showing one-half, would display one-third as the next logical choice. Thus, the erroneous answer of one-third, prevalent among non-LD students as well, made sense from her frame of reference (Hunt, et al., 2016a; Tzur & Hunt, 2015). For us,

Fig. 3.4 Tasks and their role as platforms to bring forward diverse ways of knowing

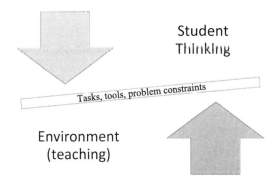

the student's thinking within this and other tasks served as a basis to infer into what they understood at the time (Hunt, 2015; Tzur & Hunt, 2015).

An instructional method suitable for engaging students in actively working on mathematics as a means to expand their available concepts and engender new, intended ones, termed *variation*, was also explicated in research on mathematics teaching in China (Gu et al., 2006; Huang et al., 2015; Jin & Tzur, 2011b). Teaching with variation promotes conceptual change by sequencing tasks in ways that support the student's engagement in goal-directed activities while promoting either or both types of reflection. Teaching with variation (of tasks) could involve solving different problems using the same method or solving the same problem using different methods/strategies. An example of the former is the sequence of French fry tasks in which we engaged Lia after moving from two to three (no ruler, no folding) people, then to four, five, six, eleven, eight, etc. Initially, she resolved a challenge (no ruler, no folding) by developing the method of iterating just one person's share for sharing a given whole (paper strip) among three people. Then, she continued using the same method while we essentially varied the number of people chosen for the task. In general, teachers learn to greatly appreciate, and use, mindfully choosing numbers for tasks as an effective instructional tool. An example of the latter (one problem solved with different methods) could be seen in Lia's use of guessing-and-checking a small piece to add (or reduce) from one person's share vs. "eyeballing" the gap between her iterated whole and the given whole so she could partition it according to the number of people sharing the fry (shown by the third yellow strip from the top in Fig. 3.2).

We note that there are no "safe procedures" for how a sequence of variation tasks could nurture the intended conceptual change. Yet, the theoretical lens we presented in this chapter give some guidance. We further illustrate such guidance after considering the core tenant to foster the mechanism for conceptual change, namely, reflection.

Core tenant #3: Promote noticing and reflection As we have articulated, the heart of learning as conceptual change is the reflection on activity-effect relationships, including two types of reflection and two stages (Simon et al., 2004; Tzur & Simon, 2004). This theoretical lens implies that direct teaching is not likely to promote conceptual learning (Hunt & Silva, 2020). Direct teaching methods, such as telling and/or modeling for students how to solve problems (Geary, 2004), may yield changes in performance (e.g., mastery of some procedures to order unit fractions by comparing the denominators as whole numbers; see Fuchs et al., 2013). Students with LD or MD may even retain some of what they have been taught for some time – as assessment of future performance may show (Fuchs et al., 2016). But, we argue, direct instruction hardly ever advances their conceptual understandings (Woodward & Tzur, 2017).

Instead, teaching methods that draw on a constructivist theory of learning entail that a core tenant for fostering students' conceptual learning is to promote noticing of and reflection on relationships between *their* goal-directed activities and effects of those activities (Tzur, 2008). Promoting noticing and reflection is indirect (or implicit), because a teacher cannot model, determine, or manage what a student's mental system sets as its goal, triggers as an activity to accomplish that goal, takes notice of as effect(s) of that activity, compares these effects to one's anticipated effects, or link across instances of similar experiences. Simply put, noticing and reflecting reside solely within the student. What a teacher can and should do is learn to occasion those mental processes in students (Pirie & Kieren, 1994).

Tzur (2008) and Hunt & Tzur (2017) have discussed specific ways by which teachers can orient (occasion) intended noticing and reflections in their students. We further explain those in Chap. 5. Here, we point out that for students to construct the participatory (prompt-dependent, provisional) stage of a new concept, a teacher can foster Reflection Type-I by orienting students to use their anticipated effect(s) as a gauge against which to compare actual effects of their activity. Questions we asked of Lia, such as, "Would you make the share of one-in-four people longer or shorter than for one-in-three?" were used to nurture her establishment of an anticipated effect. Questions that followed her completion of the activity, such as, "Did making the share for one-in-four longer than one-in-three got you closer to your goal?" then oriented her Reflection Type-I. Questions such as, "Would you need to make the next attempt for one-in-four longer/shorter than the one-in-three?" would further orient her reflection across instances, which include her initial, erroneous adjustment and the next one.

For students to construct the anticipatory stage of a new concept, a teacher can foster Reflection Type-II by orienting students to compare several instances in which their goal-directed activity yielded and should be anticipated to continue yielding similar effects. For example, we asked Lia, "In what ways were your adjustment of one person's share when moving from three to four, to five, and to six people, were similar?" or "Would you anticipate this direction of adjustment be the same when moving from eleven people to eight people and why?" Such questions promoted her abstraction of the invariant, logical necessity of inverse relation among unit fractions. We remind our reader that teachers' questions to foster Reflection Type-II would be critical for any student, let alone those with LD, to construct the intended concept. The reason is twofold: (a) such reflection does not occur automatically and (b) it is postulated to engender the anticipatory stage – without which a student is likely to be left behind.

3.4 Concluding Remarks

In this chapter, we have presented a theoretical lens that can serve teachers of students with LD (or MD) as a tool for reasoning about and promoting mathematics learning as a conceptual change. Using an example of a highly successful, teaching-learning process in which one such student (Lia) constructed a concept of unit fractions, we articulated:

(a) *Two types of mental comparison* – between anticipated and actual effects (Type-I) or across instances of activity-effect dyads (Type-II), which constitute the mechanism of reflection that engenders this change;
(b) *Two stages in this process of change*, from available concepts, through a participatory (prompt-dependent, provisional) stage, to an anticipatory (independent) stage of a new concept;
(c) *Three core tenants for teaching* implied by this theoretical lens, including (a) bringing forth students' available concepts (e.g., bridging tasks), (b) engaging students in variation tasks, and (c) orienting students' noticing and reflection on relationships between their goal-directed activities and effects of those activities.

Having seen how this theory fostered similar learning in hundreds of students with LD or MD, we argued that for mathematics teaching to effectively advance conceptual learning in those students it is imperative to steer away from teacher-directed instruction. Instead, we encouraged teachers to use individualization of instruction by assessing a student's available concepts and use the theory to foster reorganization of those into new, intended concepts (Tzur, 2008; Tzur & Lambert, 2011). Knowledge of research-based progressions can then augment the teacher's reasoning by guiding a program of instruction and assessment for students with LD and MD (Empson, 1999; Mack, 2001; Olive & Vomvoridi, 2006). Chapters 4 and 5 expound upon this idea of creating instructional trajectories that begin with students' available concepts (Simon, 1995) to improve access and opportunities for constructing mathematical ideas as they grapple with problematic situations in their experience (Clements & Sarama, 2009).

References

Biddlecomb, B., Olive, J., & Sutherland, P. (2013). *JavaBars 5.3*. Retrieved from http://math.coe. uga.edu/olive/welcome.html#Software%20developed%20throgh%20the%20Fractions%20 Project

Clements, D. H., & Sarama, J. (2009). *Learning and teaching early math: The learning trajectories approach*. Routledge.

Empson, S. B. (1999). Equal sharing and shared meaning: The development of fraction concepts in a first-grade classroom. *Cognition and Instruction, 17*(3), 283–342.

Fuchs, D., & Deshler, D. D. (2007). What we need to know about responsiveness to intervention (and shouldn't be afraid to ask). *Learning Disabilities Research and Practice, 22*(2), 129–136.

Fuchs, L. S., Fuchs, D., & Compton, D. L. (2010). Rethinking response to intervention at middle and high school. *School Psychology Review, 39*(1), 22–28.

Fuchs, L. S., Malone, A. S., Schumacher, R. F., Namkung, J., Hamlett, C. L., Jordan, N. C., … Changas, P. (2016). Supported self-explaining during fraction intervention. *Journal of Education & Psychology, 108*(4), 493–508.

Fuchs, L. S., Schumacher, R. F., Long, J., Namkung, M., Hamlett, C. L., Jordan, N. C., … Changas, P. (2013). Improving at-risk learners' understanding of fractions. *Journal of Education & Psychology, 105*(3), 683–700. https://doi.org/10.1037/a0032446

Geary, D. C. (2004). Mathematics and learning disabilities. *Journal of Learning Disabilities, 37*, 4–15. https://doi.org/10.1177/00222194040370010201

Gersten, R., Chard, D. J., Jayanthi, M., Baker, S. K., Morphy, P., & Flojo, J. (2009). Mathematics instruction for students with learning disabilities: A meta-analysis of instructional components. *Review of Educational Research, 79*(3), 1202–1242.

Gu, L., Huang, R., & Marton, F. (2006). Teaching with variation: A Chinese way of promoting effective mathematics learning. In L. Fan, N.-Y. Wong, J. Cai, & L. Shiqi (Eds.), *How Chinese learn mathematics: Perspectives from insiders* (pp. 309–347). World Scientific Publishing.

Hackenberg, A. J., Norton, A., & Wright, R. J. (2016). *Developing fractions knowledge*. Sage.

Harrington, C., Hodkowski, N. M., Wei, B., & Tzur, R. (in press). A teacher's conceptualization of the distributive property for continuous (fractional) units. In NNN & MMM (Ed.), *Proceedings of the 44th conference of the international group for the psychology of mathematics education*. PME.

Hord, C., Tzur, R., Xin, Y. P., Si, L., Kenney, R. H., & Woodward, J. (2016). Overcoming a 4th grader's challenges with working-memory via constructivist-based pedagogy and strategic scaffolds: Tia's solutions to challenging multiplicative tasks. *The Journal of Mathematical Behavior, 44*, 13–33.

Huang, R., Miller, D. L., & Tzur, R. (2015). Mathematics teaching in a Chinese classroom: A hybrid-model analysis of opportunities for students' learning. In L. Fan, N.-Y. Wong, J. Cai, & S. Li (Eds.), *How Chinese teach mathematics: Perspectives from insiders* (pp. 73–110). World Scientific.

Hunt, J. H. (2015). How to better understand the diverse mathematical thinking of learners. *Australian Primary Mathematics Classroom, 20*(2), 15–21.

Hunt, J. H., & Empson, S. B. (2015). Exploratory study of informal strategies for equal sharing problems of students with learning disabilities. *Learning Disabilities Quarterly, 38*(4), 208–220.

Hunt, J. H., MacDonald, B. L., & Silva, J. (2019a). Gina's mathematics: Thinking, tricks, or "teaching"? *The Journal of Mathematical Behavior, 56*, 100707. https://doi.org/10.1016/j.jmathb.2019.05.001

Hunt, J. H., & Silva, J. (2020). Emma's negotiation of number: Implicit intensive intervention. *Journal for Research in Mathematics Education, 51*(3), 334–360. https://doi.org/10.5951/jresemtheduc-2019-0067

Hunt, J. H., Silva, J., & Lambert, R. (2019b). Empowering students with specific learning disabilities: Jim's concept of unit fraction. *The Journal of Mathematical Behavior, 56*, 100738. https://doi.org/10.1016/j.jmathb.2019.100738

Hunt, J. H., & Tzur, R. (2017). Where is difference? Processes of mathematical remediation through a constructivist lens. *The Journal of Mathematical Behavior, 47*, 62–76. https://doi.org/10.1016/j.jmathb.2017.06.007

Hunt, J. H., Tzur, R., & Westenskow, A. (2016a). Evolution of unit fraction conceptions in two fifth-graders with a learning disability: An exploratory study. *Mathematical Thinking and Learning, 18*(3), 182–208. https://doi.org/10.1080/10986065.2016.1183089

Hunt, J. H., Welch-Ptak, J. J., & Silva, J. M. (2016b). Initial understandings of fraction concepts evidenced by students with mathematics learning disabilities and difficulties: A framework. *Learning Disability Quarterly, 39*(4), 213–225. https://doi.org/10.1177/0731948716653101

Jin, X., & Tzur, R. (2011a). *'Bridging': An assimilation- and ZPD-enhancing practice in Chinese pedagogy*. Paper presented at the 91st Annual Meeting of the National Council of Teachers of Mathematics (Research Pre-Session).

Jin, X., & Tzur, R. (2011b). *Progressive incorporation of new into known: A perspective on and practice of mathematics learning and teaching in China*. Paper presented at the Annual Conference of the Association of Mathematics Teacher Educators, Irvine, CA.

Kilpatrick, J., Swafford, J., & Findell, B. (Eds.). (2001). *Adding it up: Helping children learn mathematics*. National Academy.

Lockhart, P. (2009). *A mathematician's lament: How school cheats us out of our most fascinating and imaginative art form*. Bellevue Literary Press.

Mack, N. K. (2001). Building on informal knowledge through instruction in a complex content domain: Partitioning, units, and understanding multiplication of fractions. *Journal for Research in Mathematics Education, 32*(3), 267–295.

Ministry of Education of the People's Republic of China. (2011). *Mathematics curriculum standards for compulsory education*. Beijing Normal University.

National Assessment of Educational Progress (NAEP). (2019). Retrieved from https://www.nationsreportcard.gov/math_2019/nation/achievement?grade=4. Retrieved September 30, 2019, from National Center for Education Statistics https://www.nationsreportcard.gov/math_2019/nation/achievement?grade=4

National Governors Association Center for Best Practices. (2010). *Common Core State Standards Initiative*. Retrieved from Washington, DC: http://www.corestandards.org/the-standards

National Mathematics Advisory Panel. (2008). *Foundations for success: The final report of the National Mathematics Advisory Panel*. US Department of Education.

Olive, J., & Vomvoridi, E. (2006). Making sense of instruction on fractions when a student lacks necessary fractional schemes: The case of Tim. *The Journal of Mathematical Behavior, 25*(1), 18–45. https://doi.org/10.1016/j.jmathb.2005.11.003

Piaget, J. (1985). *The equilibration of cognitive structures: The central problem of intellectual development*. (T. Brown & K. J. Thampy, Trans. The University of Chicago.

Pirie, S. E. B., & Kieren, T. E. (1994). Growth in mathematical understanding: How can we characterize it and how can we represent it? *Educational Studies in Mathematics, 26*(2–3), 165–190.

Siegler, R. S., Thompson, C. A., & Schneider, M. (2011). An integrated theory of whole number and fractions development. *Cognitive Psychology, 62*, 273–296.

Simon, M. A. (1995). Reconstructing mathematics pedagogy from a constructivist perspective. *Journal for Research in Mathematics Education, 26*(2), 114–145.

Simon, M. A., Kara, M., Norton, A., & Placa, N. (2018a). Fostering construction of a meaning for multiplication that subsumes whole-number and fraction multiplication: A study of the Learning Through Activity research program. *The Journal of Mathematical Behavior, 52*, 151–173. https://doi.org/10.1016/j.jmathb.2018.03.002

Simon, M. A., Kara, M., Placa, N., & Avitzur, A. (2018b). Towards an integrated theory of mathematics conceptual learning and instructional design: The Learning Through Activity theoretical framework. *The Journal of Mathematical Behavior, 52*, 95–112. https://doi.org/10.1016/j.jmathb.2018.04.002

Simon, M. A., Placa, N., & Avitzur, A. (2016). Participatory and anticipatory stages of mathematical concept learning: Further empirical and theoretical development. *Journal for Research in Mathematics Education, 47*(1), 63–93.

Simon, M. A., & Tzur, R. (2004). Explicating the role of mathematical tasks in conceptual learning: An elaboration of the hypothetical learning trajectory. *Mathematical Thinking and Learning, 6*(2), 91–104.

Simon, M. A., Tzur, R., Heinz, K., & Kinzel, M. (2004). Explicating a mechanism for conceptual learning: Elaborating the construct of reflective abstraction. *Journal for Research in Mathematics Education, 35*(3), 305–329.

Steffe, L. P. (1990). Adaptive mathematics teaching. In T. J. Cooney & C. R. Hirsch (Eds.), *Teaching and learning mathematics in the 1990s* (pp. 41–51). National Council of Teachers of Mathematics.

Steffe, L. P. (2001). A new hypothesis concerning children's fractional knowledge. *The Journal of Mathematical Behavior, 20*(3), 267–307. https://doi.org/10.1016/S0732-3123(02)00075-5

Steffe, L. P., & Olive, J. (2010). *Children's fractional knowledge*. Springer.

Sullivan, P., Askew, M., Cheeseman, J., Clarke, D., Mornane, A., Roche, A., & Walker, N. (2015). Supporting teachers in structuring mathematics lessons involving challenging tasks. *Journal of Mathematics Teacher Education, 18*(2), 123–140.

Tzur, R. (2000). An integrated research on children's construction of meaningful, symbolic, partitioning-related conceptions, and the teacher's role in fostering that learning. *The Journal of Mathematical Behavior, 18*(2), 123–147.

Tzur, R. (2004). Teacher and students' joint production of a reversible fraction conception. *The Journal of Mathematical Behavior, 23*, 93–114.

Tzur, R. (2007). Fine grain assessment of students' mathematical understanding: Participatory and anticipatory stages in learning a new mathematical conception. *Educational Studies in Mathematics, 66*(3), 273–291. https://doi.org/10.1007/s10649-007-9082-4

Tzur, R. (2008). Profound awareness of the learning paradox (PALP): A journey towards epistemologically regulated pedagogy in mathematics teaching and teacher education. In B. Jaworski & T. Wood (Eds.), *The international handbook of mathematics teacher education: The mathematics teacher educator as a developing professional* (Vol. 4, pp. 137–156). Sense.

Tzur, R. (2011). Can dual processing theories of thinking inform conceptual learning in mathematics? *The Mathematics Enthusiast, 8*(3), 597–636.

Tzur, R. (2018). Simon's team's contributions to scientific progress in mathematics education: A commentary on the Learning Through Activity (LTA) research program. *The Journal of Mathematical Behavior, 52*, 208–215. https://doi.org/10.1016/j.jmathb.2018.02.005

Tzur, R. (2019a). Developing fractions as multiplicative relations: A model of cognitive reorganization. In A. Norton & M. W. Alibali (Eds.), *Constructing number: Merging perspectives from psychology and mathematics education* (pp. 163–191). Springer Nature.

Tzur, R. (2019b). Elementary conceptual progressions: Reality check + implications. In J. Novotná & H. Moraová (Eds.), *Proceedings of the international symposium on elementary mathematics teaching* (Vol. 1, pp. 29–40). Charles University.

Tzur, R. (2019c). Hypothetical learning trajectory (HLT): A lens on conceptual transition between mathematical "markers". In D. Siemon, T. Barkatsas, & R. Seah (Eds.), *Researching and using progressios (trajectories) in mathematics education* (pp. 56–74). Sense.

Tzur, R. (in press). Two-stage changes in anticipation: Cognitive sources of Aha!Moments. In B. Czarnocha & W. Baker (Eds.), *Creativity of Aha!Moment and mathematics education*. Brill.

Tzur, R., & Hunt, J. H. (2015). Iteration: unit fraction knowledge and the French fry task. *Teaching Children Mathematics, 22*(3), 148–157.

Tzur, R., Johnson, H. L., McClintock, E., Kenney, R. H., Xin, Y. P., Si, L., … Jin, X. (2013). Distinguishing schemes and tasks in children's development of multiplicative reasoning. *PNA, 7*(3), 85–101.

Tzur, R., & Lambert, M. A. (2011). Intermediate participatory stages as Zone of Proximal Development correlate in constructing counting-on: A plausible conceptual source for children's transitory 'regress' to counting-all. *Journal for Research in Mathematics Education, 42*(5), 418–450.

Tzur, R., & Simon, M. A. (2004). Distinguishing two stages of mathematics conceptual learning. *International Journal of Science and Mathematics Education, 2*, 287–304. https://doi.org/10.1007/s10763-004-7479-4

Tzur, R., Xin, Y. P., Si, L., Woodward, J., & Jin, X. (2009). Promoting transition from participatory to anticipatory stage: Chad's case of multiplicative mixed-unit coordination (MMUC). In M. Tzekaki, M. Kaldrimidou, & H. Sakonidis (Eds.), *Proceedings of the 33rd conference of the international group for the psychology of mathematics education* (Vol. 5, pp. 249–256). PME.

von Glasersfeld, E. (1995). *Radical constructivism: A way of knowing and learning*. Falmer.

Woodward, J., & Tzur, R. (2017). Final commentary to the cross-disciplinary thematic special series: Special education and mathematics education. *Learning Disability Quarterly, 40*(3), 146–151. https://doi.org/10.1177/0731948717690117

Xin, Y. P., Chiu, M. M., Tzur, R., Ma, X., Park, J. Y., & Yang, X. (2019). Linking teacher–learner discourse with mathematical reasoning of students with learning disabilities: An exploratory study. *Learning Disabilities Quarterly, 00*(0), 1–14. https://doi.org/10.1177/0731948719858707

Xin, Y. P., Park, J. Y., Tzur, R., & Si, L. (2020). The impact of a conceptual model-based mathematics computer tutor on multiplicative reasoning and problem-solving of students with learning disabilities. *The Journal of Mathematical Behavior, 58*(online first). https://doi.org/10.1016/j.jmathb.2020.100762

Zaslavsky, O. (2008). Meeting the challenges of mathematics teacher education through design and use of tasks that facilitate teacher learning. In B. Jaworski & T. Wood (Eds.), *The international handbook of mathematics teacher education: The mathematics teacher educator as a developing professional* (Vol. 4, pp. 93–114). Sense.

Chapter 4
Commentary on Part I

Ann Dowker

Abstract The author first provides a summary of the first three chapters in this book, which are focused on theoretical issues relevant to mathematical difficulties. The chapter progresses to discuss the componential nature of arithmetic and the impact this has on assessment and interventions for children with mathematical difficulties. Then the chapter discusses the impact cultural practices have on mathematics cognition.

Keywords Mathematics difficulties · Arithmetic · Assessment · Cultural practices

The first three chapters in this book deal with mathematical difficulties, and how they influence and are influenced by considerations of diversity and equity in education. The first chapter is 'Considerations of Equity for Learners Experiencing Mathematics Difficulties: At the Nexus Between Mathematics and Special Education' by Robyn Ruttenberg-Rozen, Brenda Jacobs, and Brianne Brady. They discuss the narratives that are constructed around learners experiencing difficulties and the consequences of those narratives. They argue that many pupils' abilities are discounted because learners learn about mathematics in a variety of ways, and those that are not regarded by society as the 'norm' may not be recognized at all.

The second chapter is 'Characteristics of the Learners' by Kay Owens and Shirley Yates. They discuss the diversity of learners, and summarize work by Reisman & Kauffman on characteristics that may affect learning. These include differences and difficulties in cognitive characteristics such as general speed of learning; speed of learning of specific content; memory; verbal skills; symbolic learning; attention; and ability to form concepts, relationships and generalisations. They also include differences and difficulties in physical and psychomotor characteristics, such as visual perception, visual discrimination, figure-ground perception, visual

A. Dowker (✉)
Department of Experimental Psychology, Oxford University, Oxford, UK
e-mail: ann.dowker@psy.ox.ac.uk

and auditory figure-ground distractibility, form constancy, visual-sequential memory, spatial relationships, auditory perception and auditory discrimination. They also include differences and difficulties in social and emotional development.

The third chapter is 'Discerning Learning as Conceptual Change: A Vital Reasoning Tool for Teachers' by Ron Tzur and Jessica Hunt. This article argues that learning can be seen as a process of change in learners' concepts; and that this should be addressed by starting from the concepts that children do have (assets) rather than those which they lack (deficits). They postulate two types of comparison that are important in such conceptual change: comparing anticipated and actual effects, and comparing different instances of activity-effect dyads. They also propose two stages in conceptual change: 'from available concepts, through a participatory (prompt-dependent, provisional) stage, to an anticipatory (independent) stage of a new concept'. They conclude by proposing the tenets for teachers: (1) 'Bring forward students' prior knowledge'. This may involve *bridging*: the use of initial problems that the children are capable of solving independently (rather than those that need to be modelled or assisted by teachers), that bring forward concepts which are reorganized to form the basis for solving more advanced problems. (2) 'Engage students in tasks that engender goal-directed activities relevant for the intended conceptual change'. This may involve *variation*: encouraging flexible problem-solving either by solving different problems using the same method or solving the same problem using different methods. (3) 'Encourage noticing and reflection'. This may be promoted by orienting students to compare anticipated effects of their activity with actual effects. They may first be asked to anticipate the likely effects and then asked whether their activities did result in these effects.

These recommendations, and the emphasis on diversity in learning, are consistent with, and indeed have important and fascinating implications for, theories about the componential nature of arithmetic. There is by now overwhelming evidence (e.g. Dowker, 2015; Cowan et al., 2011; Gifford & Rockcliffe, 2012; Jordan et al., 2009; Pieters et al., 2015) that arithmetical ability is not unitary, but includes a variety of components, ranging from counting to word problem solving to memory for arithmetical facts to understanding of arithmetical principles. Moreover, though the different components often correlate with one another, weaknesses in any one of them can occur relatively independently of weaknesses in the others. Weakness in even one component can ultimately take its toll on performance in other components, partly because difficulty with one component may increase the risk of the individual relying exclusively on another component, and failing to perceive and use relationships between different arithmetical processes and problems; and partly because when children fail at certain tasks, they may come to perceive themselves as 'no good at maths' and develop a negative attitude to the subject.

Awareness of the componential nature of arithmetical performance has led to attempts to develop interventions that are targeted to individually assess strengths and weaknesses in arithmetic. Such individualized, component-based techniques of assessing and remediating mathematical difficulties have surprisingly early origins. They have been in existence at least since the 1920s (Buswell & John, 1926;

Brownell, 1929; Greene & Buswell, 1930; Williams & Whitaker, 1937; Tilton, 1947). The educationist Weaver (1954) argued that:

Arithmetic competence is not a unitary thing but a composite of several types of quantitative ability: e.g. computational ability, problem-solving ability, etc. These abilities overlap to varying degrees, but most are sufficiently independent to warrant separate evaluations…Children exhibit considerable variation in their profiles or patterns of ability in the various patterns of arithmetic instruction. . .(E)xcept for extreme cases of disability, which demand the aid of clinicians and special services, remedial teaching is basically good teaching, differentiated to meet specific instructional needs (pp. 300–301).

On the other hand, component-based techniques of assessment and intervention have rarely been used extensively. This appears to be mainly due to practical problems: In under-resourced classrooms, it is difficult to provide individualized instruction. Also, especially before the internet, there was limited communication of findings between different locations and between researchers, educators, and policy-makers.

There are increasing numbers of intervention programmes involving individualized assessment and targeted interventions (Chodura et al., 2015; Dowker, 2017). A recent and initially government-sponsored example in the UK is the Numbers Count programme (Dunn et al., 2010; Torgerson et al., 2011; Torgerson et al., 2012). It involves careful assessment of individual children's strengths and weaknesses, followed by intensive individualized intervention targeted to address specific weaknesses, and emphasise the development of number concepts through multisensory teaching. Other individualized interventions include Mathematics Recovery (Willey et al., 2007; Wright et al., 2006); Catch Up Numeracy (Dowker, 2016; Dowker & Sigley, 2010; Holmes & Dowker, 2013); Calcularis (Kaeser et al., 2013); Dynamo Maths (Esmail, 2020); and Pirate Math (Fuchs et al., 2010). In particular, recent development of mathematics intervention programs emerging from collaborative work of scholars from the fields of mathematics and special education are presented later in this book (e.g., Chaps. 12, 13, 14, 15, and 16). These all appear to be giving positive results; nevertheless much more research, in far more parts of the world, still needs to be done on interventions and their development and implementation. Far more systematic research is also needed in order to compare different programmes. But even when such research is done, it is likely that there will prove to be no single best intervention programme, and that different programmes would be suitable for different groups of children.

As well as evidence that arithmetical ability is in general made up of various components, there is much evidence that different components and different ways of expressing and demonstrating them may be emphasized in different cultures and that this will influence the educational practices, assessments, and interventions that may be carried out. Language, including the level of correspondence between the linguistic practices of home and school, is likely to influence responses to both assessment and school instruction; and mathematics assessments that do not depend strongly on language may prove useful (Greisen et al., 2018).

    More generally, cultural practices have a strong impact on mathematical cogni-
tion. For example, Posner (1982) found that the Dioula, a mercantile group of people from the Ivory Coast, learned to use rather complex calculation strategies for
trading and selling purposes. Even those merchants who had never been to school
were adept at calculation. Baoule people in the same region, who were farmers
rather than merchants, did not demonstrate such high-level calculation abilities.

    Carraher et al. (1985) studied Brazilian child street traders between 9 and
15 years. All attended school, though many attended somewhat irregularly. They
were given the same arithmetic problems in three different contexts: (1) a 'street'
context, where the researchers approached them as customers and asked them about
prices and change; (2) a 'word problem' context where they were given school-type
word problems dealing with prices and change in hypothetical vending situations;
and (3) a numerical context, where they were given the problems in the form of
written sums. The children performed much better in the street context than on word
problems and much better on word problems than on written sums. They solved
almost all – 98% – of the problems correctly in the street context; 74% of the same
problems were solved correctly when presented in the form of word problems; but
only 37% were solved correctly when presented in the form of written sums. By
contrast, middle-class children, who attended school regularly but had no street
market experience, performed better in a numerical context than in a market-type
context.

    Other studies of the effects of schooling versus street trading experience were
carried out by Saxe (1985, 2012; Saxe & Esmonde, 2005). Saxe studied the arithmetical strategies of Oksapmin children in Papua New Guinea. Some were street
vendors with little or no schooling; some attended school but had no vending experience; and some had both types of experience. They were all given word problems
based on the prices and profits for selling sweets. Those with more schooling relied
more on written numbers and place value notation. Those with little or no schooling
relied more on the specific features of the currency.

    Among children with equal amounts of schooling, children with vending experi-
ence used more derived fact strategies. Those without vending experience relied
more on well-learned, school-taught algorithms. Those with more schooling relied
more on written numbers and place value notation. Those with little or no schooling
referred more to the specific features of the currency.

    The majority of the research on the effects of specific cultural practices on learn-
ing was carried out in the late twentieth century, and there appears to have been
comparatively little in the twenty-first century. Hodge & Cobb (2019) have suggested that part of the reason for this may be that different cultures are less isolated
and separable than in the past. There has, however, been extensive research on the
effects of the language and counting system on number sense and counting, both in
the late twentieth century (e.g. Miura et al., 1993) and more recently (e.g.
Bahnmueller et al., 2018; Butterworth et al., 2008; Dowker & Nuerk, 2016; Göbel
et al., 2011; Pica et al., 2004). There have also been discussions, from various points
of views, about adapting educational practice to cultural backgrounds: for example,
Bal & Traimor (2014); Hodge & Cobb (2019); Hollins (2015); Howard & Terry

(2011); Strekalova-Hughes et al. (2021). But there is clearly room for much more research on how educational practices can be adapted to children's experiences of cultural practices, as well as their individual strengths and weaknesses. The chapters discussed here will contribute significantly to such research.

References

Bahnmueller, J., Nuerk, H. C., & Moeller, K. (2018). A taxonomy proposal for types of interactions of language and place-value processing in multi-digit numbers. *Frontiers in Psychology, 9*, 1024. https://doi.org/10.3389/fpsyg.2018.01024

Bal, A., & Trainor, A. A. (2014). Development and preliminary analysis of a rubric for culturally responsive research. *Journal of Special Education, 47*, 203–216. https://doi.org/10.1177/0022466912436397

Brownell, W. (1929). Remedial cases in arithmetic. *Peabody Journal of Education, 7*(2), 100–107.

Buswell, G. T., & John, L. (1926). Diagnostic studies in arithmetic. In *Supplementary Educational Monographs, 30*. University of Chicago Press.

Butterworth, B., Reeve, R., Reynolds, F., & Lloyd, D. (2008). Numerical thought with and without words: evidence from indigenous Australian children. *Proceedings of the National Academy of Sciences, 105*, 13179–13184. https://doi.org/10.1073/pnas.0806045105

Carraher, T. N., Carraher, D. W., & Schliemann, A.D. (1985). Mathematics in the streets and in the schools. *British Journal of Developmental Psychology, 3*, 21–29. https://doi.org/10.1111/j.2044-835X.1985.tb00951.X

Chodura, S., Kuhn, J.-T., & Holling, H. (2015). Interventions for children with mathematical difficulties: a meta-analysis. *Zeitschrift für Psychologie, 223*, 129–144. https://doi.org/10.1027/2151-2604/a000211

Cowan, R., Donlan, C., Shepherd, D. L., Cole-Fletcher, R., Saxton, M., & Hurry, J. (2011). Basic calculation proficiency and mathematics achievement in elementary school children. *Journal of Education & Psychology, 103*, 786–803. https://doi.org/10.1037/a0024556

Dowker, A. (2015). Individual differences in arithmetical abilities: the componential nature of arithmetic. In R. Cohen Kadosh & A. Dowker (Eds.), *Oxford Handbook of Mathematical Cognition* (pp. 878–894). Oxford University Press.

Dowker, A. (2016). Factors that influence improvement in numeracy, reading, and comprehension in the context of a numeracy intervention. *Frontiers in Psychology, 7*, 1929. https://doi.org/10.3389/fpsyg.2016.01929

Dowker, A. (2017). Interventions for primary school children with difficulties in mathematics. *Advances in Child Development and Behavior, 53*, 255–287. https://doi.org/10.1016/bs.acdb.2017.04.004

Dowker A., & Nuerk H. C. (2016). Linguistic influences on mathematics. *Frontiers in. Psychology, 7*, 1035. https://doi.org/10.3389/fpsyg.2016.01035

Dowker, A. D., & Sigley, G. (2010). Targeted interventions for children with arithmetical difficulties. *British Journal of Educational Psychology, II. 7*, 65–81. https://doi.org/10.1348/97818543370009x1258369933249

Dunn, S., Matthews, L., & Dowrick, N. (2010). Numbers count: developing a national approach to intervention. In I. Thompson (Ed.), *Issues in teaching numeracy in primary schools* (pp. 224–234). Open University Press.

Esmail, K. K. (2020). *Dynamo maths developmental dyscalculia assessment: standardisation and validation*. JellyJames Publishing.

Fuchs, L. S., Powell, S. R., Seethaler, P. R., Cirino, P. T., Fletcher, J. M., Fuchs, D., & Hamlett, C. L. (2010). The effects of strategic counting instruction, with and without deliberate practice, on number combinations skill among students with mathematics difficulties. *Learning and Individual Differences, 20*, 89–100. https://doi.org/10.1177/001440291007600201

Gifford, S., & Rockliffe, F. (2012). Mathematics difficulties: does one approach fit all? *Research in Mathematics Education, 14*, 1–16. https://doi.org/10.1080/14794802.2012.657436

Göbel, S. M., Shaki, S., & Fischer, M. H. (2011). The cultural number line: a review of cultural and linguistic influences on the development of number processing. *Journal of Cross-Cultural Psychology, 42*, 543–565.

Greene, C. E., & Buswell, G. T. (1930). Testing, diagnosis, and remedial work in arithmetic. In *Yearbook of the National Society for the Study of Education, 1930, Part 1* (pp. 269–319). Public School Publishing Company.

Greisen, M., Hornung, C., Baudson, T. G., Muller, C., Martin, R., & Schiltz, C. (2018). Taking language out of the equation: the assessment of basic math competence without language. *Frontiers in Psychology, 9*, 1076. https://doi.org/10.3389/fpsyg.2018.01076

Hodge, L. L., & Cobb, P. (2019). Two views of culture and their implications for mathematics teaching and learning. *Urban Education, 54*, 860–884. 1177/0042085916641173.

Hollins, E. R. (2015). Culture in School Learning. In *Revealing the deep meaning*. Routledge. https://doi.org/10.4324/9781315813615

Holmes, W., & Dowker, A. (2013). Catch up numeracy: a targeted intervention for children who are low-attaining in mathematics. *Research in Mathematics Education, 15*, 249–265. https://doi.org/10.2307/749455

Howard, T., & Terry, C. L. (2011). Culturally responsive pedagogy for African American students: promising programs and practices for enhanced academic performance. *Teaching Education, 22*, 345–362. https://doi.org/10.1080/10476210.2011.608424

Jordan, J. A., Mulhern, G., & Wylie, J. (2009). Individual differences in trajectories of arithmetical development in typically achieving 5- to 7-year-olds. *Journal of Experimental Child Psychology, 103*, 455–468. https://doi.org/10.1016/j.jecp.2009.01.011

Kaeser, T., Baschera, G. M., Kohn, J., Kucian, K., Richtmann, V, Grond, U., Gross, M., & Von Aster, M. (2013). Design and evaluation of the computer-based training program Calcularis for enhancing numerical cognition. *Frontiers in Psychology, 4*: 489. https://doi.org/10.3389/fpsyg.2013.00489

Miura, I., Okamoto, Y., Kim, C., Steere, M., & Fayol, M. (1993). First graders' cognitive representation of number and understanding of place value: cross-cultural comparisons - France, Japan, Korea, Sweden and the United States. *Journal of Educational Psychology, 85*, 24–30. https://doi.org/10.1037/0022-0663.85.1.24

Pica, P., Lerner, C., Izard, V., & Dehaene, S. (2004).. Exact and approximate arithmetic in an Amazonian indigene group. *Science, 306*(5695), 499–503. https://doi.org/10.1126/science.1102085f

Pieters, S., Roeyers, H., Rosseel, Y., Van Waelvelde, H., & Desoete, A. (2015). Identifying subtypes among children with developmental coordination disorder and mathematical learning disabilities, using model-based clustering. *Journal of Learning Disabilities, 48*, 83–95. https://doi.org/10.1177/0022219413491288

Posner, J. K. (1982). The development of mathematical knowledge in two west African societies. *Child Development, 53*, 200–208. https://doi.org/10.1111/j.1467-8624.1982.tb01309.X

Saxe, G., & Esmonde, I. (2005). Studying cognition in flux: a historical treatment of fu in the shifting structure of Oksapmin mathematics. *Mind, Culture, and Activity, 12*, 171–225.

Saxe, G. B. (1985). Effects of schooling on arithmetical understandings: studies with Oksapmin children in Papua New Guinea. *Journal of Educational Psychology, 77*, 503–513. https://doi.org/10.1037/0022-0663.77.5.503

Saxe, G. B. (2012). *Cultural development of mathematical ideas: Papua new guinea studies*. Cambridge University Press.

Strekalova-Hughes, E., Nash, K. T., Schmer, B., & Caldwell, K. (2021). Meeting the needs of all cultureless learners: culture discourse and quality assumptions in personalized learning research. *Review of Research in Education, 45*, 372–407. https://doi.org/10.3102/0091732X20985081

Tilton, J. W. (1947). Individualized and meaningful instruction in arithmetic. *Journal of Education & Psychology, 38*, 83–88.

Torgerson, C. J., Wiggins, A., Torgerson, D.J., Ainsworth, H., Barmby, P., Hewitt, C., Jones, K., Hendry, V., Askew, M., Bland, M., Coe, R., Higgins, S., Hodgen, J., Hulme, C., & Tymms, P. (2011) *The every child counts independent evaluation report to DfE, 31st mar.,2011.* Department for Education.

Torgerson, C. J., Wiggins, A., Torgerson, D. J. et al. (2012). The effectiveness of an intensive individual tutoring programme (numbers count) delivered individually or to small groups of children: a randomised controlled trial. *Effective Education, 4*, 73–86. https://doi.org/10.108 0/19415532.2013.778591

Weaver, J. (1954). Differentiated instruction in arithmetic: an overview and a promising trend. *Education, 74*, 300–305.

Willey, R., Holliday, A., & Martland, J. (2007). Achieving new heights in Cumbria: raising standards through mathematics recovery. *Educational and Child Psychology, 24*, 108–118.

Williams, C., & Whitaker, R. (1937). Diagnosis of arithmetic difficulties. *The Elementary School Journal, 37*(8), 592–600.

Wright, R. J., Martland, J., & Stafford, A. (2006). *Early numeracy: assessment for teaching and intervention* (2nd ed.). Paul Chapman.

Part II

Chapter 5
Connecting Theory to Concept Building: Designing Instruction for Learning

Jessica H. Hunt and Ron Tzur

Abstract Supporting students to develop conceptual understanding can be challenging. In this chapter, we address how teachers can promote students' conceptual growth. Our aim is to illustrate connections between the framework for knowing and learning presented in Chap. 3 and instructional design considerations. We address the essential question, "What tools do teachers utilize to promote students' conceptual change?" First, we compare and contrast interventions designed for remediation versus interventions designed for learning, making connections between designing for learning and mathematical proficiency. Next, we review core features of student learning: students' noticing of relationships between goals, actions, and their effects and unpack two key design considerations teachers can use to support growth in students' concepts. Specifically, we unpack (1) bridging, variation, and reinstating types of tasks and (2) interactive prompting and gesturing. Finally, we use the fundamental concept of ten as a unit to illustrate how teachers might use such design moves to inform their teaching.

Keywords Conceptual teaching · Bridging tasks · Variation tasks · Reinstating tasks · Prompting and gesturing

In Chap. 3, we discussed what it means for a student to know mathematics well, core tenets of a strengths-based theory of student learning, and mechanisms that students use to build upon their strengths and advance their mathematics proficiency. In this chapter, we encourage teachers to consider how they can use the mechanisms that bring about conceptual change to promote students to grow their mathematical

J. H. Hunt (✉)
North Carolina State University, Raleigh, NC, USA

R. Tzur
University of Colorado Denver, Westminster, CO, USA

© The Author(s), under exclusive license to Springer Nature Switzerland AG 2022   83
Y. P. Xin et al. (eds.), *Enabling Mathematics Learning of Struggling Students*,
Research in Mathematics Education,
https://doi.org/10.1007/978-3-030-95216-7_5

conceptions. Specifically, our aim is to support the reader to make connections between the framework for knowing and learning presented in Chap. 3 and instructional design considerations, so that instruction can be designed for *student learning*.

We present this chapter with the full realization that systems are often put into place in schools to improve students' mathematics *performance* as opposed to promoting learning and more robust notions of mathematics proficiency. As we noted in Chap. 3, a common goal in such systems seems to rest in the desire to address students' difficulties that emerge in the regular mathematics classroom and, through remediation, bring students' mathematics performance to a desired level. By *remediation*, we are referring to the instructional process by which teachers work to change students' mathematics performance. By *intervention*, we are referring to the system or structures through which the process occurs.

Intervention systems in the USA are conceptualized by research and practice as a three-tiered structural approach, where intensity of supplemental instruction increases gradually in each tier (Bryant et al., 2008). The first tier is usually understood as the mathematics classroom, where whole-class mathematics teaching is employed and procedures are put into place to monitor children's learning. The second tier is conceptualized as supplemental to the first and involves small group sessions ranging in time from 20 to 35 minutes a day, for several days a week, for several weeks in settings inside or outside of the mathematics classroom (Bryant et al., 2008; Fuchs et al., 2008). The third tier is, by far, the most intensive. It often involves individualized interaction between children and teachers (Fuchs & Fuchs, 2006). Instruction at the third tier can be in addition to instruction taking place in the second tier, or it can involve an enhanced version of tier 2 instruction (e.g., increased number of sessions or amount of time spent in each session; additional area of instructional emphasis). Overall, the additional supports in the second and third tiers are meant to work with instruction happening in the mathematics classroom, with a focus on improving children's mathematical performance to meet a norm expected of their peers.

A wealth of research documents increased mathematical *performance* when remediation is used to support mathematics intervention (Brophy & Good, 1986; Ellis & Worthington, 1994). Indeed, if the goal of intervention is to increase children's mathematical performance, then remediation can be an effective instructional technique (Hiebert & Grouws, 2007). Yet, we challenge you to consider that remediation is limited in terms of its ability to promote meaningful and lasting change in students' conceptions and, ultimately, students' sense of themselves as mathematically competent people.

5.1 Rethinking the "Issue" of Student Performance

Mathematically competent people do more than simply perform mathematics. Recall from Chap. 3 that Kilpatrick et al. (2001) explained mathematical proficiency as encompassing five interrelated strands – *conceptual understanding*, *procedural*

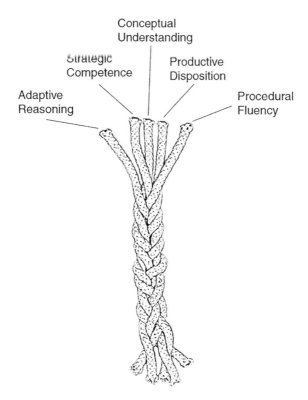

Fig. 5.1 NRC (2001) five interconnected strands of mathematics proficiency

fluency, strategic competence, adaptive reasoning, and *productive disposition*. As shown in Fig. 5.1, these strands are interconnected and inter-reliant. As a result, intervention should work to promote all five strands to build students who are mathematically competent as opposed to remediating students to perform at an expected mathematical level.

Accordingly, effective teaching builds on students' own ways of sense making and promotes student learning by asking the fundamental question, "How is the student thinking about the mathematics?" This question is asked in reference to a clear goal for student learning that drives the planning of activities, preparation of materials, and planning of conversation and interaction with students, all of which become the drivers of student learning.

When mathematical understanding [and learning] begins with the student engaging in mathematical situations that both use and challenge their ways of knowing, it allows a space to "grapple with key mathematical ideas that are comprehendible but not yet well formed" (Hiebert & Grouws, 2007, pp. 387). When teachers utilize a student's conceptual strengths, the students are positioned as already possessing a way of knowing that they use to understand and to learn, building more productive dispositions and powerful conceptions in mathematics. In this way, we encourage teachers to use systems of intervention not mostly or merely as spaces for mathematics remediation but as spaces for mathematics conceptual learning. In the next sections, we connect pedagogy to the mechanisms that promote student conceptual learning outlined in Chap. 3.

5.2 Designing for Learning: Three Considerations to Promote Students' Mathematics

Interactions, including conversations, between a student and a teacher in a well-designed instructional space constitutes a powerful pedagogy because it builds on the mechanisms of learning. Said another way, instruction can become optimized when we pair it with the knowledge of how children already think and how they learn more advanced concepts. Below, we revisit learning, its core mechanisms, and connect the mechanisms to pedagogy.

Revisiting Learning and Its Core Mechanisms In Chap. 3, we framed learning as changes in a child's concepts via reflection on the effects of their activities (Simon et al., 2004). Learning happens when children use their current ways of knowing to think about and solve a task. The learners' own concepts set their goal and the activity they carry out on some object(s).

We also explained core mechanisms of learning, such as noticing and reflection. Reflection consists of two types (Tzur & Simon, 2004). In Type I, the child compares between their anticipated effect (informed by their goal) and the noticed (actual) effects of their activity. In Type II, the child compares across instances in which the activity has been used and yielded similar effects, to abstract such an activity-effect linkage as a new concept. For example, when engaged in the French Fry activity described in Lia's case in Chap. 3 (Tzur & Hunt, 2015), a student may reflect on attempts to create one person's share that were shorter/longer than previous attempts for a different number of sharers. This typically leads to realizing the inverse relation between the number and the size of fractional units that fit in a given whole (e.g., $1/5 > 1/6$ *because* $6 > 5$). A *concept* grows out of those two types of reflection and refers to the abstract relationship a student forms between his or her activity and its effects.

Finally, we explained that Tzur and Simon (2004) equate the results of reflection with novel mathematical understandings constructed at two stages: participatory and anticipatory. Students who have *participatory understanding* of a concept need to engage in an activity to notice a result and reflect on it. That is, students' understandings in the participatory stage are provisional in that they depend on some form of being prompted (whether internally or by another person). In contrast, students who have an *anticipatory understanding* of a concept have abstracted the understanding of the mathematical necessity resulting from prior, goal-directed activity. They can draw upon this understanding in situations that are similar to, yet distinct from, those in which the understanding was constructed in the first place (i.e., new "knowing," von Glasersfeld, 1995).

When a new concept is constructed at the participatory stage, calling upon a concept depends on students being prompted for the actions that lead to the anticipated effect. If prompted, students perform like anyone who already established the concept; otherwise, they revert to using prior concepts. Later, the same new concept may be established at the spontaneous, anticipatory stage. Here, students can independently bring forth and use the concept to find/explain solutions to tasks and begin to build new participatory ways of reasoning (Tzur & Simon, 2004).

The distinction between the stages is crucial for teaching. A participatory stage of a new concept, while important, is not likely to support learning the next, more advanced concept(s) on their own. Anticipatory concepts must be present (in tandem with participatory concepts) for learning to occur (Hunt & Silva, 2020; Tzur & Simon, 2004; Siegler, 2007). As a result, these two stages of conceptual understanding are extremely important for teaching in that they ground every move that we discuss in the subsequent sections. Within, we discuss two design moves teachers can use as they work with students: (a) using tasks to bridge, promote, and reinstate thinking and (b) using gesturing and prompting to support students' noticing and reflection.

First Design Move: Use Tasks to Bridge, Promote, and Reinstate Thinking
A strengths-based approach to intervention uses what the student already knows as a starting point for instructional decision making (Hunt, 2015). In Chap. 3, we highlighted the role of tasks and materials in supporting students to think and be active within an environment. Recall that tasks and materials in and of themselves are not the same as the thinking that students generate within them. We also said that tasks are designed to bring forward a student's *prior knowledge*, or available conceptions (e.g., Hunt & Empson, 2015), so that they can be accessed and later supported to building *new, more advanced concepts*. We will illustrate the role of tasks in each stage of students' reasoning and use the prior case of Lia (Chap. 3) to discuss each move.

Bridging Tasks that act as platforms for students to access what they already know can be said to be a bridge to their prior knowledge (Jin & Tzur, 2011; Huang et al., 2015). In *bridging*, students solve tasks that activate their available concepts while also holds potential for linking to the intended concepts for learning. For example, in Chap. 3, we discussed Lia's thinking in a series of tasks involving a "French fry" (Tzur, 2007). For Lia, the task of sharing a French fry among two people served as a bridge to her current ways of reasoning. That is, she had prior ideas about fair shares and a concept of two that she could use to easily share the French fry among two people. Because the task bridged to what she already knew, she could also bring forth her concept of number as a unit made of smaller units and thus reason about the resulting quantity and coordinate the length of that quantity with the length of the whole (e.g., quantify the result as "one-half" of the French fry).

Importantly, different prior knowledge makes certain tasks a bridging task for some students but not for others. Here is an example. We once worked with a young student in the fourth grade on a task very similar to the one we posed to Lia. In this task, we used breadsticks as the items being equally shared. The student was puzzled, looked down and away several times before saying that he did not know how to solve the problem. Waiting several minutes for the student to think, he suddenly asked, "Breadsticks…are those like, are those like in lunch?" After affirming the student's thought and asking if he ate breadsticks at lunch in school or home, the student commented, "Yeah, I've had them… they're OK. I don't know. I don't like them very much. They tasted like cardboard!" After agreeing it would not be good to share something that did not taste good, the child chose a new context and solved the problem thoughtfully.

We tell this story to emphasize the importance of considering prior knowledge in bridging tasks from a wide stance. Bridging tasks should bring forward *much* of a student's prior knowledge – and this prior knowledge is not always mathematical in nature (Civil, 2007). The task about breadsticks presented to the student in the above example was not a bridge because it ignored the impact of the context in which the problem was presented on the child's reasoning. That is, teachers should take into consideration that students' life experiences shape their understanding of mathematics (Civil, 2007). In this way, bridging tasks are idiosyncratic in nature. What acts as a bridge to one student's ways of reasoning may not work for another's (e.g., Hunt et al., 2019).

Variation After a student's prior knowledge is brought forth, another tool that teachers have in terms of tasks is variation (Gu et al., 2006). In *variation*, we engage students in solving gradually more challenging tasks, either one task in different ways or different tasks in the same way. These kinds of tasks help students notice commonalities within and across tasks presented in learning situations. Doing so supports students to realize and abstract what, mathematically, must remain the same in related situations, hence building and solidifying a concept (Simon et al., 2004). For example, in the case of Lia, variation tasks could be to share the same-size paper strip among three, four, and five people. As described in Chap. 3, we surmised that Lia might make an estimate of the size of one person's share and iterate that piece the number of times needed for people who share the entire French fry. We designed this variation with an intention that comparing her iterated whole to the given one (to be shared) would bring about the need to adjust the length of the estimated, single piece if it proved to be too short. The action of partitioning via the iteration helps students coordinate the length of unit fractions with a given sized whole (Tzur & Hunt, 2015), eventually conceptualizing for themselves the inverse relation between the number and size of unit fractions.

We call these types of tasks "variation" because they support the same mathematical goal yet changes in terms of constraint(s) on solution method. For example, additional tasks can then engage the child in using the same strategy to equally share the same-size paper strip among 7, 10, or even 13 people. Along the way, we impose more constraints and challenges (e.g., find the correct share length in as few attempts as possible). The goal of the tasks remains consistent: to find the correct size share for *n* people such that *n* iterations reproduce the length of the referent whole being considered as a unit of one (1). The role of the constraints is to promote the need for more sophisticated reasoning that connects to how the child is already operating in the task.

Variation tasks in the context of the French fry activities serve as a platform for the student to construct and abstract a twofold, invariant relationship: (a) there is only a single unit that uniquely fits *n* times within a given whole (symbolized as $1/n$) and (b) a larger number of iterations to produce more shares necessarily yields smaller-size units – and vice versa (symbolized as $1/7 > 1/10$). Lia can then use this newly formed concept in related situations that deepen the concept, such as a second type of variation that involves different tasks in which the same solution method can be applied across the tasks. For example, we might challenge Lia to use the known length of different unit fractions to recreate the length of the referent whole.

Reinstating Bridging and variation tasks work with a child's participatory and anticipatory stages to support concept learning. Yet, these tasks do not work completely on their own within instruction designed to promote student learning. One obvious addition is the role of interaction between the student(s) and the teacher as students are immersed in tasks. Another, not so obvious addition is the use of a task that can *reinstate* noticing and reflection on the part of the student. This means that there are times in which, despite the best design and/or sequencing of tasks, constraints, and interactions, students may not notice the effects of their actions and reflect on them. This means that the core mechanisms needed to progress from participatory to anticipatory stages of a concept become "shut off."

The most straightforward way to explain reinstating tasks is as a mechanism to reorient the child back to a point that they were noticing and reflecting upon actions, such that the core mechanisms that promote learning can be "turned back on." We illustrate the use of a reinstating task in a case study at the conclusion of this chapter.

Second Design Move: Use Prompting and Gesturing to Support Students' Noticing and Reflection Having discussed how to use tasks to support students to bring forth their prior knowledge and advance it, we next outline how teachers can use prompting and gesturing to assess student thinking and help support students to notice and reflect upon their reasoning (i.e., their goal-directed actions and effects). We begin with assessing. As we did with tasks, we ground our discussion of prompting and gesturing in the case of Lia and include a synopsis of how teachers can use gesturing and prompting shown in Table 5.1.

Before we begin to describe the tools teachers can use in interacting with students, we emphasize that interactions are always based in a response to something a student does or explains about what *they* do within the confines of a task. Said differently, interactions within strengths-based interventions always take the form of a teacher's response to a student's way of reasoning. You might think of a basic unit

Table 5.1 Interactions to support participatory and anticipatory stages of a concept (Hunt & Tzur, 2017)

Prompt or gesture	Example (using "Lia's" case)
Prompt: Assess *(anticipatory or participatory)*	
Clarify	Say more Remind me again …
Justification	So, you are saying…. Convince me this is true
Critique/counterargument	Wait- I thought….
Prompts/gestures: Promote noticing *(participatory)*	
Prompt: Revoice	You said…. Is that right?
Prompt: Activity -> Effect	What happened? Is that what you thought would happen?
Gesture: Reshow	Reshowing the student's exact actions in a task
Prompt: Promote reflection type I *(participatory)*	
Gesture: Reshow	Any time a teacher reshows a student's actions in a problem
Prompt: Activity -> Effect	What happened? Is that what you thought would happen?
Prompt: Extend reasoning	Convince me how we know that….*(directly following a cause and effect situation)*
Prompt: Promote reflection type II *(participatory)*	
Prompt: Make prediction	How much longer the next estimate should be to get it exactly? Tell me first, before you do it.

of interaction as a communication exchange that originated with the child's utterance or action, followed by the teacher's response to the child's utterance or action, and led to the child's next utterance or action.

Prompting to Assess Evolving Understanding Assess interactions can generally be described as responding to a student's (in this case, "Lia's") ways of acting and speaking, with the goal of digging (inferring) into the student's reasoning. In our collective work (Hunt & Tzur, 2017), prompts that assess students' understanding usually take one of three forms: (a) clarification, (b) press for justification, and (c) critique. Using prompting to assess student reasoning can happen at any point (realtime) in intervention when teachers wish to explore how a student thinks. Seen through a student-adaptive pedagogy lens, the non-particular use of assess interactions makes sense, because a teacher who works to respond to a student's mathematical activity must constantly make inferences into the thinking underneath the student's activity.

Returning to our case of Lia, we might assess how Lia is thinking as she works within a variations task involving sharing a French fry equally among four people after successfully sharing it among three people. Suppose that we asked Lia to state to us, before creating the share size for four people if it would be shorter or longer than the share size for three people. Lia confidently says, "Longer." One way that we might assess her response is to use critique and counterarguments. We may say something like, "Wait; I thought shorter because four is a bigger number than three" and see how she responds. If Lia seems unsure or changes hers answer, then that tells us something different about how Lia is reasoning than if she sticks with her response and elaborates on it (e.g., "Four is larger than three, so I need to make it longer").

Prompting and Gesturing to Promote Noticing and Type I Reflection Prompting and gesturing interactions to promote Type I reflection sounds self-explanatory, yet its application can be quite complex. Interactions at this stage include (a) *Revoice*, (b) *Cause and effect*, (c) *Reshow*, and (d) *Extend reasoning*. Teachers can use these kinds of interactions at any point in an intervention when the goal is to support student's Type I reflection.

Prompts to promote *Cause-and-effect* (goal -> activity -> effect) on the part of the student occur when the teacher encourages the student to notice what happened in her activity, or the relationship between the student's activity in a task and its actual effects. The interaction is related to Type I reflection as well, because it supports the student to reflect on what they thought might happen, what actually did occur, and the alignment (or misalignment) between the two. *Gesturing* in the form of reshowing is also related to this prompt, because teachers can reshow a student's exact actions in a problem before asking them to evaluate cause and effect. Doing so can help students with working memory or attention differences reflect on their actions successfully (Hunt & Silva, 2020). An example of these types of interactions would be when the teacher responds to Lia's thinking and actions within a task by asking her to state what she noticed happened after her actions (e.g., "What happened when you repeated that [points to child's estimated share size and then reshows his repeating actions]? Is that what you thought would happen?").

Teachers can pair cause-and-effect prompts with a request to students to defend their reasoning. Asking students to defend their reasoning can help them extend and build upon reflection. A similar example to our work with Lia shows how prompts and gestures may come together in a cohesive way. Much like Lia, this student is considering how to adjust an estimate of one person's share for three people. Her estimate was too short (we use "S" for student and "T" for teacher):

S: Make it longer [proceeds to construct another estimate that adds the entire length of the shortage to her estimate].

T: Before you start there, you used your last guess [points to part of whole] and you went one, two, three [*reshows* students' repeating of part]. What happened [*cause and effect*]?

S: It didn't make it.

T: It didn't make it. So, if we do that much more onto each, will the three of those make it, do you think [*justification*]?

S: [pauses for four seconds] It would make it but probably more than that.

T: Where do you think it might stop [*make a prediction*]?

S: [taps the table beyond the fry] Over there maybe.

T: Beyond the end of the Fry. OK. (Hunt & Tzur, 2017, p. 71)

This example shows how prompts can coincide and work together with gestures to support students as they build concepts.

Prompting and Gesturing to Promote Type II Reflection   Prompting and gesturing interactions to promote Type II reflection have a core purpose of supporting and extending students' reflection upon the logical necessity of the effects of their actions within a task or commonalities across their actions across tasks. Teachers can use these kinds of interactions in an intervention when the goal is to support student's Type II reflection. This is typically done when the teacher has assessed that students already have a participatory stage of that concept (Tzur, 1996).

Comparison/prediction prompts support the child's comparison of outcomes of past activity with anticipated results of current activity. Essentially, the role of the teacher when using these kinds of prompts is to challenge the student to use their past activity and apply it (i.e., could you predict before acting?). Often, we have found that these kinds of prompts are the most consistently and heavily used interaction at work in intervention programs, because Type II reflection is what seems necessary to support anticipated actions (i.e., new concepts at the anticipatory stage; see Tzur, 2011).

Examples of the teacher using comparison/prediction prompts include asking students to use their past activity to support activity in the current situation. For example, after a student created an estimate, the teacher may ask: "Last time, you used a little bit of the leftover amount to make your next guess a little longer [*Revoice*]. You said it was too short. Now, you said you will make your next guess a little longer again [*Revoice*]. How much more will you have to put onto each to cover it all? Convince me of the amount before you do it." Such prompts also involve responding to the children by asking them to use a past activity as a way to defend a statement. For example, the teacher may say: "Can you tell what you might estimate if we share between eight people if this was what we got when we shared between seven people?"

We conclude this section with another example similar to that of Lia. The example again shows how prompts come together in a cohesive way. This student is considering how to use the length of one-eighth to produce an estimate length for one-ninth:

T: Can you predict what you might estimate if we share between nine people – if this was what we got when we shared between eight people?

S: There are more; and nine is bigger than eight; so, it would be shorter for the nine people [constructs an estimate and iterates it across the Fry. The estimate is too short]. Nine of these didn't go to the end so I need a longer one.

T: Oh. So, you made an estimate and when you repeated it for the nine sharers, it didn't go to the end? [*restate*]

S: [nods]

T: And you said you needed a longer one. [*restate*] How much longer to get it exactly? Tell me first, before you do it. [*make a prediction*]

S: You have to take the leftover part and split it up among the nine people. So, that much longer for each of the nine parts [aligns her previous estimate to the ninth iteration. Partitions the shortage amount into nine parts. Makes next estimate the length of her previous estimate plus one of the nine parts she made in the shortage amount]. (Hunt & Tzur, 2017, p. 71)

5.3 Supporting Reflection: Emma's Numbers and the Intersection of Tasks and Prompts

Another student, "Emma," was in fifth grade when we worked to advance her conception of number (Hunt & Silva, 2020). Like Lia, Emma was not confident in her mathematical abilities and was initially very reluctant to engage with us and with mathematics. She often asked us to "show her the answer" or resorted to memorized procedures, often applied incorrectly or in the wrong context. For example, across several missing-addend problems (e.g., "I have 12 buttons in total. Four are showing and the rest are hidden. How many buttons are hidden?"), Emma would request a sheet of paper so that she could add the numbers together in a vertical fashion, apparently using a procedure she had learned in school. When asked about her reasoning, Emma's explanation revealed that her understanding of the place value involved with 12 was not fully developed. That is, she explained that you could think of the "1" in "12" as one unit of one as opposed to one unit of ten. In other problems, (e.g., adding 13 cubes and nine cubes), Emma would represent 13 by using linking cubes (counting and arranging them one by one), then display nine additional cubes, and, finally, determine the total number of cubes by counting all the cubes again, starting from 1 (the first cube). Importantly, when asked if there were additional ways to determine the total, Emma did not yet conceive of any.

The procedural nature of Emma's reasoning seemed to remove her from the mental actions of noticing and reflection. This removal proved to have serious repercussions for her mathematics learning and conceptual growth. Much like Lia, we worked from what we knew of Emma's prior knowledge to understand how to plan instruction that could build on it. We learned that Emma utilized procedures to add two-digit numbers yet believed that the numerals in the tens place represented single digits. Emma also utilized a count-all strategy to find sums and was comfortable and confident doing so. Emma also attempted to use her fingers as tools with which to count, although it was evident this strategy had been discouraged somewhere in her school experience (Hunt & Silva, 2020). We were curious to see if Emma would utilize connected strategies to conceive of addition and subtraction in more sophisticated ways using her current ways of making sense, including finger counting.

Specifically, we decided to utilize tasks that could support Emma to see a need to adapt her count-all strategy into count-on and, eventually, breaking apart larger numbers into their place values. One task involved rolling a dice and moving a marker across a linear number game board (Tzur & Lambert, 2011). The game was called "How Far from the Start," with the goal of the gameplay to figure out how far away the game marker was after two rolls of a dice and two moves of the game board markers. The game moves mimicked the actions of having to generate and add together two addends.

Many strategies to combine the two addends were possible. For example, Emma could use a count-all strategy by counting visible spaces on the gameboard. She might also begin noticing that she could count by more than one space at a time. Another strategy Emma could use was count-on. Here, Emma could use the result of one of her dice rolls and subsequent moves of the game board marker as a starting point for her count. In our tasks, the initial numbers for the two addends ranged from three to nine. Playing many rounds of this game supported Emma to solve several problems that involved using multiple addends to figure a total. Covering of addends, using larger numbers, and moving the location of the unknown (e.g., from an unknown finish value to an unknown start or change value) were our designed constraints in the learning situations (Tzur & Lambert, 2011; also see Hunt & Silva, 2020). Specifically, our aim was to promote reflection on the double count involved with Emma seeing number not as the result of a counted sequence but as a composite unit, or a sequence contained inside of another sequence (Olive, 2000).

Emma brought forth a count-on relatively early in the intervention. That is, she used one of the addends in the problems as an input from which to begin her count (instead of from 1) to determine a solution to the question, "How Far from the Start Are You?" Her shift to count-on was especially supported by us covering one of the addends on the game board so that the spaces were no longer visible. Yet, we noticed that, across several turns of the board game, Emma began to procedurally figure the solution and utilize her fingers to reshow the steps of what she already used the procedure to determine. For example, with a problem involving adding three to 13, Emma figured the answer (16) by using her finger to count up to the solution. Specifically, she described beginning from three, counting four, five, and six, and then adding the "one" to arrive at the total of 16 that she already figured out procedurally.

Just as we had done when we first started working with Emma, we asked her about the meaning of the "one" in 16. To justify her reasoning, she wrote the algorithm out on paper vertically and stated that she simply added back on the "one" 13 to obtain her answer. We asked Emma about the value of the "one" in 13:

Emma: That's the one in the 13.
Teacher: But is that like, ONE [shows one finger]?
Emma: Yeah - only one.
Teacher: So, if this is one, how come I can't just count one more…seven?
Emma: Because…three doesn't have a one. (Hunt & Silva, 2020, p. 21–22)

Emma began the task by using an algorithm to find the total. She names the one in 13 as "one" as opposed to "10." We decided to next promote Emma's use of her figurative counting (fingers) to think more about combining the addends. Through questioning, our aim was to reinstate Emma's noticing and reflecting on the action of counting on from a start value other than 1. To do this, we invited Emma to consider counting on not from the larger addend of 13 but from the smaller addend of 3:

Teacher: Suppose you had started counting from this three [points to the three underneath 13 and references third space on the game board]? How would you do that? [removes paper and game board]

Emma: [puts up hand] 3…, 4 [raises one finger], 5 [raises 2nd finger], 6 [3rd finger], 7 [4th finger], … 8 [5th finger], 9 [6th finger], 10 [7th finger], 11 [8th finger], 12 [9th finger], 13 [10th finger; stares at ten fingers and pauses for 5 seconds].

Teacher: 13…. How many have you counted so far?

E: Ten.

Teacher: How many more do you need to count?

Emma: I think… [sticks out lower lip; pauses for 3 seconds]. I think…. [frowns, looks down].

Teacher: [grabs a paper and pen] So you started at three [writes three and an empty number line] …and you did 4 [makes hop on number line], 5 [makes hop on number line], 6 [makes hop on number line], 7 [makes hop on number line], 8 [makes hop on number line], 9 [makes hop on number line], 10 [makes hop on number line], 11 [makes hop on number line], 12 [makes hop on number line], 13 [makes hop on number line]. Then you stopped and did this [holds up all ten fingers and wiggles them, covers up the number line].

Emma: Oh! 13… 14 [raises a finger], 15 [raises a finger], 16 [raises a finger]. I need three more. I needed three more to get 16. (Hunt & Silva, 2020, p. 21–22).

In this excerpt, we can see that Emma finally noticed thirteen to be ten and some more. We asked Emma to begin counting from the three to promote her to see a connection between counting ten fingers and the "1" in the number 13. In starting to count from three, Emma noticed "10" in her figurative counting (her use of 10 fingers at that point).

Emma made a connection between "13," "10," and "three" in her figurative counting. Yet, she did not yet seem to connect her activity to the problem solution. As opposed to decomposing the second addend into 10 and three, reflecting upon and coordinating her figurative counting, and arriving at the solution of 16, here, she *counted up to* a known answer that she previously arrived at procedurally. Arguably, Emma used this reasoning to compensate for demands on working memory during the coordination of counting. Yet, we argue that her use of procedures took away from both *Reflection Type-I* and *Reflection Type-II*. As a result, her conceptions of 10 remained as they were, despite our attempts to promote advances in her concept of number as sequences contained inside of other sequences.

In fact, over the next several teaching sessions, Emma continued to procedurally figure out a stop value to count-up-to, using her fingers to represent her counting. This activity continued despite our attempts to ask for alternate ways in which to consider the problem. This use of algorithms to first figure out the answer and then count-up-to eclipsed Emma's propensity to notice and reflect upon her figurative activity to advance her reasoning (3... 4 [that is 1], 5 [2], 6 [3] ... 16 [13, or ten and three]). As opposed to an awareness of two number sequences at once, Emma kept the two number sequences separate when considering addends larger than 10 whose result crosses decades, leaving her potential to construct a concept of ten as a unit and use of break apart strategies untapped.

We point out the focus on noticing and reflecting across goal-directed activity and its effects as critical conceptual change mechanisms. For example, past attempts to increase performance in number through activities such as building numbers with cubes and base ten materials did not seem to be effective for Emma. She was taught to circle the tens digit to aid in ordering and comparing numbers and was also provided with a number line at her desk, and a hundreds chart in her math journal, to help her see the order of numbers. She was also taught procedures for adding numbers and heavily relied on them to figure out sums.

Conversations with Emma's teacher revealed that the extra instruction was not helpful. For instance, if Emma saw a written number (e.g., 17), she could point to the tens and ones yet could not state the value of these numbers. Together with the teacher, we talked about a way to support Emma to not only access but also advance her knowledge by supporting her to notice and reflect upon her activity to promote conceptual change.

Promoting Learning We used bridging, variation, and reinstating tasks alongside interactive prompting and gesturing to support the Emma's growth. We planned prompts to help her notice what happened in her activity as she works in and across relevant situations (e.g., "Counting all of the objects helped me find 'how many'"). For example, a task that served to both bridge to her current reasoning (count-all) and extend it (to count-on) through variation was the "How Far From Start" game (see Chap. 3). The task has the same mathematical goal throughout, yet it changes in terms of the constraint(s) on the child's solution methods.

When we planned this task for Emma, we thought she might engage in noticing and the first type of reflection. For example, because she could use count-all on the boardgame spaces at first, we thought Emma might notice that "Counting all helped [her] find out how many." We also thought Emma might notice an activity that took her away from her goal (e.g., "Counting an object twice does not help me find 'how many'"). This noticing and reflection could help her become aware of and reorganize her ways of knowing (e.g., "If I use the first addend as a starting point, I can also reach my goal and I don't have to count all"), especially if we covered up the gameboard spaces that comprised the first addend in the problem.

Such constraints also could help Emma engage in a consistent activity in terms of her goal (e.g., "When I use the first addend to count on from, I use the second addend to know when to stop counting"). We planned additional constraints within the task to help Emma use reflection to expand upon her goal or make it more sophisticated. For example, we then covered the first and second addend, which may have helped Emma consider more sophisticated ways of counting on (e.g., "I need to find another way to keep track of the second addend").

Reflection across consistent activity attempts seemed to support Emma to begin anticipate connected activity in future situations ("This addend is more than ten; another way to keep track is breaking the addend into ten and some more"). Here, we describe the additional tasks and prompts that supported Emma to build upon her reasoning and move past using a procedural strategy that removed her from reasoning and making sense of her actions. Specifically, we created a new kind of task – *Reinstating* – in which Emma was taken back to a period in the intervention where she was still making sense, noticing, and reflecting upon her actions. We did so because she continually used a prior compensatory strategy (i.e., an algorithm) across many tasks, arguably because the new actions and reflections on those actions she was building did not trump her current compensations for "getting through" such problems.

As earlier stated, there are times that, despite the best design and/or sequencing of tasks, interactions, and representations, children become removed from noticing the effects of their actions and reflecting on the result. This means that the core mechanisms needed to progress from participatory to anticipatory stages of a concept become "shut off." Our reinstating tasks for Emma were the tool we used to restore her noticing and reflecting upon actions such that the core mechanisms that promote learning can be "turned back on."

Given Emma's activity, we made the instructional decision to return to problems where Emma was successful and reinstate her noticing and reflection. Specifically, for Emma, that meant returning to problems where the result crossed a decade yet used smaller numbers. Within these tasks, we began with our current constraints (both addends covered) and eventually utilized additional problem constraints (e.g., we removed the gameboard and asked Emma to imagine her activity as if the board spaces were still visible). Throughout, we paired the variation tasks with core prompting and gesturing to support Emma to engage in both types of reflection and advance her reasoning. Here are Emma and the teacher's interactions as she solved the task 8 + 6:

T: [covers spaces Emma counted out earlier] How far from the start are you?
S: 6, 7-8-9-10-11; 12, ..., 13, 14 [looks at her fingers].
T: Why? [*Assess- Justification*]
S: I had six and added eight more.
T: Ok; so did you know it would be 14 first or did you need to count it up?
S: I had to count it up.
T: And you knew when to stop because? [*Cause and Effect; Extend Reasoning*]
S: Umm [stares at five fingers raised, then looks over at three fingers raised]. Eight.
T: You saw eight fingers, so you knew you counted eight? [*Assess: Clarify*]
S: [Nods "Yes"]
T: I wonder if there is a way to do that with bigger numbers. [*Make a Prediction*]
S: [Smiles slightly]

And later in a task involving adding 13 and 14:

S: [Counting] 14-15-16-17-18 [has five fingers raised]; 19-20-21-22-23 [has 10 fingers raised; puts all fingers back down]; 24-25-26-27 [looks at teacher–researcher].
T: Did you know to stop at 27? [*Assess- Clarify*]
S: No [confidently].

T: Then how did you know when to stop? [*Assess- justification*]

S: Because it says 14 [looks at her fingers].

T: But I only see four [points to fingers]. [*Assess- counterargument*]

S: It's because I went 10. [Shows fingers again] 14, 15, 16, 17, 18, 19, 20, 21, 22, 23… [has all 10 fingers up, then closes them into two fists which she shows to the researcher], then 24-25-26-27 [shakes her four fingers at the researcher].

T: You did 10… and four more. Is 14 like 10 and four?

S: [smiles broadly and nods] (Hunt & Silva, 2020, p. 21–22).

5.4 Conclusion

As we stated in Chap. 3, teachers who view "intervention" as a platform for students to gain access to the opportunity to use their own mathematics and reflect upon it as empowered learners, can promote their students' conceptual learning. The *second* part of this work involves employing tools within interventions that help students bring about their conceptual change. We have argued that key design and pedagogical moves available to teachers include (a) *bridging, variation, and reinstating* tasks and *(b) interactive prompting and gesturing* that support students' noticing and reflection upon their actions toward logical necessity and mathematical connections. By utilizing these moves, teachers can effectively support students' development of conceptual understanding, reasoning, sense making, productive disposition, and adaptive expertise.

References

Brophy, J., & Good, T. (1986). Teacher behavior and student achievement. In *Handbook of research on teaching* (pp. 238–375).

Bryant, D. P., Bryant, B. R., Gersten, R., Scammacca, N., & Chavez, M. M. (2008). Mathematics intervention for first-and second-grade students with mathematics difficulties: The effects of tier 2 intervention delivered as booster lessons. *Remedial and Special Education, 29*(1), 20–32.

Civil, M. (2007). Building on community knowledge: An avenue to equity in mathematics education. In *Improving access to mathematics: Diversity and equity in the classroom* (pp. 105–117).

Ellis, E. S., & Worthington, L. A. (1994). *Research synthesis on effective teaching principles and the design of quality tools for educators.* National Center to Improve the Tools of Educators, College of Education, University of Oregon.

Fuchs, L. S., Fuchs, D., & Zumeta, R. O. (2008). A curricular-sampling approach to progress monitoring: Mathematics concepts and applications. *Assessment for Effective Intervention, 33*(4), 225–233.

Fuchs, D., & Fuchs, L. S. (2006). Introduction to response to intervention: What, why, and how valid is it? *Reading Research Quarterly, 41*(1), 93–99.

Gu, L., Huang, R., & Marton, F. (2006). Teaching with variation: A Chinese way of promoting effective mathematics learning. In L. Fan, N.-Y. Wong, J. Cai, & L. Shiqi (Eds.), *How Chinese learn mathematics: Perspectives from insiders* (pp. 309–347). World Scientific.

Hiebert, J., & Grouws, D. A. (2007). The effects of classroom mathematics teaching on students' learning. In *Second handbook of research on mathematics teaching and learning* (Vol. 1, pp. 371–404).

Hunt, J. H., & Empson, S. B. (2015). Exploratory study of informal strategies for equal sharing problems of students with learning disabilities. *Learning Disability Quarterly, 38*(4), 208–220.

Hunt, J. (2015). How to better understand the diverse mathematical thinking of learners. *Australian Primary Mathematics Classroom, 20*(2), 15–21.

Hunt, J., & Silva, J. (2020). Emma's negotiation of number: Implicit intensive intervention. *Journal for Research in Mathematics Education, 51*(3), 334–360.

Hunt, J. H., Silva, J., & Lambert, R. (2019). Empowering students with specific learning disabilities: Jim's concept of unit fraction. *The Journal of Mathematical Behavior, 56*, 100738.

Hunt, J., & Tzur, R. (2017). Where is difference? Processes of mathematical remediation through a constructivist lens. *The Journal of Mathematical Behavior, 48*, 62–76.

Huang, R., Miller, D. L., & Tzur, R. (2015). Mathematics teaching in a Chinese classroom: A hybrid-model analysis of opportunities for students' learning. In L. Fan, N.-Y. Wong, J. Cai, & S. Li (Eds.), *How Chinese teach mathematics: Perspectives from insiders* (pp. 73–110). World Scientific.

Jin, X., & Tzur, R. (2011). *'Bridging': An assimilation- and ZPD-enhancing practice in Chinese pedagogy.* A presentation at the 91st Annual Meeting of the National Council of Teachers of Mathematics.

Kilpatrick, J., Swafford, J., & Findell, B. (2001). In National research council (Ed.), *Adding it up: Helping children learn mathematics* (Vol. 2101). National Academy Press.

National Research Council (U.S.). (2001). *Adding it up: Helping children learn mathematics.* Washington, DC: National Academy Press.

Olive, J. (2000). Children's number sequences: An explanation of Steffe's constructs and an extrapolation to rational numbers of arithmetic. *The Mathematics Educator, 11*(1), 4.

Tzur, R. (1996). *Interaction and children's fraction learning.* UMI Dissertation Services (Bell & Howell).

Tzur, R. (2007). Fine grain assessment of students' mathematical understanding: Participatory and anticipatory stages in learning a new mathematical conception. *Educational Studies in Mathematics, 66*(3), 273–291.

Tzur, R. (2011). Can dual processing theories of thinking inform conceptual learning in mathematics? *The Mathematics Enthusiast, 8*(3), 597–636.

Tzur, R., & Lambert, M. A. (2011). Intermediate participatory stages as zone of proximal development correlate in constructing counting-on: A plausible conceptual source for children's transitory "regress" to counting-all. *Journal for Research in Mathematics Education, 42*(5), 418–450.

Tzur, R., & Hunt, J. (2015). Iteration: Unit fraction knowledge and the French fry tasks. *Teaching Children Mathematics, 22*(3), 148–157.

Tzur, R., & Simon, M. (2004). Distinguishing two stages of mathematics conceptual learning. *International Journal of Science and Mathematics Education, 2*(2), 287–304.

Siegler, R. S. (2007). Cognitive variability. *Developmental Science, 10*(1), 104–109.

Simon, M. A., Tzur, R., Heinz, K., & Kinzel, M. (2004). Explicating a mechanism for conceptual learning: Elaborating the construct of reflective abstraction. *Journal for Research in Mathematics Education, 35*(3), 305–329.

Von Glasersfeld, E. (1995). *Radical constructivism: A way of knowing and learning. Studies in mathematics education series: 6.* Falmer Press, Taylor & Francis.

Chapter 6
Meaningful Assessments of Students Who Struggle to Learn Mathematics

Dake Zhang, Carolyn A. Maher, and Louise C. Wilkinson

Abstract This chapter presents the results of collaboration among special education, mathematics education, and psycholinguistic researchers, with a focus on meaningful assessments for students who have difficulties with learning mathematics. Gaining knowledge of the mathematical abilities of students who struggle or who have particular learning disabilities is often limited in traditional assessments. Access to struggling students' ability to reason and solve problems may be obscured because of inadequate approaches and limitations in assessing the knowledge of these students using traditional test items. Our goal is to address these concerns and other obstacles to assessment and suggest additional and alternate ways of measuring students' knowledge. It is noteworthy for researchers to attend to what are termed "learning disabilities" in mathematics and the characteristics of goal requirements in students' individualized education programs (IEPs). IEP requirements are understood in relation to assessment approaches that are used to identify students who struggle to learn mathematics. In particular, it is essential to consider former, existing, and emerging theoretical views of mathematics and special education educators regarding what might be considered the "ability" of students to learn mathematics.

Keywords Mathematics · Assessment · Students with disabilities/difficulties

6.1 Introduction

Effective instruction for students with mathematics difficulties/disabilities (MD) requires accurate and meaningful assessment to understand students' knowledge, skills, and abilities in order to identify where improvement is needed. Additionally,

D. Zhang (✉) · C. A. Maher
Graduate School of Education, Rutgers University, New Brunswick, NJ, USA
e-mail: Dake.zhang@gse.rutgers.edu

L. C. Wilkinson
School of Education, Syracuse University, Syracuse, NY, USA

© The Author(s), under exclusive license to Springer Nature Switzerland AG 2022   101
Y. P. Xin et al. (eds.), *Enabling Mathematics Learning of Struggling Students*,
Research in Mathematics Education,
https://doi.org/10.1007/978-3-030-95216-7_6

assessment is important in evaluating the effectiveness of teaching or other interventions. Historic approaches to assessing mathematics are primarily score-based, with a focus on assessing students' procedural mathematics knowledge. The conditions of standardized testing, regardless of the specific format, are such that success depends on the accuracy in providing the standard answers to the questions. Testing procedures are standardized, and an individual student's status can be compared with a population or a sub-population (based on responses) to determine the individual's position or rank in that population. Students who fall below a certain percentile rank are considered to have some learning problems or be at risk for not learning adequately. For example, often in psychological research, students with MD are identified as those whose achievement scores in mathematics fall below a specific cutoff, which may range from the 5th to the 30th percentile (Geary et al., 2007; Swanson & Hoskyn, 1998). However, students with MD are often disadvantaged in these score-based approaches, as their special needs can distort and mask their abilities to respond quickly and accurately. More recently, a focus on assessment has moved from measuring mathematics content knowledge to assessing students' reasoning abilities. Assessments that engage struggling students in problem-solving activities, along with interviews and error analysis with a microgenetic approach, offer the potential to examine and evaluate how these students represent, communicate, discuss, and support solutions to problems in contexts that promote thoughtfulness. These assessments are addressed in this chapter.

In this chapter, we (a) review traditional assessment methods (i.e., norm-referenced and curriculum-based assessments) that are widely used in special education and school psychology fields; (b) discuss the advantages and disadvantages of traditional score-based assessments and their limitations in revealing the mathematical abilities of students; (c) address accommodations in standardized assessment and alternative approaches; and (d) recommend multiple methods, including clinical interviews, error patterning analysis, strategy analysis, and a microgenetic approach, both in formative and summative ways, to help teachers and researchers provide meaningful assessments to understand and evaluate students' abilities to reason and solve mathematical problems.

6.2 Part I: Traditional Mathematics Assessment Methods in Current Education Practices

In this section, we review current and widely used mathematics assessment approaches, including large-scale norm-referenced assessments and curriculum-based assessments and measurements. Both types of assessment, in which the test is administered and scored in a predetermined, standard manner, are considered standardized assessments. In this case, when a predetermined testing protocol is applied, all test takers are directed to follow the same administration procedures and answer questions in the same format. Additionally, a standardized test is scored in a "standard" or consistent manner, which makes it possible to compare the relative performance of individual students or groups of students. While different types of tests and assessments may be standardized in this way, the term "standardized test" is primarily associated with large-scale tests administered to large populations of students. In special education, another type of commonly used standardized assessment is the curriculum-based assessment or curriculum-based measurement. These standardized assessments are typically used as summative assessments conducted at the end of a learning unit or after a period of time and are used primarily by educators and school administrators to determine learning acquisition.

6.2.1 Large-Scale Norm-Referenced Assessment

Norm-referenced tests in mathematics include many high-stakes achievement assessments, such as state standard tests. Most states have their own state standard tests that reflect the state mathematics standards, for example, State of Texas Assessments of Academic Readiness (STAAR), Pennsylvania System of School Assessment (PSSA), Rhode Island Comprehensive Assessment System (RICAS), Nevada Proficiency Examination Program (NPEP), and so forth. There are also several standardized assessments that are aligned with Common Core Standards and are adopted by multiple states, such as the Partnership for Assessment of Readiness for College and Careers (PARCC) used by District of Columbia, Louisiana, Massachusetts, and Maryland, and Smarter Balanced, used by California, Connecticut, Delaware, Washington, Oregon, Vermont, and South Dakota (Gewertz, n.d.). These assessments are designed to evaluate the extent to which the state standards for mathematics have been met at the level of individual students, schools, and school districts. Students' scores can be standardized to determine a student's position in a population of the same age or grade level. Although these state standard assessments are rarely used by school psychologists as instruments to identify a student's learning disabilities, each student is assigned a scaled score for their level of achievement. Students who are identified as not meeting the basic level in these state standard tests are typically identified for remediation so that intensive interventions can be created to address ways of improving achievement levels.

These high-stakes, state standard assessments have resulted in increased school accountability for developing plans to increase the academic achievement of students with disabilities. Since the enactment of No Child Left Behind [NCLB] (2002), students with disabilities have been required to participate in assessments of reading/language arts and mathematics. Schools that do not demonstrate adequate yearly progress (AYP) for these students are held accountable and are subject to corrective action. After 2015, the Every Student Succeeds Act (ESSA), reauthorized by the Obama administration, maintained the expectation of school accountability for students' progress. Both NCLB and ESSA mandate that schools report the performance of subgroups of students.

Standardized tests also include national or international academic assessments, such as the National Assessment of Educational Progress (NAEP, 2017), important international comparison studies such as Trends in International Mathematics and Science Study (TIMSS, 2011), and the Programme for International Student Assessment (PISA, 2015). NAEP, known as The Nation's Report Card, provides important information about how well students are performing academically and is administered to a representative sample of students in the United States (National Center for Education Statistics [NCES], n.d.). Both TIMSS and PISA are large-scale assessments designed to inform educational policy and practice by providing an international perspective on teaching and learning. The TIMSS assessment evaluates performance of fourth and eighth graders in mathematics and science. The mathematics assessment involves three domains for fourth graders (i.e., number, measurement and geometry, and data) and four domains for eighth graders (i.e., number, algebra, geometry, and data and probability). PISA targets 15-year-old students and includes two tests that are related to mathematics (i.e., mathematics literacy and problem solving) among six subjects. PISA has a greater focus on students' mathematical reasoning as it includes content, cognitive, and application components, whereas TIMSS focuses mainly on mathematical content (NCES, n.d.). In contrast to the state high-stakes standard tests, NAEP, TIMSS, and PISA do not report scores to individual students or to schools. Rather, the results of NAEP are reported for groups of students with similar characteristics (e.g., gender, race and ethnicity, and school location; NCES, n.d.), and the results of TIMSS and PISA are reported for nations and states. While scores cannot be used to identify, diagnose, or evaluate students with learning disabilities, they provide average scores, the percentage of students below the low-performance benchmark, and the achievement gap between high and low achievers. For example, TIMSS 2019 data suggested that the percentage of US fourth graders below the low international benchmark (7%) was not significantly different from the international median (8%) and that the score gap between top and bottom performing was larger than the top and bottom gap in most countries (NCES, n.d.). Additionally, educators are able to investigate how certain non-cognitive factors, such as culture, instruction, curriculum, and family SES, affect students' achievement in mathematics, a perspective that supports the social-cultural aspect of mathematics learning disabilities.

Some large-scale, norm-referenced standardized assessments can be used as diagnostic tests. These tests are used to obtain further information about a specific skill, area of academic achievement, or certain cognitive capacities. Some diagnostic assessments specifically focus on mathematics knowledge. These diagnostic tests are closely related to standard special education practices, as school psychologists or the Child Study Team usually administer diagnostic cognitive or achievement assessments as part of the comprehensive evaluation to identify a student's disability type and to determine the eligibility for special education services. For instance, the KeyMath-3 Diagnostic Assessment (Connolly, 2007) provides a comprehensive assessment of three domains of mathematics (basic concepts, operations, and applications) in 12 content subtests (i.e., numeration, algebra, geometry, data analysis and probability, measurement, mental computation and estimation, addition and subtraction, multiplication and division, foundations of problem solving, and applied problem solving). Test items and goals are aligned with the NCTM *Standards* and may be used for benchmarking and for monitoring student progress. Another example of a diagnostic assessment that specifically measures mathematical abilities is the Test of Mathematical Abilities (TOMA-3; Brown et al., 2013). This test seeks to assess areas of mathematics functioning that are not typically addressed with other instruments. For example, the TOMA-3 addresses topics such as a student's expressed attitude toward mathematics and general vocabulary level when used in a mathematical sense, and how knowledgeable the student is regarding the functional use of mathematics facts and concepts in the general culture. Many other norm-referenced achievement tests include one or more subtests in mathematics. For example, the Woodcock-Johnson IV Tests of Achievement includes three subtests (i.e., Applied Problems, Calculation, and Math Facts Fluency); the Peabody Individual Achievement Test-4 includes a mathematics subtest; the Kaufman Test of Educational Achievement-III includes subtests of Math Concepts & Applications and Math Computation; and the Wechsler Individual Achievement Test-Third Edition includes Numerical Operations, Math Problem Solving, and Math Fluency.

Large-scale, norm-referenced assessments serve a different purpose when compared with other assessments. For example, these assessments can be used as a snapshot for international comparisons, as a portrait of an individual student's cognitive profile to identify obstacles and strengths, or to position a student in comparison to a larger population. While large-scale, norm-referenced standardized assessments may fulfill the above purposes, they fail to reach many other purposes of assessments, such as to precisely capture a student's individual growth in an area. Also, problems related to large-scale norm-referenced assessments have been documented by researchers (Deno, 1985), with three primary areas of concern: (a) The tests need personnel with specialized training to be administered and interpreted, (b) the content may not represent material taught in the actual curriculum, and (c) the results are not sufficiently sensitive to measure students' short-term growth and small gains. Large-scale, norm-referenced assessments are widely used in the traditional IQ-achievement-discrepancy approach to identify students with learning disabilities. Specifically, students who are referred for a full evaluation are tested with norm-referenced assessments such as the Woodcock-Johnson Test of Achievement, and along with their IQ scores, they are used to identify whether discrepancies exist between cognitive capacity and actual mathematics achievement. For example, if a student achieves a standard score of 100 on an IQ test and a score of 85 on a mathematics achievement test, the student may be diagnosed as having a learning disability.

These large-scale assessments are usually used as summative assessments and differ from formative assessments that can be used to inform instructional decisions. The summative assessments are primarily used to assign students a grade for accountability purposes. However, with the wide implementation of the response to intervention (RtI) model, students' learning disabilities are defined as non-responsiveness to interventions, especially non-responsiveness to small-group and intensive interventions. The large-scale interventions cannot meet the needs of the RtI model because of its lack of sensitivity to student gains in a short period of time.

Further, they lack consideration of the impact of the instructional materials used during the interventions. In contrast, curriculum-based assessment measures, which we will discuss in the next section, are a preferred choice to avoid the shortcomings of large-scale assessments.

6.2.2   Curriculum-Based Measurement

Curriculum-based measurement (CBM) is another type of standardized and systematic assessment method that has been used for assessing the progress of students with MD (Deno, 2003; Overton, 2011). CBM uses content from the curriculum to assess student progress within a given topic in the curriculum. A CBM includes multiple parallel probes. With carefully designed sampling, each probe selects items that represent aspects of the curriculum for an entire school year. While each probe is different, each assesses the same types of skills at the same level of difficulty. Teachers administer the CBM probes biweekly or weekly to monitor student progress by visual analysis of students' growth trends and by comparing a student's progress with an expected rate of progress.

CBM is considered both a standardized assessment and an informal assessment (Overton, 2011). Validated CBMs have standardized directions for administration and interpretation, and some of the tests are norm referenced in order to compare the performance of an individual student with a group of the same age or grade. An example of a well-known CBM in mathematics is Early Math Fluency (Hosp et al., 2007), which includes three measures: Quantity Discrimination, Missing Number, and Number Identification. All Early Math Fluency assessments have an administration time of 1 min. Research-based norms are included in Hosp et al.'s document, with fluency benchmarks researched by Chard et al. (2005) that provide screening norms for each of the tests of the three areas. Another example of a well-validated mathematics CBM is the CBM-Computation Fluency (https://www.interventioncentral.org/curriculum-based-measurement-reading-math-assesment-tests). This is a brief, timed assessment designed to inform teachers whether a student is developing computation fluency and remains on track to master grade-appropriate mathematics facts (basic computation problems). Each student is given a worksheet containing mathematics facts and is given 2 min to answer as many problems as possible. The worksheet is then collected and scored, with the student receiving credit for each correct digit in his or her answers. Teachers can then compare any student's performance to research norms and evaluate whether the student is making adequate progress.

CBM may be considered a *formative* type of evaluation because teachers use the results to inform their instructional decisions. In particular, with the implementation of RtI models, CBMs are widely used to monitor student learning outcomes, to identify at-risk students, to evaluate intervention effectiveness, and to inform instructional decisions (Fuchs & Fuchs, 2004; Fuchs et al., 1990; Hosp & Hosp, 2003). This monitoring of progress allows the teacher to know if the child is making adequate progress under the current educational conditions and to make adjustments if necessary.

In comparison to large-scale, norm-referenced tests, CBMs have some obvious advantages. First, probes are brief and easy to administer; they can be administered frequently such as weekly or bi-weekly; and teachers do not need special trainings to be qualified to administer the probes. Second, probes are aligned with topics in the school curriculum and can sensitively reflect a student's small gains or growth during a shorter period of time. Third, the probes are inexpensive to implement in contrast with large-scale norm-referenced assessments. Further, the results yielded by the measures are easy to explain to others, such as to parents and other school personnel. In general, CBMs are useful and widely recommended for screening students at risk and to monitor progress in the RtI model for identifying students with MD and providing early interventions.

However, some challenges are posed by CBMs. While a review of the literature suggested the existing CBMs in mathematics have acceptable levels of reliability, the criterion validity of mathematics CBMs is not as good as that for reading CBMs (Foegen et al., 2007). A lack of evidence in validity is also suggested in another literature review (Christ et al., 2008). More importantly, similar to the large-scale standard test, CBMs also determine students' grades based on students' problem-solving accuracy and speed, whereas students' problem-solving or reasoning process is neglected.

6.2.3 Accommodations Provided in Standardized Assessment

Based on the Americans with Disabilities Act (ADA) and the Individuals with Disabilities Education Act (IDEA), individuals with disabilities are eligible to receive necessary testing accommodations while being assessed. Testing accommodations are "changes to the regular testing environment and auxiliary aids and services that allow individuals with disabilities to demonstrate their true aptitude or achievement level on standardized exams or other high-stakes test" (ADA.org). This act enables modifications to be made to standardized test conditions in order to level the playing field of students with and without disabilities. A purpose of identifying appropriate accommodations is to achieve valid testing results demonstrating students' true competence (Fuchs & Fuchs, n.d.). The most frequently used accommodations allowed in high-stakes assessment by most states include dictated response, Braille, large print, extended testing time, and use of interpreters (Bolt & Thurlow, 2004). There are also a variety of testing accommodations that are not mentioned in education laws in many states but are widely used in special education classroom practices. For example, representation accommodations address students' comprehension barriers and attention problems in mathematics testing (e.g., color coding, highlighting key information, simplified language interpretations). Equipment and materials accommodations include providing amplification equipment, audio-videocassettes, calculators, and magnification equipment. Response accommodations include providing computers, scribes, spell checkers, and writing in the test booklet. Scheduling timing accommodations include providing extended time, testing over multiple days, testing at a time beneficial to the student, and providing resting time between problems. Setting accommodations include individual administration, separate rooms, small-group administration, and administration of assessments in a student's home.

Accommodation supports are meant to compensate for a student's documented disability and are designed to eliminate obstacles to responding to test items. They are intended to provide equal opportunities to students by adequately measuring the target knowledge or skill (Thurlow, 2002). Different from modifications which denote changes in practices that lower expectations to compensate for a disability, accommodations are changes in practices that hold a student to the same standard as students without disabilities (i.e., grade-level academic content standard) but provide differential supports so that the student can access the general education curriculum (Harrison et al., 2013).

In a classical education or psychology assessment, a standardized protocol to implement the assessment is commonly employed. That is, a pre-determined standardized testing procedure and scoring procedure is used to assess all participating students. Accommodation supports are usually also standardized. For example, they may be designed to provide simplified explanations to students who have comprehension difficulties when solving mathematics problems. In classic psychology research, such a standardized assessment is required and necessary to ensure that all participants receive the same treatment and to rule out any confounding variables that could cause threats to internal validity (i.e., the extent to which a study establishes a cause-and-effect relationship between an independent variable and a dependent variable). However, a recent study by Zhang and Rivera (in press) raised questions about the assessment validity of a commonly used standardized test with a predetermined accommodation method for assessing students with MD. In that study, the researchers observed substantial variations, both interpersonally and intrapersonally, in the cognitive barriers that hindered students with MD from accessing the mathematics problems. These variations made it difficult, or almost impossible, to use a standardized testing protocol with predetermined, "one-size-fits-all" accommodation support to meet the complex and specific needs of students with MD. For example, some students with difficulties comprehending the content of the mathematics problems benefited by being provided with simplified interpretations that enabled the removal of comprehension barriers. This could be explained, at least in part, because each student had different obstacles that led to comprehension difficulties.

6.3 Part II: Assessing Students' Mathematical Reasoning

In this section, we will introduce several methods to assess students' reasoning. These assessment methods are typically used as formative assessments. Formative assessments are usually brief and frequent check-ins throughout teaching the content and should be used to identify where students are successful and where they are struggling (Black & Wiliam, 2010; Dann, 2014). Formative assessments serve as an ongoing evaluation of student learning as a means of providing continual feedback about performance to both learners and instructors. The value of offering both frequent feedback and feedback embedded throughout the lessons is that teachers can learn from the information they are collecting and use formative assessment results to make ongoing judgments, thus making use of the data they collect to inform instruction and make adjustments to their lessons. This information serves as an aid in determining what material needs to be revisited, what interventions can be used, and what materials may provide greater insight into students' cognitive obstacles. Formative assessments are dynamic, ongoing, and useful for revealing important details about where students are in their learning progressions. In this section, we will introduce and describe three commonly used, informal, formative assessment methods that provide in-depth analysis of student mathematical reasoning.

6.3.1 What Do We Want to Learn about Students' Knowledge? Knowledge Outcome Versus Learning Process

What is the objective of a meaningful assessment? There is no agreed-upon conclusion among researchers in special education, mathematics education, and psychology fields regarding what we need to assess: students' knowledge and skills, students' reasoning, or some combination (Crooks & Alibali, 2014; Star, 2005). Reviewing the history of mathematics education as described in the following paragraphs reveals an evolution from a focus on the outcome (i.e., the use of skill and accuracy in calculation) to a shift in the conceptual understanding of mathematical behaviors (Ellis & Berry, 2005).

These different perspectives can be traced back to debates in the 1920s between behaviorism and constructivism in mathematics education (Thorndike, 1923). Behavioral psychologists, such as Edward Thorndike and B. F. Skinner, proposed the association theory, which justified drill as a means of forming and strengthening the stimulus-response (S-R) bonds that are viewed as constituting the S-R bonds of arithmetic. Accordingly, changes in behavior (R) could be obtained through programmed instruction (S). These perspectives had a wide application in classroom mathematics instruction in the 1950s (Skinner, 1954). Consequently, students were taught mathematics by (a) identifying the stimuli, (b) identifying the response, and (c) making connections between the stimuli and the response by repetitive practice. Thus, the objective of an assessment was to evaluate how accurately and quickly a student demonstrated the R as the desired outcome of learning. And thus, the problems in students with MD could be interpreted as their difficulties with establishing an S-R bond between a problem prompt and an accurate answer.

However, one controversy evolved regarding whether accuracy and speed are valid indicators of a student's knowledge and true mathematical ability, especially a student's ability to reason successfully in justifying a solution to a problem (Brownell, 1944/2007). Cognitive psychologists, including W. A. Brownell, argued that mathematics suffered from the general application or misapplication of association theory. Brownell upheld that a meaningful learning theory focused on the process of learning rather than merely the outcome. For Brownell, the criterion for arithmetic skill was the ability to think quantitatively rather than to perform with accuracy and speed. Brownell found that children used a variety of procedures and strategies other than direct recall to complete their arithmetic tasks, which provided evidence that children's learning in mathematics was not merely the result of drill and practice and that meaningful assessment should focus on students' problem-solving processes and not only on outcomes. From this theoretical perspective, the difficulties LD students encounter are in building conceptual understanding and reasoning.

Subsequent to Brownell, other cognitive psychologists whose work was applied extensively to mathematics education included Piaget's cognitive developmental stages and Vygotsky's social constructivism that called for a mathematical community in which children could learn mathematics with understanding. Based on a constructivist perspective, knowledge cannot simply be transferred directly from instructors to learners but has to be actively built by learners based on their personal experiences (Confrey, 1990; von Glasersfeld, 1990, 2000; Noddings, 1990). From an observer's perspective, students' mathematical behaviors are viewed as schemata of actions and operations (Steffe, 1992; von Glasersfeld, 1995/2013). By observing and analyzing students' mathematical behaviors, including the physical models that students build, the pictures that they draw, and their written symbolic statements of the problem situation, we can gain insight into students' mental representations and infer their mental actions and operations (Maher & Sigley, 2020; Maher & Davis, 1990). From this perspective, students' problem-solving behaviors, regardless of whether they lead to the "correct answer," are meaningful explorations rooted in their experiences.

Much of the existing research on students with MD focuses on success in calculation or on procedural fluency (Lewis & Fisher, 2016). Traditionally, students who lack procedural fluency might be labeled as poor mathematics students. Assessment that only values procedural fluency is unlikely to attend to conceptual reasoning, and students' potential for reasoning in problem solving may not be recognized. For students in special education, a meaningful assessment should consider the evaluation of student engagement in contexts in which they are comfortable sharing their ideas and representing their knowledge.

6.3.2 Importance of Assessing Students' Reasoning and Argumentation

Research has documented that in a natural way, children – even young children – are able to justify solutions to problems with arguments that are "proof like" (Maher & Davis, 1995; Maher & Martino, 2000). In presenting their solutions to problems, children provide convincing arguments that take proof-like forms: cases, induction, contradiction, and upper/lower bounds. Their justifications are driven by an effort to make sense of the problem situation, notice patterns, and pose theories (Mueller et al., 2011). Their solutions are refined through discussions and arguments as they negotiate meaning with classmates and structure their investigations (Weber et al., 2008).

Inspired by the theory of "meaningful learning," we propose that a "meaningful assessment" is based on the view that students should have many opportunities to display their conceptual understanding of mathematics through their explorations, explanations, and justifications as expressed with oral and written language, symbols, models, and other constructions. Using a variety of tools to represent understanding, students are encouraged to share and support their ideas, question task requirements and solutions that are unclear to them, revisit earlier ideas, and work together in pairs or small groups to exchange and share ideas and to resolve differences in solutions to strands of well-defined, mathematical tasks. In 1986, Freudenthal (p. 47) wrote, "Learners should be allowed to find their own levels and explore the paths leading there with as much and as little guidance as each particular case requires." Freudenthal expressed this view over three decades ago, and subsequent research (Maher, 2010) by careful observation and analyses of students working on strands of rich, open-ended, and engaging mathematical tasks provides strong support for this view. Maher's long-term study focused on discovering what mathematical understandings students could build with minimal intervention. In these NSF-funded studies, classrooms were organized so that students could work collaboratively, using each other and available resources and tools to revisit tasks and discuss strategies and arguments for solutions, enabling researchers to follow a cohort of students from Grade 1 through high school and beyond. Two cross-sectional studies in other districts provided replication for the interventions. These studies have contributed to our understanding of how mathematical ideas develop in learners who are engaged in a range of mathematical practices. Some results from the earlier research are presented in two volumes (see Maher, 2010; Maher & Yankelewitz, 2017).

Prior research suggested that students naturally provide justifications for their problem solutions as they are encouraged by facilitators to persevere and make sense of the problem task (see, for example, Maher & Martino, 1998). Building on earlier work with students, some of whom were identified as struggling (Maher & Martino, 2000), we accepted the challenge of attending to the mathematical reasoning of struggling students (Zhang et al., 2018). Too often these students are dismissed as not having the knowledge or ability to make sense of problems and to reason, perhaps because they lack computational fluency, have low summative test results, have low expectations in their ability to think and reason, or are hesitant or have not had the opportunity to express their knowledge. Too often these students wait to be told what rule or algorithm to use, which is then applied without understanding.

6.3.3   Language Communication and Mathematical Reasoning in Struggling Students

To fully understand students' construction of mathematical knowledge requires examining their language use, both oral and written (Bailey et al., 2018), and how that usage changes over time. Mathematical knowledge can be represented in multiple semiotic ways. Thus, representations of mathematical knowledge do not depend uniquely on language representations. Representing one's mathematical knowledge can include a blend of mathematical symbolism and disciplinary language, everyday language, and gestural and visual displays. Consequently, mathematics learning and teaching involves mediating complex relationships among language, symbolic, and visual forms (Wilkinson et al., 2020), so that students are offered multiple opportunities to make connections among semiotic systems, each having its own conventions and each involving specific challenges (O'Halloran, 2015). However, we note that language representations are often privileged as forms of representation in standardized assessments of mathematical knowledge (Bailey et al., 2021).

The process of using multi-semiotic representations of a mathematical understanding is detailed in Sigley and Wilkinson's (2015) study of a bilingual upper middle-grade student, Ariel. Their analysis revealed the interdependency between his developing mathematical understandings and his deploying of the specialized language unique to that discipline. Initially, Ariel displayed his mathematical understanding and conveyed it to others while attempting to solve an algebraic problem by identifying patterns. Sigley and Wilkinson's analysis revealed Ariel's mathematical and language learning from this initial conjecture to his reformulation and generalization of an algebraic rule. Over time, Ariel gradually adopted the formal, more conventional expression of mathematical representation, the mathematical register (Wilkinson, 2015, 2018, 2019; Wilkinson et al., 2020). Ariel represented his understandings by employing that specialized form of natural English language as well as by using non-verbal representations such as graphics, tables, and more informal language. Thus, one significant aspect of examining the interaction of students during mathematical learning activities is attending to how they represent and communicate their understandings with language, both oral and written, during problem solving. As noted, Sigley and Wilkinson's (2015) study illustrated how one student used language, both informal and formal to explain, justify, and argue his position when communicating with other students during a problem-solving process.

Sigley and Wilkinson (2015) concluded that, during instruction, it may not be necessary for all teachers in every situation to require their students to convey their understandings in "perfect English." The focus for instruction should be on how best to support each student's constructive process, including deploying language representations, while at the same time teachers should be modeling and encouraging multi-semiotic representations of knowledge by all students (Wilkinson, 2018).

Thus, we propose that students should be provided with multiple opportunities to construct ideas by thinking through and testing ideas in interaction with other students and teachers via language and non-language representations (Bailey et al., 2021). These opportunities create the optimal circumstances for students to construct their own mathematical understandings, and subsequently or even simultaneously, they may then build a more complete knowledge of the mathematics register and how to use it. The analysis of Ariel's problem solving illustrates how mathematical knowledge and language knowledge are both interconnected and integrated in an interactive learning activity.

In many cases, the challenges of mathematical reasoning for students with MD are related to their language problems (Riccomini et al., 2015; Rubenstein & Thompson, 2002). There is a medium to high rate (30–70%) of co-occurrence of reading disabilities and MD (Willcutt et al., 2013). This suggests that students with language difficulties fairly frequently encounter difficulties learning mathematics. Students with MD particularly struggle with not-everyday vocabulary (e.g., rhombus, polygon, parallelogram, imaginary number), context-dependent vocabulary (e.g., foot as in 12 inches versus a body part), and phrases that are contradictory to daily language (e.g., A is divided into B).

There are several aspects of language that challenge struggling students' mathematical learning. First, some students with MD may be unable or reluctant to engage in mathematical learning activities because they do not understand the question being asked or the problem posed. In such circumstances, students are presented with word problems with assumed background knowledge (Jitendra et al., 2015). In these cases, some students from underrepresented minority groups and varied cultural backgrounds, such as English language (EL) students, may have the capacity to understand the problems as stated in words but may not interpret the problems as majority native English-speaking middle-class students do. Second, students with language problems, such as EL students, may become frustrated in their efforts to "think aloud," because they do not have the command of English to produce the requisite mathematical formulations. Think-aloud studies are often used by mathematics education researchers to infer students' thinking and reasoning (van der Walt, 2009; Zhang et al., 2014). Due to language difficulties, students may fail in responding to prompts, which is often perceived by teachers as inactivity and disengagement. In classroom activities, these factors may render a student-centered discussion into a teacher-centered activity. Third, limited language skills might interfere with struggling students' verbal reasoning that is required in explaining, contradicting, and justifying arguments (Kurkul & Corriveau, 2018).

In addition to the problems associated with children being able to effectively process language input and generate explanatory outputs, there is interest in how struggling students process information. Issues with working memory (Geary, 2011), attention (Maynard et al., 1999), and executive function (Toll et al., 2011) may interfere with the reasoning process. Working memory refers to the capacity to store information for short periods of time when engaging in cognitively demanding activities (Baddeley, 1986). Working memory is a strong contributor to students' mathematics achievement across various mathematics domains (Peng et al., 2016), and a poor working memory is found in many children with MD (Geary, 2004; McLean & Hitch, 1999). In psychology, attention refers to the process of selectively focusing on specific information in the environment (APA, n.d.). Learning disability and attention-deficit disorder (ADD) frequently co-occur (Maynard et al., 1999), with estimates of co-morbidity of ADD and learning disabilities ranging from approximately 20% (Javorsky, 1996) to 50% (Riccio & Jemison, 1998). Executive functions refer to a set of cognitive skills required to maintain and hold relevant information in the face of interference or distraction in order to reach a goal (Engle et al., 1999; Jacob & Parkinson, 2015). Students with MD are found to differ from their normal-achieving peers in all executive function skills (Toll et al., 2011).

Research findings suggest that these cognitive problems should not be viewed as learning deficits but as learning differences (Lewis, 2014; Lewis & Fisher, 2016). A deficit label suggests that the learner lacks the ability to learn, whereas the perspective of a learning difference signifies that learners take alternative pathways to learning and require, in some cases, supports to do so. Although students with MD demonstrate poorer scores in testing, this does not mean that being unable to reason mathematically is a deficit. The unique ways of cognitive processing during learning make a standard protocol ineffective and inaccessible to students with MD by obscuring their potential mathematical reasoning. Therefore, providing modified protocols and other relevant accommodations for these students may be warranted to make visible the reasoning potential of students with MD.

6.4 Part III: Alternative Assessment Methods to Measuring Mathematical Reasoning of Students with MD

A dominant form of summative assessment is the standardized academic achievement test used for assigning students a grade for accountability purposes. In contrast, formative assessments are brief and frequent check-ins throughout teaching the content and should be used to identify where students are successful and where they are struggling (Black & Wiliam, 2010; Dann, 2014). Formative assessments serve as an ongoing evaluation of student learning as a means of providing continual feedback about performance to both learners and instructors. The value of being both frequent and embedded throughout the lessons is that teachers can learn from the information they are collecting and use formative assessments to make ongoing judgments, thus making use of the data they collect to inform instruction and make adjustments to their lessons. The information aids in determining what material needs to be revisited, what interventions can be used, and what materials may provide them with greater insight into cognitive obstacles that students may encounter. Formative assessments are dynamic, ongoing, and useful for revealing important details about where students are in their learning progressions. In this section, we introduce and describe three commonly used informal, formative, assessment methods that provide in-depth analyses of student mathematical reasoning. Bailey and colleagues (Bailey et al., in press) offer an example of how to conduct such formative assessments with students learning mathematical knowledge and how to represent that knowledge with language.

6.4.1   Clinical Interviews: Informal Assessment Focusing on Students' Mathematical Reasoning

Interviewing students as an assessment method has been widely used in mathematics education research and practice (Hunting, 1997). This method echoes the recent orientations to psychological testing that encourage more qualitative approaches in which students are active participants and the examiner is not just an observer but also a participator; the purpose is not to identify deficits but to describe modifiability and individualized assessment (Hunting & Doig, 1997). Piaget (1965/2006) is credited with pioneering the use of clinical interviews to gain a deeper understanding of children's cognitive development. Interviews can be used for research purposes or for classroom teachers to assess and understand their students' reasoning abilities. The clinical interview is a widely used approach in mathematics education research in which the researcher typically records the interview process. A constructivist approach to conducting clinical interviews seeks to assess the knowledge the student brings to the task. To do so, interviewers need to provide a variety of tools/opportunities/pathways/environments for students to represent their knowledge: physical tools for building models, paper/pencil/crayons for making drawings, tables, and charts; and opportunities to observe the notations and symbols used to represent student knowledge. Interviewers, then, can attend to student language, oral and written, and can capture obstacles that impede the process of learning. They can monitor students' gains by revisiting the same or similar tasks and, as needed, provide accommodations. By creating an environment in which students work in pairs and/or small groups, opportunities for talking aloud and sharing knowledge with others evolve. Student's language representations of their mathematical knowledge as it is evolving provide evidence of a deeper understanding (Freudenthal, 1991; Sigley & Wilkinson, 2015).

Table 6.1 presents examples of tasks that were used in other studies to monitor and indicate students' growth in mathematical reasoning. These tasks, with appropriate modifications, can be used to monitor the construction of knowledge of students with MD (Bailey et al., 2021).

A significant characteristic of a constructivist student interview is that assessment administrators (e.g., teachers, researchers) determine which tasks to give an individual student based on an ongoing formative assessment of the student's specific progress and obstacles. This suggests that each student may receive a different task, an approach that differs from many standardized tests in which every student receives the same items or tasks from the same pool and follows the same procedures. Another significant characteristic of the interview is the choice of tasks available. Interviewers rarely utilize tasks that focus on routine performance, such as

Table 6.1 Example tasks used to assess students' mathematical reasoning (Maher et al., 2010)

Tasks	Structure
Outfits	Making outfits from shirts and pants. For example, Alice has 3 shirts (e.g., a white shirt, a blue shirt, and a yellow shirt), and 2 pairs of pants (e.g., blue pants, white pants). How many different outfits can Alice have? Modifications as needed: 2 shirts × 2 pants; 2 shirts × 3 pants; 2 shirts × 2 pants × 2 hats.
Towers	Building models of towers with Unifix cubes available. How many different towers of a certain height can be built selecting from cubes of 2 colors (starting with four cubes tall)?
	Modifications as needed: Folding back to 3 and 2 cubes tall; moving forward to 5 cubes tall).
Taxicab geometry	A student is given a specific section of a neighborhood represented by a grid. All trips originate at the home. Three locations of intersections are indicated on the map (school, library, store). Student is asked to show all possible routes that could have taken to each location. Student is asked if there is a shorter route, and if so, why? Also, is there more than one shortest route to each location? If not, why not? If so, how many? Justify your answers.
Pizzas	A customer can then select from 4 different toppings (e.g., peppers, sausage, mushrooms, pepperoni). How many different choices for pizza does a customer have?
	Modifications, as needed: Folding back to 3 and 2 topping choices.

calculation, facts, and fluency. A constructivist perspective suggests that students should be challenged at an appropriate level and engage in solving non-routine tasks so that they are challenged to demonstrate and extend their cognitive abilities by reflecting on the relationship between their goals and the actual effects of their current activities or strategies (Davis, 1990; Simon & Tzur, 2004; Simon et al., 2004). The selection of high-level cognitively complex tasks provides opportunities to promote children's capacity to think, reason, and solve problems (Smith & Stein, 1998; Steffe & Thompson, 2000a, 2000b).

Clinical interviews used in research and practice involve a high level of interaction between the test administrator and the student who is interviewed. In a standardized assessment, the test administrator typically follows the standard directions, such as reading the instructions and then waiting for the student to complete the item. In contrast, during a student interview, both the interviewee and the interviewer are interactive as they engage in the assessment activity. The interviewer, using questioning or probing, may seek to trigger justifications and explanations from the interviewee who is engaged in the solution or justification of a problem task. This format is particularly beneficial for students with MD because it grants test administrators opportunities to provide individualized accommodations to help students with MD to gain access to solving the problem tasks and to provide individualized interpretations to make sure that a student understands the task requirements.

An advantage of the use of interviews is that a student's ability to reason mathematically is not only measured as "right" or "wrong" but rather by an in-depth analysis of how the problem is solved (Confrey & Harel, 1994). Qualitative analysis clearly suggests that a response that is different from a standard answer also shows students' exploration and reasoning attempts based on their experiences. From a constructivist perspective, confusion might occur when students with MD fail to assimilate a way of thinking in response to some prompt, or alternatively, when they are unable to interpret the question. An error is a window into a cognitive obstacle that may be impeding performance. With a clinical interview, teachers or researchers capture the patterns expressed by students as they explore solutions to problems, thus providing an opportunity to reveal how students represent their knowledge with gestures, words, models, diagrams, drawings, and symbols. Interviewers have opportunities to document their explorations, strategies, and obstacles encountered as students work on strands of tasks. This method enables the interviewer to capture students' individual voices in representing knowledge, often silenced by a lack of confidence or inadequate verbal skills.

Although clinical interviews can provide opportunities to explore a student's thinking in great detail and depth, they are time intensive (Maher & Martino, 1998). Such interviews require the interviewers to have relatively high-level skills in scaffolding students to construct mathematical ideas, prompting students to explain their reasoning or showing their solutions, and making on-the-spot decisions about what task to assign to the student next based on the ongoing interpretation of students' mathematical behaviors. Unfortunately, these higher-level skills are not mastered by less experienced teachers, which is one of the major reasons why the interview assessment is not used as frequently as traditional assessment methods (i.e., large-scale standardized assessment and CBM), in which the test administrator only needs to follow a standardized and specified protocol. Special challenges arise when interviewing students with MD. Some students with MD have barriers to accessing problem tasks due to a lack of adequate mathematics vocabulary, difficulties with comprehending the problem context, and lack of familiarity with the story context due to cultural differences. For some students, a history of lack of success can discourage further effort to persist in the task (Zhang & Rivera, 2021). Clinical interviews for research purposes are typically recorded using videos. Because privacy concerns have made it increasingly difficult to obtain parental consent for video recordings of student interviews, the camera can just focus on the actions and words of the student, thereby preserving anonymity.

The Case of Brandon We will use a case study as an example of a constructivist clinical interview to reveal how researchers or teachers can use this approach to assess the mathematical reasoning of a student identified with MD. Brandon was in a mathematics class identified as the group with lowest mathematical success in the school. In a classroom research session led by Visiting Researcher Martino, Brandon and his partner Justin were videotaped working separately on the 4-tall Tower Problem. In this task, the children were given Unifix cubes available in two colors and asked to build all possible 4-tall towers and convince themselves that they found them all with no duplicates. The boys began by generating new towers using trial and error. They recognized and eliminated duplicate towers and reorganized their towers by creating eight partner pairs, that is, a tower with a corresponding partner with opposite colors in each height position. The students concluded that they had solved the problem with eight pairs of towers. Approximately 1 month later, Researcher Martino provided the students with paper and pencil and asked them to provide a written justification for finding all 3-tall towers, selecting from cubes available in two colors. Brandon produced a written justification using his earlier organizing strategy of pairs of towers, 4-tall, to justify his solution of eight, 3-tall towers. Four months after the administration of the first Tower Problems, the students were given another problem, the 4-topping Pizza Problem, a task whose solution has the same structure as the 4-tall Tower Problem. In this session, Brandon was paired with a new partner, Colin. The boys were asked to find all possible pizzas that could be made, selecting from four different topping choices. The two boys worked separately, producing charts to record their pizza outcomes. Brandon's chart made use of a notation for representing each topping choice. He inserted the numeral "0" to represent the absence of a topping and the numeral "1" to represent its presence. He used a "guess and check" strategy to account for all choices. Researcher Martino asked Brandon, "What are you doing?" and Brandon responded, "Making a graph, just like Colin. I put peppers, sausage, mushrooms, and pepperoni down and have them like 1, 0, 1, 0 and put…make a graph." The researcher probed further, asking Brandon what 1, 0, 1, 0 meant, and invited him to explain certain organizations, giving Brandon an opportunity to explain his representation choice and justification. Brandon, using his system of coding the different topping choices, offered an inclusive organization by cases, accounting for all possible pizza choices (see Maher & Martino, 1998, pp. 77–84, for a transcript of the videotaped interview). Three weeks later, in a one-on-one interview by Researcher Martino, Brandon was asked about his Pizza Problem solution. During the interview, he provided a valid justification for his solution, responding in detail to probing questions. He was then asked if the problem reminded him of any others he had worked on in the past, and Brandon spontaneously introduced the 4-tall Tower Problem (see Maher & Martino, 1998, pp. 84–101, for full transcript and images of student work; see https://doi. org/10.7282/t3-bv69-dj55 for video narrative (VMCAnalytic, Abadir, 2021) for Brandon's interview by Martino justifying his solution to the 4-Topping Pizza Problem and recognition of isomorphism with the 4-Tall Tower Problem, selecting from cubes of two colors).

What is particularly remarkable in Brandon's reconstruction of a justification for the 4-tall Tower Problem is that rather than offering eight tower pairs for his solution as he did previously, he built a new model using Unifix cubes to provide a justification by cases, demonstrating the structural equivalence of the solutions to the two problems. In so doing, he reorganized the towers into a pattern that showed a case justification for his solution, similar to his solution of the 4-Topping Pizza Problem. Brandon was able to recognize the equivalent structure of solutions, that is, identifying an isomorphism between the solutions for the two tasks, to that of the great mathematician Poincare (Greer & Harel, 1998).

Without the use of interviews and the opportunities they provide to probe deeper into student reasoning, it is unlikely a more traditional assessment would have captured Brandon's mathematical thinking. Brandon had an opportunity to represent his problem solutions in a number of ways: a table, a chart, a physical model, and verbal explanations. Alongside a partner, he offered an explanation about what he was doing and explained his problem solving later to the interviewer. Brandon had sufficient time to reflect on his earlier work and revisit his earlier ideas by being afforded the opportunity to revisit the earlier tasks. For a video study of Brandon's problem solving, see Private Universe Project Math Project Brandon interview [video]. Retrieved from https://doi.org/10.7282/T3VX0FRDj. The use of video-recording of these learning activities enabled subsequent analysis, shared and co-constructed by the researcher (see, for example, Wilkinson et al., 2018.)

6.4.2 Error Pattern and Strategy Analysis

Another approach that teachers frequently use to examine students' reasoning process is a fine-grained analysis of student error patterns and strategy uses in their responses. We define strategies as students' attempted methods to solve a mathematics problem, including both strategies that lead to a correct answer and faulty strategies that represent students' misconceptions and lead to an incorrect answer. An examination of students' error patterns and faulty strategies can help teachers to understand students' reasoning and misconceptions, which then can lead to improved instruction. Student errors provide a window through which glimpses of mental functioning can be obtained (Fisher & Lipson, 1986). Error analysis allows teachers to assess student understanding or misunderstanding and to identify and analyze errors that a student repeatedly makes when solving a mathematical problem. Specifically, with systematic coding and examination of students' work, researchers and teachers have been able to identify different types of faulty strategies and errors, which represent major types of common misconceptions and operational mistakes.

Zhang et al. (2016) examined strategies that students with MD used for solving fraction comparison problems. Three major strategies were identified. First, a common intuitive strategy among many students was the "whole-number strategy," in which students overgeneralized their existing whole number strategies to solve fraction problems, such as comparing denominators or numerators to decide the magnitude of a fraction (e.g., 1/4 > 1/3 because 4 is greater than 3, or 3/7 > 1/2 because 3 and 7 are greater than 1 and 2, or 1/7 > 2/3 because the largest number 7 is in 1/7). A common misconception for solving fraction comparison problems is the notion that the larger the parts (i.e., denominators or numerators) the greater the fraction (National Mathematics Advisory Panel, 2008). This misconception is rooted in students' prior mathematical experiences with whole numbers (Schneider & Siegler, 2010). Second, using concrete materials (e.g., a manipulative pie or fraction tiles) or visual representations (e.g., number lines, graphs) to present the fraction magnitude is another common strategy recommended for elementary students, and this is considered the first step for understanding fractions (Piaget et al., 1960). A third type of

Fig. 6.1 Students' comparison of 3/9 and 3/5

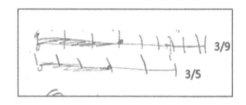

strategy can be called numerical transformation, such as rounding a fraction to a decimal number or a whole number (1/3 is greater than 0 and less than half), converting a fraction to a decimal number (1/4 = 0.25), and using the cross-multiplication algorithm to find the common denominator to compare two fractions (e.g., 1/4 = 3/12; 1/3 = 4/12).

Zhang et al. (2016) also examined the error patterns when students solved the problems with their attempted strategies. Figure 6.1 includes two common error types that students made when they attempted to use the visual representation strategy. First, when solving arithmetic problems or fraction comparison problems, students with MD may draw two 0–1 number line segments of different lengths to represent the unit of 1 in the same problem context (e.g., to compare 3/9 and 3/5). The length of the whole or the unit 1 for identifying 3/9 is much longer than that for identifying 3/5, resulting in an invalid conclusion that 3/9 was greater than 3/5. This type of obstacle suggests that students tried to use a visual representation strategy, yet lacked understanding of the equal-division concept and failed to recognize that 3/9 and 3/5 refer to the 3/9 and 3/5 of the same unit. Second, as we can see, these students were having difficulties with spatially dividing a number line equally into nine or three identical parts, even if they were attempting to do so.

An advantage of error pattern and strategy analysis, in comparison to the classic clinical interviews, is that it does not always require video recording to track every behavior or dialogue; it primarily relies on students' permanent products such as worksheets, test sheets, and so on. Frequently, researchers require students to write down their explanations on the test sheet or they help students to write down their explanations. Behavior notes are typically taken to record students' problem-solving behaviors. This advantage makes the method more acceptable for obtaining the consent of parents and schools for children with learning problems; as such, it also makes feasible a larger sample than the clinical interview and thus increases the generalizability of the findings.

6.4.3   Strategy Analysis with the Microgenetic Approach

The microgenetic approach is a series of concentrated observations designed to capture changes in student thinking and to identify what gave rise to those changes. The microgenetic approach is defined by three characteristics: "Observations span the whole period of rapidly changing competence; the density of observations within this period is high, relative to the rate of change; observations of changing performance are analyzed intensively to indicate the processes that gave rise to them" (Siegler & Svetina, 2002, p. 793). The microgenetic approach not only answers whether the changes happen after an intervention but also answers how changes happen during the intervention (Siegler, 2006).

Siegler and his group have conducted numerous research studies (Fazio et al., 2016; Opfer & Siegler, 2004; Ramani et al., 2012; Siegler & Svetina, 2006) with the microgenetic approach to investigate children's strategic development during learning, reasoning, and problem solving (Siegler, 2006), including research on students with MD. Siegler's microgenetic approach is relatively standardized: The teacher or researcher first presents a pre-determined task to the student, asks the student to solve it on their own and explain how they obtained the answer, and then the researcher provides feedback by saying, "yes, you are correct," or "no, the correct answer should be…," and asks "could you tell me why the correct answer is …?" Students' responses and explanations are coded as different types of strategies, and the frequency of each strategy use is counted to indicate students' strategic development or reasoning process. Such testing or data collection takes place with a very high frequency and density, for example, during every learning session. In this way, researchers reveal the nuanced changes of reasoning development throughout the whole learning process. The microgenetic approach creates a unique frame for assessing and analyzing how students' reasoning happens from five perspectives (Siegler, 2006): source of change (what leads children to adopt new strategies), path of change (the sequence of strategies children use while gaining competence), rate of change (the amount of time or experience from initial use to consistent use of a strategy), breadth of change (how widely the new strategy is generalized to other problems), and variability of change (differences among children in the previous four dimensions). This fine-grained and multiple-angled assessment can provide rich information for understanding the underlying development of mathematical reasoning.

A particular advantage of the microgenetic approach is its revealing of the strategies that children develop in response to instruction, and thus this approach helps understand how instruction exercises its effects. In other words, it can serve as a meaningful evaluation of program effectiveness in terms of facilitating students' mathematical reasoning development. The microgenetic approach not only answers whether the changes happen after an intervention but also answers how changes happen during the intervention (Siegler, 2006). Echoing the clinical interview, microgenetic studies have successfully revealed how task assignment facilitates children's reasoning and helps children adopt new strategies. Moreover, numerous studies have successfully employed this approach to investigate the mathematics learning of regular-achieving students, and there have been multiple studies (Zhang et al., 2013; Zhang et al., 2014; Zhang et al., 2016) using this method to assess students' progress during an intervention for students with MD. Zhang et al. (2013) utilized the microgenetic approach to evaluate how students with MD develop their multiplicative reasoning skills. The participants were two fifth graders with MD and one at risk. Investigators coded and analyzed four strategies the children used. Results showed that unitary counting (i.e., count by 1s to solve a multiplicative problem) was dominant during baseline before the intervention. During the teaching sessions, a noticeable increase in the use of double counting and a decrease in the use of unitary counting were observed. This study suggests that the monitoring of students' strategic changes can be used as a measure for assessing and progress monitoring students' dynamic growth of mathematical thinking and problem-solving skills.

When applied to students with special education or students with MD, there are also a few disadvantages with this microgenetic approach. A characteristic of this approach is intensive observations throughout the learning process, typically in a one-on-one format; in other words, this approach is time consuming for teachers. As such, it is possible to use this approach for a small number of students, but it would be very difficult to implement for every student in a class. Second, similar to all assessment methods that attempt to understand students' reasoning process, the development of a coding scheme is subjective, and some strategies are not exclusive from each other. In particular, because the assessment procedures in the microgenetic approach are standardized, fewer individualized prompts and accommodations are available in microgenetic analysis than in clinical interviews and this may result in a lack of responsiveness.

6.5   Conclusions

In this chapter, we reviewed five types of assessment that are widely used in special education and mathematics education research and classroom practices for students who have learning disabilities or are struggling to learn mathematics. We described the features, advantages, and disadvantages of each assessment. We began with the two types of standardized assessments: the large-scale, norm-referenced assessment and the curriculum-based assessment and measurement; we also reviewed commonly used testing accommodations for these standardized tests. Then we reviewed alternative assessment methods to identify what students know; to monitor their learning trajectories; and to evaluate their paths to deeper understanding, including clinical interviews, error and strategy analysis, and a microgenetic approach.

While it is acknowledged that standardized, norm-referenced, large-scale assessments have been criticized for not addressing students' reasoning, it is not our intention in this chapter to rank whether one assessment is more or less meaningful than the other. Rather, our purpose is to provide a broad range of knowledge about these important assessment techniques for both special education and mathematics education teachers who work with students with MD. Each method of assessment has unique advantages and disadvantages for gaining understanding of what students know and are capable of learning in mathematics. As we considered different kinds of performance by students, we selected a lens that enables tracking what students are capable of learning in contrast to viewing failures in performance from a deficit perspective. The selection of assessments in research and/or practice depends on the goals and purposes for assessment. It is noteworthy that a promising new trend in building new items for standardized testing is inclusion of behavioral notes as supplementary records in reporting scores; some newly developed psychometrics techniques, such as cognitive diagnosis models (CDM; de la Torre & Chiu, 2016), are used to diagnose students' missing strategies or flawed concepts based on students' responses to tasks in standardized, large-scaled tests. It seems a promising direction to integrate different assessment techniques to maximize teachers' understanding of students' mathematics problem-solving performance.

Appendix A

A Summary of Assessment Categories

Assessment types	Large-scale standardized assessment			Curriculum-based assessment	Assessment of student reasoning		
	National and international comparison assessment	High stakes state standard assessment	Diagnostics assessments		Clinical interviews / teacher classroom interviews	Student work error analysis and strategy analysis	Strategy analysis with microgenetic approach
Purposes	TIMSS & PISA: To provide a snapshot of international comparison. NAEP: The Nation's Report Card, provides important information about how students are performing academically national-wide.	Test students with standardized assessment aligned to state standards in mathematics.	Compare an individual student to a large population; identify student's weaknesses or strengths.	To monitor student's progress, to evaluate a student's response to intervention, and to inform intervention decisions.	To examine how students construct mathematical ideas in problem solving.	To examine student's error patterns and strategy uses in responses to test items.	To capture changes in student thinking and identify what gave rise to those changes.

	Large-scale standardized assessment		Curriculum-based assessment		Assessment of student reasoning		
General benefits for students with mathematics learning difficulties or disabilities.	International benchmarks of different levels of mathematics achievement. Provide the average score, percentage of students who score below the low international benchmark, and the achievement gap between high and low achievers in each country.	Students who fail to pass are considered as at risk. Schools held accountable for improving the progress of students with disabilities.	Determine the percentile rank of individual students. Identify students with learning disabilities with the traditional IQ-achievement discrepancy model.	Widely used in the RtI model to identify students with learning disabilities and provide early intervention. Sensitive to students' short-term growth and reflects the classroom curriculum. Easy to administer and can be administered frequently.	Not graded based on scores but based on students' reasoning. Examiners may provide individualized accommodations to help students with LD during the one-on-one interviews.	A student's ability to reason mathematically is not only measured as "right" or "wrong" but rather by an in-depth analysis of students' written solutions and explanations. When researching students with learning disabilities, researchers do not need parental consent for video recording.	To reveal the nuanced strategic changes that children with learning disabilities develop in response to an intervention.

Constraints for students	Large-scale standardized assessment			Curriculum-based assessment	Assessment of student reasoning		
	No individual-level data available.	Graded with accuracy. Standardized and pre-determined accommodations which may be insufficient to meet students' individualized special needs.	Not sensitive to students' short-term growth. Do not closely align to math curriculum.	Focuses on accuracy and does not reflect students' reasoning process.	Some students with MD have difficulties explaining their math ideas. Some students with disabilities do not make any responses or respond with nonsensical answers. Difficult to implement clinical interviews for every student in a class.	Students with disabilities may skip questions or refuse to provide any responses or explanations. Only student's final written product is analyzed; information of detailed changes during the learning process is omitted. Researchers/teachers may have to make inferences to interpret students' errors or strategies.	Standardized protocol to ask for students' explanations of strategic use; few opportunities for individualized accommodations. Difficult to implement for every student in a class. Some students with disabilities do not make any responses or respond with nonsensical answers.

	Large-scale standardized assessment		Curriculum-based assessment		Assessment of student reasoning		
Constraints for teachers	Only selected states or districts participate in the assessment. No individual-level data available.	Teachers do not have the assessment items except the released items. Only assessed once per year. Administered once per year. "Teach to Test" in school practice.	These assessments need to be administered by professionals with training. Cannot be assessed frequently.	Focusing on accuracy. Some CBM instruments do not have strong validity data support. Not many validated CBM instruments available in mathematics.	Time intensive. Require the interviewers to have advanced skills in scaffolding students to construct mathematical ideas, prompting students to explain their reasoning or showing their solutions, and making on-the-spot decisions about what task to assign next to the student. Require video recording if the interview is for research purposes.	Teachers need training to identify and categorize students' error patterns and strategies.	Teachers need training to identify and categorize students' strategic developmental levels. Video recording is recommended for research purposes.
Grained level	Global comparison at the national, state, and district level. No individual-level data.	Graded upon accuracy data. Only reflect percent correct but not student reasoning.	Graded upon accuracy data. Only reflect percent correct but not student's reasoning.	Graded upon accuracy data. Only reflect percent correct but not student's reasoning.	Fine-grained analysis on students' reasoning.	Fine-grained analysis of student error patterns and strategies used for problem solving.	Fine-grained analysis of student strategic development during learning.

References

Abadir, R. (2021). *Nine-year old Brandon's problem solving for accounting for all possible pizzas choosing from 4 toppings and recognition of a connection to the towers-4 tall selecting from 2 colors,* https://doi.org/10.7282/t3-bv69-dj55

Baddeley, A. D. (1986). *Working memory.* Oxford University Press.

Bailey, A., Maher, C., & Wilkinson, L. C. (2018). Introduction: Language, literacy, and learning in the STEM disciplines. In A. Bailey, C. Maher, & L. Wilkinson (Eds.), *Language, literacy, and learning in the STEM disciplines: How language counts for English learners* (pp. 1–10). Routledge Taylor Francis.

Bailey, A., Maher, C., Wilkinson, L. C., & Nyakoojo, U. (2021). The role of assessment in teaching mathematics with English-speaking and EL students. In D. Varier & S. Nichols (Eds.), *American education research association theory to practice series: educational psychology for teachers and teaching: Teaching on assessment* (pp. 151–172). Information Age Publishers.

Black, P., & Wiliam, D. (2010). Inside the black box: Raising standards through classroom assessment. *Phi Delta Kappan, 92*(1), 81–90.

Bolt, S. E., & Thurlow, M. L. (2004). Five of the most frequently allowed testing accommodations in state policy: Synthesis of research. *Remedial and Special Education, 25*(3), 141–152.

Brown, V. L., Cronin, M. E., & Bryant, D. P. (2013). *Test of mathematical abilities* (3rd ed.). Pro-Ed.

Brownell, W. A. (2007). The progressive nature of learning in mathematics [Special issue]. *Mathematics Teacher, 100,* 26–34. (Reprinted from *Mathematics Teacher,* 1944, *37*[4], 147-157).

Chard, D. J., Clarke, B., Baker, S., Otterstedt, J., Braun, D., & Katz, R. (2005). Using measures of number sense to screen for difficulties in mathematics: Preliminary findings. *Assessment for Effective Intervention, 30*(3), 3–14.

Christ, T. J., Sculin, S., Tolbize, A., & Jiban, C. L. (2008). Implications of recent research: Curriculum-based measurement of math computation. *Assessment for Effective Intervention, 33,* 198–205.

Confrey, J. (1990). What constructivism implies for teaching. In R. B. Davis, C. A. Maher, & N. Noddings (Eds.), *Journal for Research in Mathematics Education, Monograph No. 4: Constructivist views on the teaching and learning of mathematics* (pp. 107–122). National Council of Teachers of Mathematics.

Confrey, J., & Harel, G. (1994). Introduction. In G. Harel & J. Confrey (Eds.), *The development of multiplicative reasoning in the learning of mathematics* (pp. vii–xxviii). State University of New York Press.

Connolly, A. J. (2007). *KeyMath-3 diagnostic assessment: Manual forms A and B.* Pearson.

Crooks, N. M., & Alibali, M. W. (2014). Defining and measuring conceptual knowledge in mathematics. *Developmental Review, 34*(4), 344–377.

Deno, S. L. (1985). Curriculum-based measurement: The emerging alternative. *Exceptional Children, 52*(3), 219–232.

Deno, S. L. (2003). Developments in curriculum-based measurement. *The Journal of Special Education, 37*(3), 184–192. https://nces.ed.gov/timss/pdf/comparing_timss_naep_%20pisa.pdf

Dann, R. (2014). Assessment as learning: Blurring the boundaries of assessment and learning for theory, policy and practice. *Assessment in Education: Principles, Policy & Practice, 21*(2), 149–166.

Davis, R. B. (1990). Discovery learning and constructivism. In R. B. Davis, C. A. Maher, & N. Noddings (Eds.), *Journal for Research in Mathematics Education, Monograph 4: Constructivist views on the teaching and learning of mathematics* (pp. 93–106). National Council of Teachers of Mathematics.

de la Torre, J., & Chiu, C. Y. (2016). A general method of empirical Q-matrix validation. *Psychometrika, 81*(2), 253–273.

Ellis, M. W., & Berry, R. Q. (2005). The paradigm shift in mathematics education: Explanations and implications of reforming conceptions of teaching and learning. *The Mathematics Educator, 15*(1), 7–17.

Engle, R. W., Kane, M. J., & Tuholski, S. W. (1999). Individual differences in working memory capacity and what they tell us about controlled attention, general fluid intelligence, and functions of the prefrontal cortex. In A. Miyake & P. Shah (Eds.), *Models of working memory: Mechanisms of active maintenance and executive control* (pp. 102–134). Cambridge University Press.

Fazio, L. K., DeWolf, M., & Siegler, R. S. (2016). Strategy use and strategy choice in fraction magnitude comparison. *Journal of Experimental Psychology: Learning, Memory, and Cognition, 42*, 1–16. https://doi.org/10.1037/xlm0000153

Fisher, K. M., & Lipson, J. I. (1986). Twenty questions about student errors. *Journal of Research in Science Teaching, 23*(9), 783–803.

Foegen, A., Jiban, C., & Deno, C. (2007). Progress monitoring measures in mathematics. A review of the literature. *The Journal of Special Education, 41*(2), 121–139.

Freudenthal, H. (1991). *Revisiting mathematics education: China lectures.* Kluwer Academic Publishing.

Freudenthal, H. (1986). *Didactical phenomenology of mathematical structures* (Vol. 1). Springer Science & Business Media.

Fuchs, L. S., & Fuchs, D. (2004). *Using CBM for progress monitoring.* National Center on Student Progress Monitoring.

Fuchs, L. S., & Fuchs, D. (n.d.). *Fair and unfair testing accommodations: What's considered appropriate when assessing the academic performance of students with disabilities?* https://www.aasa.org/SchoolAdministratorArticle.aspx?id=14932

Fuchs, L. S., Fuchs, D., Hamlett, C. L., & Stecker, P. M. (1990). The role of skills analysis in curriculum-based measurement in math. *School Psychology Review, 19*(1), 6–22.

Geary, D. C. (2004). Mathematics and learning disabilities. *Journal of Learning Disabilities, 37*(1), 4–15.

Geary, D. C., Hoard, M. K., Byrd-Craven, J., Nugent, L., & Numtee, C. (2007). Cognitive mechanisms underlying achievement deficits in children with mathematical learning disability. Child Development, 78, 1343–1359.

Geary, D. C. (2011). Consequences, characteristics, and causes of mathematical learning disabilities and persistent low achievement in mathematics. *Journal of Developmental and Behavioral Pediatrics, 32*(3), 250–263.

Gewertz, C. (n.d.). What tests does each state require? *Education week.* https://www.edweek.org/teaching-learning/what-tests-does-each-state-require

Greer, B., & Harel, G. (1998). The role of isomorphisms in mathematical cognition. *The Journal of Mathematical Behavior, 17*(1), 5–24.

Harrison, J. R., Bunford, N., Evans, S. W., & Owens, J. S. (2013). Educational accommodations for students with behavioral challenges: A systematic review of the literature. *Review of Educational Research, 83*(4), 551–597.

Hosp, M. K., & Hosp, J. L. (2003). Curriculum-based measurement for reading, spelling, and math: How to do it and why. *Preventing School Failure, 48*(1), 10–17.

Hosp, M. K., Hosp, J. L., & Howell, K. W. (2007). *The ABCs of CBM: A practical guide to curriculum-based instruction.* The Guilford Press.

Hunting, R. P. (1997). Clinical interview methods in mathematics education research and practice. *The Journal of Mathematical Behavior, 16*(2), 145–165.

Hunting, R. P., & Doig, B. A. (1997). Clinical assessment in mathematics: Learning the craft. *Focus on Learning Problems in Mathematics, 19*(3), 29–48.

Jacob, R., & Parkinson, J. (2015). The potential for school-based interventions that target executive function to improve academic achievement: A review. *Review of Educational Research, 85*(4), 512–552.

Javorsky, J. (1996). An examination of youth with attention-deficit/hyperactivity disorder and language learning disabilities: A clinical study. *Journal of Learning Disabilities, 29*(3), 247–258.

Jitendra, A. K., Petersen-Brown, S., Lein, A. E., Zaslofsky, A. F., Kunkel, A. K., Jung, P. G., & Egan, A. M. (2015). Teaching mathematical word problem solving: The quality of evidence for strategy instruction priming the problem structure. *Journal of Learning Disabilities, 48*(1), 51–72.

Kurkul, K. E., & Corriveau, K. H. (2018). Question, explanation, follow-up: A mechanism for learning from others? *Child Development, 89*(1), 280–294.

Lewis, K. E. (2014). Difference not deficit: Reconceptualizing mathematical learning disabilities. *Journal for Research in Mathematics Education, 45*(3), 351–396.

Lewis, K. E., & Fisher, M. B. (2016). Taking stock of 40 years of research on mathematical learning disability: Methodological issues and future directions. *Journal for Research in Mathematics Education, 47*(4), 338–371.

Maher, C. A. (2010). The longitudinal study. In C. A. Maher, A. B. Powell, & E. B. Uptegrove (Eds.), *Combinatorics and reasoning: Representing, justifying and building isomorphisms* (pp. 3–8). Springer Publishers.

Maher, C. A., & Davis, R. B. (1990). Building representations of children's meanings. In R. B. Davis, C. A. Maher, & N. Noddings (Eds.), *Journal for Research in Mathematics Education, Monograph 4: Constructivist views on the teaching and learning of mathematics* (pp. 79–90). National Council of Teachers of Mathematics.

Maher, C. A., & Davis, R. B. (1995). Children's explorations leading to proof. In C. Hoyles & L. Healy (Eds.), *Justifying and proving in school mathematics* (pp. 87–105). Mathematical Sciences Group, Institute of Education, University of London.

Maher, C. A., & Martino, A. (1998). Brandon's proof and isomorphism. In C. A. Maher (Ed.), *Can teachers help children make convincing arguments? A glimpse into the process* (pp. 77–101). Universidade Santa Ursula, ISSN 0104-9720 (in Portuguese and English).

Maher, C. A., & Martino, A. M. (2000). From patterns to theories: Conditions for conceptual change. *The Journal of Mathematical Behavior, 19*(2), 247–271.

Maher, C. A., Powell, A. B., & Uptegrove, E. B. (Eds.). (2010). *Combinatorics and reasoning: Representing, justifying and building isomorphisms* (Vol. 47). Springer Science & Business Media.

Maher, C. A., & Yankelewitz, D. (Eds.). (2017). *Children's reasoning while building fraction ideas.* Sense Publishers.

Maynard, J., Tyler, J. L., & Arnold, M. (1999). Co-occurrence of attention-deficit disorder and learning disability: An overview of research. *Journal of Instructional Psychology, 26*, 183–187.

McLean, J. F., & Hitch, G. J. (1999). Working memory impairments in children with specific arithmetic learning difficulties. *Journal of Experimental Child Psychology, 74*(3), 240–260.

Mueller, M., Yankelewitz, D., & Maher, C. (2011). Sense making as motivation in doing mathematics: Results from two studies. *The Mathematics Educator, 20*(2), 33–43.

National Center for Education Statistics. (n.d.). *TIMSS 2019 U.S. results.* https://nces.ed.gov/timss/results19/doc/TIMSS2019_compiled.pdf

National Mathematics Advisory Panel. (2008). *Foundation for success: The final report of the National Mathematics Advisory Panel.* U.S. Department of Education.

NAEP. (2017). *National Assessment of Educational Progress: Mathematics assessment.* Retrieved November 1, 2018, from https://nces.ed.gov/nationsreportcard/mathematics/

No Child Left Behind Act. (2002). NCLB Act of 2001, Pub. L. No. 107-110, § 115. Stat, 1425.

Noddings, N. (1990). Constructivism in mathematics education. In R. B. Davis, C. A. Maher, & N. Noddings (Eds.), *Journal for Research in Mathematics Education, Monograph 4: Constructivist views on the teaching and learning of mathematics* (pp. 7–29). National Council of Teachers of Mathematics.

O'Halloran, K. (2015). The language of learning mathematics: A multimodal perspective. *The Journal of Mathematical Behavior, 20-A*, 63–74.

Opfer, J. E., & Siegler, R. S. (2004). Revisiting preschoolers' living things concept: A microgenetic analysis of conceptual change in basic biology. *Cognitive Psychology, 49*, 301–332.

Overton, T. (2011). *Assessing learners with special needs.* Pearson Education.

Ramani, G. B., Siegler, R. S., & Hitti, A. (2012). Taking it to the classroom: Number board games as a small group learning activity. *Journal of Educational Psychology, 104*, 661–672.

Peng, P., Namkung, J., Barnes, M., & Sun, C. (2016). A meta-analysis of mathematics and working memory: Moderating effects of working memory domain, type of mathematics skill, and sample characteristics. *Journal of Educational Psychology, 108*(4), 455–473.

Piaget, J. (2006). The stages of the intellectual development of the child. In B. A. Marlowe & A. S. Canestrari (Eds.), *Educational psychology in context: Readings for future teachers* (pp. 98–106). SAGE Publications. (Original work published 1965).

Piaget, J., Inhelder, B., & Szeminska, A. (1960). *The child's conception of geometry*. Basic Books.

PISA. (2015). *Mathematics literacy: Average scores*. https://nces.ed.gov/surveys/pisa/pisa2015/pisa2015highlights_5.asp

Riccio, C. A., & Jemison, S. J. (1998). ADHD and emergent literacy: Influence of language factors. *Reading & Writing Quarterly Overcoming Learning Difficulties, 14*(1), 43–58.

Riccomini, P. J., Smith, G. W., Hughes, E. M., & Fries, K. M. (2015). The language of mathematics: The importance of teaching and learning mathematical vocabulary. *Reading & Writing Quarterly, 31*(3), 235–252.

Rubenstein, R. N., & Thompson, D. R. (2002). Understanding and supporting children's mathematical vocabulary development. *Teaching Children Mathematics, 9*, 107–112.

Star, J. R. (2005). Reconceptualizing procedural knowledge. Journal for Research in Mathematics Education, 36(5), 404–411.

Schneider, M., & Siegler, R. S. (2010). Representations of the magnitudes of fractions. *Journal of Experimental Psychology: Human Perception and Performance, 36*, 1227–1238.

Sigley, R., & Wilkinson, L. C. (2015). Ariel's cycles of algebraic problem solving: A bilingual adolescent acquires the mathematics register. *The Journal of Mathematical Behavior, 40*, 75–87.

Simon, M. A., & Tzur, R. (2004). Explicating the role of mathematical tasks in conceptual learning: An elaboration of the hypothetical learning trajectory. *Mathematical Thinking and Learning, 6*, 91–104.

Simon, M. A., Tzur, R., Heinz, K., & Kinzel, M. (2004). Explicating a mechanism for conceptual learning: Elaborating the construct of reflective abstraction. *Journal for Research in Mathematics Education, 35*, 305–329.

Siegler, R. S. (2006). Microgenetic analyses of learning. In W. Damon & R. M. Lerner. (Series Eds.), D. Kuhn & R. S. Siegler. (Vol. Eds.), Handbook of child psychology: Vol. 2. Cognition, perception, and language (6th ed., pp. 464–510). Wiley.

Siegler, R. S., & Svetina, M. (2002). A microgenetic/cross-sectional study of matrix completion: Comparing short-term and long-term change. *Child Development, 73*, 793–809.

Siegler, R. S., & Svetina, M. (2006). What leads children to adopt new strategies? A microgenetic/cross sectional study of class inclusion. *Child Development, 77*, 997–1015.

Skinner, B. F. (1954). The science of learning and the art of teaching. *Harvard Educational Review, 24*, 86–89.

Smith, M. S., & Stein, M. (1998). Selecting and creating mathematical tasks: From research to practice. *Mathematics Teaching in the Middle School, 3*(5), 344–350.

Steffe, L. P. (1992). Schemes of action and operation involving composite units. *Learning and Individual Differences, 4*(3), 259–309.

Steffe, L. P., & Thompson, P. W. (Eds.). (2000a). *Radical constructivism in mathematics and science education: Essays in honor of Ernst von Glasersfeld*. Falmer Press.

Steffe, L. P., & Thompson, P. W. (2000b). Teaching experiment methodology: Underlying principles and essential elements. In A. E. Kelly & R. A. Lesh (Eds.), *Handbook of research design in mathematics and science education* (pp. 267–306). Erlbaum.

Swanson, H. L., & Hoskyn, M. (1998). Experimental intervention research on students with learning disabilities: A meta-analysis of treatment outcomes. Review of Educational Research, 68(3), 277–321.

Thorndike, E. L. (1923). The psychology of arithmetic. New York: The Macmillan Company.

Maher, C. A., & Sigley, R. (2020). Task-based interviews in mathematics education. *Encyclopedia of Mathematics Education,* 821–824.

Thurlow, M. (2002). Accommodations for students with disabilities in high school. *Issue brief: Examining current challenges in secondary education and transition, 1*(1). http://www.ncset.org/publications/viewdesc.asp?id=247

TIMSS. (2011). *Trends in International Mathematics and Science Study (TIMSS): 2011 results.* https://nces.ed.gov/timss/results11.asp

Toll, S. W., Van der Ven, S. H., Kroesbergen, E. H., & Van Luit, J. E. (2011). Executive functions as predictors of math learning disabilities. *Journal of Learning Disabilities, 44*(6), 521–532.

van der Walt, M. (2009). Study orientation and basic vocabulary in mathematics in primary school. *South African Journal of Science and Technology, 28*, 378–392.

von Glasersfeld, E. (1990). An exposition of constructivism. In R. B. Davis, C. A. Maher, & N. Noddings (Eds.), *Journal for Research in Mathematics Education, Monograph 4: Constructivist views on the teaching and learning of mathematics* (pp. 19–30). National Council of Teachers of Mathematics.

von Glasersfeld, E. (2000). Problems of constructivism. In L. P. Steffe & P. W. Thompson (Eds.), *Radical constructivism in action: Building on the pioneering work of Ernst von Glasersfeld* (pp. 3–9). Routledge/Falmer.

von Glasersfeld, E. (2013). *Radical constructivism: A way of knowing and learning* (Vol. 6). Routledge. (Original book published 1995).

Weber, K., Maher, C., Powell, A., & Lee, H. S. (2008). Learning opportunities from group discussions: Warrants become the objects of debate. *Educational Studies in Mathematics, 68*(3), 247–261.

Wilkinson, L. C. (2015). The language of learning mathematics. *The Journal of Mathematical Behavior, 40, 2–5.*

Wilkinson, L. C. (2018). Teaching the language of learning mathematics: What teachers need to say and do. *The Journal of Mathematical Behavior, 51*, 167–174.

Wilkinson, L. C. (2019). Learning language and mathematics: A perspective from linguistics and education. *Linguistics and Education, 49*, 86–95.

Wilkinson, L., Bailey, A., & Maher, C. (2020). Students' learning language and learning to reason mathematically. In M. Daszkiewicz & A. Dąbrowska (eds.) In search of the language educational paradigm; Strand 2.4 (pp. 211–226.) Kraków, Poland: Oficyna Wydawnicza Impuls Press. ISBN 978-83-7850-779-6.

Wilkinson, L. C., Bailey, A., & Maher, C. A. (2018). Students' mathematical reasoning, communication, and language representations: A video-narrative analysis. *ECNU Review of Education, 1*(3), 11–27.

Willcutt, E. G., Petrill, S. A., Wu, S., Boada, R., DeFries, J. C., Olson, R. K., & Pennington, B. F. (2013). Comorbidity between reading disability and math disability: Concurrent psychopathology, functional impairment, and neuropsychological functioning. *Journal of Learning Disabilities, 46*(6), 500–516.

Zhang, D., Li, P., Di Dolce, B., Corron, K., & Quirinale, E. (2018). *Effects of unit coordination instruction on learning fractions in students with math difficulties.* Presented at the American Educational Research Association Annual (AERA) Convention, NYC, NY.

Zhang, D., & Rivera, F. (2021). Predetermined accommodations with a standardized testing protocol: Examining two accommodation supports for developing fraction thinking in students with mathematical difficulties. *The Journal of Mathematical Behavior, 62*(4), 100861. https://doi.org/10.1016/j.jmathb.2021.100861

Zhang, D., Stecker, P., Huckabee, S., & Miller, R. (2016). Strategic development for middle school students struggling with fractions: Assessment and intervention. *Journal of Learning Disabilities, 49*(5), 515–531.

Zhang, D., Xin, Y. P., & Si, L. (2013). Transition from intuitive to advanced strategies in multiplicative reasoning for students with math disabilities. *Journal of Special Education, 47*, 50–64.

Zhang, D., Xin, Y., Harris, K., & Ding, Y. (2014). Improving multiplication strategic development in children with math difficulties. *Learning Disability Quarterly, 37*, 15–30.

Chapter 7
Commentary on Part II

Leanne R. Ketterlin-Geller

Abstract The purpose of this commentary is to summarize the information presented by Hunt and Tzur in Chap. 5 and Zhang, Maher, and Wilkinson in Chap. 6, identify common themes that bring these chapters together, and offer my reflection on these themes. Both chapters discuss ways in which assessment results can support decision-making for students who are experiencing difficulties in mathematics. Emerging from these chapters is the notion that when intentionally designed, assessment results can help teachers align instruction with students' learning needs to improve and accelerate student learning. Three themes that I build on are (a) the impact of teachers' attitudes toward students who are experiencing difficulty, (b) the need to carefully design assessments and tasks, and (c) the role of assessments in providing teachers with diagnostic information. Within each of these themes, I summarize the commonalities across chapters and extend the discussion to highlight additional perspectives or approaches.

Keywords Assessment · Students experiencing difficulty · Validity · Teacher decision-making · Item design · Asset-based assessment · Curriculum–instruction–assessment alignment · Fairness · Teachers' attitudes · Accommodations · Interview-based assessment · Diagnostic assessment · Learning progressions

Research conducted for this manuscript is based upon work supported by the National Science Foundation under Grant No. 1721100. Any opinions, findings, and conclusions or recommendations expressed in these materials are those of the authors and do not necessarily reflect the views of the National Science Foundation.

L. R. Ketterlin-Geller (✉)
Southern Methodist University, Dallas, TX, USA
e-mail: lkgeller@smu.edu

7.1   Designing and Implementing Assessments to Support Learning for Students Who Are Experiencing Difficulties in Learning Mathematics

In this commentary, I reflect on the information presented by Hunt and Tzur in Chap. 5 and Zhang, Maher, and Wilkinson in Chap. 6. The overarching theme that ties these chapters together is the way in which assessment results can support decision-making for students who are experiencing difficulties in learning mathematics. In the chapter by Hunt and Tzur, the authors emphasize the role tasks play in connecting teaching and learning: Tasks are used to engage students' reflective thinking during the learning process, and teachers use prompting and gesturing to deepen students' reflection. In the chapter by Zhang, Maher, and Wilkinson, the authors broaden the discussion of assessments and focus on three major topics: (1) participation of students who are experiencing mathematics difficulties in traditional assessments for accountability and formative purposes, (2) dimensions of students' thinking that are assessed, and (3) diagnostic inferences to support instructional decisions. In this commentary, I provide a brief overview of the central points highlighted in these chapters. Then, I offer my observations about themes that cut across both chapters, calling attention to (a) the impact of teachers' attitudes toward students who are experiencing difficulty, (b) the need to carefully design assessments and tasks, and (c) the role of assessments in providing teachers with diagnostic information. Within each of these themes, I summarize the commonalities across chapters and extend the discussion to highlight additional perspectives or approaches.

To help situate my commentary, it is important to clarify what I mean when I reference students who are experiencing difficulties in learning mathematics. It is difficult to put bounds on the group of students for whom I am referring. To begin, I am not talking about students who may have temporary challenges understanding a new mathematical concept, applying a newly learned procedure to solve a problem, or providing a mathematically sound rationale to justify their procedures. Mathematics is a complex discipline for which many people will experience temporary difficulties or challenges at one time or another. Instead, I am referring to students who have persistent difficulties in learning or applying mathematical knowledge and skills, especially as concepts increase in complexity within and across grades. This population of students is heterogeneous; they have various strengths and learning needs. A small percentage of these students have been diagnosed with learning disabilities. Students with a diagnosed learning disability are receiving specialized instruction as outlined on their Individual Education Program (IEP), which may or may not include instructional goals related to mathematics.

For many students who experience difficulty, the underlying cause is not a diagnosed learning disability, but instead may emerge as a result of a misalignment between instruction and their learning needs (Tzur, 2013). The implications of this misalignment can be far-reaching for students. In a series of studies conducted by Jordan et al. (2007, 2009), they observed that over 50% of the children from low-income families had low or flat growth trajectories from kindergarten through grade

1. Many of the children entered kindergarten with less developed number sense than their middle- and high-income peers, and formal instruction did not accelerate their learning during this period. In their follow-up study, the authors noted that the low mathematics performance of the children from low-income families in grade 3 was mediated by their relatively weak number sense in kindergarten. In other words, formal mathematics instruction between kindergarten and grade 3 did not result in growth or meaningful learning. To explain these findings, the authors posit a mismatch between the instruction the children received and the targeted support they actually needed to reach the learning goals and/or the curricular expectations. Had the children who had less developed number sense received appropriately intensive supplemental instruction, they may have developed more robust early mathematics knowledge and skills and experienced steeper growth trajectories that put them on course for future success.

The misalignment between instruction and students' learning needs may be caused by multiple reasons. Common causes I have seen include instruction that does not activate or draw on students' prior knowledge or conceptualizations, instruction that is not sequenced and paced in accordance with students' current level of understanding, and instruction that does not address previous misconceptions or errors, thereby allowing them to linger. When instruction is not aligned with students' learning needs, their opportunities to learn mathematics are compromised. I explicitly raise these issues because they directly relate to the chapters by Hunt and Tzur and Zhang, Maher, and Wilkinson: When designed with intentionality, assessment results can help teachers align instruction with students' learning needs to improve and accelerate student learning. After describing the unique contributions made by each chapter, I highlight areas of convergence and extensions to consider.

7.2 Assessment Considerations Presented by the Authors

In the chapter by Hunt and Tzur (Chap. 5, this volume), the authors make an explicit connection between teaching and learning and the conditions that facilitate this connection. They define learning as changes in a student's conceptualization of the content via reflection on the effects of their activities and explain how teachers' instructional moves can directly impact this process.

Hunt and Tzur define teaching in terms of instructional moves teachers make to elicit a student's current ways of knowing or understanding and then facilitate reflection during tasks to optimize the student's learning. When a student is in a participatory stage of understanding, the teacher may need to prompt reflection to support the student's progress to the anticipatory stage of understanding. Once a student moves into the anticipatory stage, the teachers' actions may shift to focus on deepening or extending reflection to subsequently deepen understanding.

Hunt and Tzur situate instructional tasks at the juncture of teaching and learning, while explicating the role tasks play in deepening understanding. Tasks connect teaching and learning through the teacher's intentional placement in the learning process and the teacher–student interactions that are facilitated during an instructional exchange. Hunt and Tzur dissect the various types of tasks and the level of interaction that encourages student reflection. In the procedures these authors describe, the teacher presents a student with tasks that are designed to bring forward the student's prior knowledge and available conceptualizations (labeled as "bridging tasks") from which they can build new conceptual understanding. Next, with student's prior knowledge primed, they engage in solving gradually more challenging tasks that include meaningful "variations" in the task design, but still focus on the key learning goal. Finally, deeper understanding is reached as the student moves into the "reinstating" phase where student learning is reinforced through reflection. Throughout this process, the teacher is interacting with the student to support their learning. The authors note the importance of the teacher's prompting and gesturing as a way of assessing student understanding and encouraging the student to clarify, justify, and critique their thinking and to deepen the student's reflection. I illustrate the connection between these components in Fig. 7.1.

Shifting to Chap. 6 by Zhang, Maher, and Wilkinson (this volume), these authors take on three major topics related to assessment. First, the authors provide an overview of various assessment approaches with an emphasis on how students with disabilities or students experiencing difficulties in mathematics participate in these programs. The authors reference the uses of large-scale assessments and curriculum-based measures (CBM) within accountability and instructional contexts, respectively. They provide examples of specific assessments and illustrate their strengths and limitations. In particular, they raise concerns about the content of many standardized mathematics assessments, noting that they may not be representative of the taught curriculum.

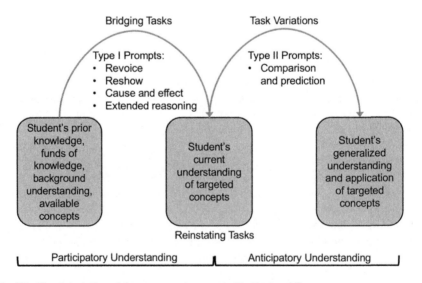

Fig. 7.1 Visual depiction of the components presented in Hunt and Tzur

To ensure that results from these assessment systems are reliable and support valid decision-making for students with disabilities or students experiencing difficulties in mathematics, Zhang, Maher, and Wilkinson discuss the provision of test accommodations and provide an important distinction from test modifications. The authors reference the lack of empirical evidence documenting the effectiveness of accommodations at supporting access and note that applying accommodations uniformly may not support greater access for students experiencing difficulty. Relatedly, when describing the role of language in communicating one's mathematical knowledge, the authors describe how language may cause an access barrier for some students, which may lead to teachers' underestimating their students' level of mathematical proficiency.

Next, Zhang, Maher, and Wilkinson consider the mathematical construct being measured and the value different constructs bring to understanding student thinking. The authors associate specific mathematical constructs with the application of theories of learning to mathematics education. For example, the authors reference the role of behaviorism in promoting instruction in fact fluency and retrieval along with the corresponding assessments designed to measure students' accuracy and computational fluency. The authors reference intentionally designing tasks to elicit students' reasoning and justification. They describe the importance of teachers attending to students' responding behaviors (e.g., how students represent and communicate their understanding) to deepen their understanding of students' mathematical thinking.

Lastly, Zhang, Maher, and Wilkinson describe the role of diagnostic assessments within the instructional design process. They provide an overview of some published diagnostic assessments and review several methods for gathering diagnostic information including clinical interviews, error pattern, and strategy analysis and integrating a microgenetic approach to strategy analysis. Through each of these approaches, the authors describe how information about students' thinking can be revealed to support future instruction and support student learning.

7.3 Convergence Across Chapters

Several themes emerged across the chapters by Hunt and Tzur and Zhang, Maher, and Wilkinson. First, both chapters emphasize the negative attribution teachers often place on students who are experiencing mathematics difficulty. Hunt and Tzur note how teachers' perceptions of their students' abilities may influence their instructional actions. Zhang, Maher, and Wilkinson similarly emphasize students' learned helplessness that sometimes accompanies low expectations. Second, both chapters describe the intentionality of assessment (or task) design as an essential component of using assessment results. Hunt and Tzur provide a detailed description of the types of tasks and accompanying prompting and gesturing that can be integrated into different phases of learning. Zhang, Maher, and Wilkinson note the important role accommodations play in administering assessments to overcome task design issues that may impact accessibility. Moreover, the authors of both chapters emphasize the importance of designing tasks to assess mathematical reasoning and justification. Third, and finally, both chapters emphasize the importance of implementing assessments and tasks to facilitate teachers' diagnostic inferences. In the remainder of this commentary, I will elaborate on each of these themes by offering some additional perspectives and extending the discussion in complementary ways.

7.3.1   Teachers' Attitudes Towards Students Experiencing Difficulties in Learning Mathematics

Highlighted in the chapters by Hunt and Tzur and Zhang, Maher, and Wilkinson are the consequences to student learning when teachers perceive students experiencing difficulties in learning mathematics as less capable of learning (e.g., deficit view). Hunt and Tzur describe how teachers' perceptions of their students' abilities may influence their instructional actions. They make a compelling case that teachers who perceive their students as lacking mathematical abilities will likely design instruction that sustains a deficit view. Similarly, Zhang, Maher, and Wilkinson recognize the role that standardized tests play in shaping teachers' perceptions of students' knowledge, skills, and abilities. They note that students may be judged as incapable of learning or applying advanced mathematics content based on their low scores.

I would like to broaden this discussion to focus on the systems in which teachers and students engage and highlight the role a school's culture plays in shaping teachers' attitudes and ultimately their actions and/or beliefs. A school's culture can have a significant impact on how students who are experiencing difficulties in learning mathematics are perceived and supported. A school's culture can be thought of as the shared values, norms, and beliefs that explicitly or implicitly guide the actions and decisions of teachers, school leaders, and the broader school community (Kilgore & Reynolds, 2011). A school's culture impacts the ways in which students experiencing difficulties in mathematics are perceived and the perception of whose responsibility it is to support their learning. As evidenced by their actions and decisions, some schools may espouse underlying beliefs about these students' mathematical abilities. For instance, these students may be thought of as "lacking" or "deficient" in their mathematical knowledge. In turn, those deficiencies may be perceived to be associated with disadvantages that are rooted in cultural, social, economic, or political differences from the normative group in the school (Healy & Powell, 2013).

Schools with a deficit view of students often hold the perception that low performance on standardized tests is caused by problems inherent to the student. Such problems may include limited parental support or involvement, insufficient preparation from their home environment or previous schooling, underlying lack of motivation, and students' insufficient background knowledge or even capacity to learn. In some instances, this deficit view permeates so deeply that it becomes a self-fulfilling prophecy. Beliefs that students have deficits or deficiencies that limit their ability to learn mathematics may fuel underlying biases that impact teachers' and/or school-level actions and decisions that harm student learning. An outcome of this perspective is where the blame for poor performance lies: In a deficit model, the blame lies with the student, their parents, or possible background or demographic characteristics, but not with teaching that takes place within the school or educational system through which the poor performance originated.

Conversely, an asset-based approach to supporting students assumes that all students have both knowledge and the capacity to learn—and deserve to learn. Accordingly, the school and related stakeholders are responsible for providing appropriately designed instruction (including assessment) that can address each student's needs. Schools in which the perception of students is grounded in an asset model believe that students' backgrounds provide a strong basis from which they can connect new knowledge and view families and communities as mathematically rich resources that can be harnessed to engage students. Actions that characterize teachers with an asset orientation include designing instruction to build new knowledge based on what students know, recognizing and honoring students' reasoning and sense making, and extending what they recognize and value as mathematical practices (Healy & Powell, 2013). At the system level, schools can examine how they incorporate families and communities into their practices, ways in which current operations may alienate people, and what additional resources are needed to engage the student, their families, and the community. Based on the actions and beliefs that are espoused and enacted, the school is establishing their culture. Each school has the authority and responsibility to establish a positive culture that values all students and views all students as capable of learning.

Intertwined with schools' and teachers' perceptions of students who are experiencing difficulties in learning mathematics is the notion of struggle. In mathematics education, this term is often used to denote how students wrestle with and persevere through problem-solving as they apply their mathematical knowledge. The field of mathematics education has coined the term "productive struggle" to indicate the experiences students face as they tackle challenging mathematical content, integrate and apply their knowledge, and persevere through finding a solution. This type of struggle is deemed productive because the content is accessible yet also challenging, and the problem-solving experience generally results in positive cognitive and/ or non-cognitive outcomes (e.g., persistence, grit). However, there is an opposite experience had by some students: unproductive struggle. When students are unproductively struggling, it is often because they do not yet have an entry point for approaching the content or problem, they have not yet gained sufficient knowledge to apply to the problem situation, or they have persistent errors or naive conceptions that impact their current thinking about the problem space. Instead of resulting in positive cognitive and/or non-cognitive outcomes, the result of unproductive struggle is often negative (e.g., unwillingness to persevere, diminished feelings of self-worth). At this point, and with repeated experiences with unproductive struggle, students may lose interest in mathematics, perceive themselves as not successful in mathematics, and cement a negative identity. Students with a negative mathematical identity who define themselves as "not a math person" may act in ways that perpetuate this self-impression, such as deferring to others who are perceived of as more mathematically capable, engaging with mathematics in low-level ways, and talking and acting with peers in a way that propagates this identity (Bishop, 2012). As such, without intervention to change the underlying causes of unproductive struggle, the lasting consequences could be dire.

Recognizing that some students experience unproductive struggle sometimes and with some mathematics content does not mean that they will continuously and persistently struggle with mathematics. Moreover, saying that someone is struggling is not akin to saying that they are failing or incapable of learning mathematics. Instead, it recognizes that all students learn differently, and some students may need additional instructional support to build and demonstrate understanding in specific mathematics concepts or procedures.

In summary, both chapters recognize the negative consequences to student learning when teachers perceive students experiencing difficulties in learning mathematics from a deficit view. I extend this discussion to recognize the role a school's culture plays in shaping teachers' attitudes, actions, and beliefs. I note that a deficit view of students may cause a misalignment between students' learning needs and their instructional opportunities. By not providing access to proximally relevant instruction and tasks, students miss the opportunity to build on their prior knowledge as a basis to deepen their understanding of mathematics concepts and may develop poor dispositions toward mathematics.

7.3.2   Intentionally Designed Assessments and Tasks

Another theme that emerged in the chapters by Hunt and Tzur and Zhang, Maher, and Wilkinson is the importance of intentionally designing assessments or tasks. Hunt and Tzur provide detailed guidance on the types of tasks and prompts that may facilitate learning across different stages of students' understanding. Zhang, Maher, and Wilkinson approach the topic from two perspectives: (1) the importance of providing accommodations to improve accessibility and (2) the need to extend the range of constructs that are measured so as to cover the breadth of mathematical practices. In this section, I will expand upon the authors' discussions by first emphasizing the importance of alignment and then underscoring the importance of fairness and equity in assessment or task design.

Alignment of Curriculum, Instruction, and Assessment: Implications for Assessment and Task Design   In 1999, the National Research Council (Bransford et al., 1999) recognized the alignment between curriculum, instruction, and assessment as a key issue impacting students' opportunities to learn. Within an aligned system, students have access to important curricular expectations during instruction, and their progress toward reaching these expectations can be monitored through an integrated and ongoing process of formative and summative assessment.

All students are provided with instructional opportunities that facilitate their learning of the curricular expectations. Assessment results are used to guide teachers' decisions about how best to facilitate learning.

For assessment results to be relevant and informative for supporting learning, formative and summative assessment practices must be clearly and carefully aligned with curriculum and instruction along multiple dimensions. Such an alignment includes the content domains, levels of cognitive engagement, and strategies and thinking processes through which students interact with the content (Ketterlin-Geller, 2016). No single test can sample the full range of curricular expectations; data combined from multiple tests should provide a comprehensive picture of student proficiency.

Within an aligned system, the curricular expectations should form a foundation of all instruction and assessment efforts. The curricular expectations to which I am referring are the state or national content standards in mathematics. In the United States, these content standards are often informed by research on how children learn mathematics (National Research Council [NRC], 2001) and may be aligned with the *Common Core State Standards in Mathematics* (CCSS-M; National Governors Association & Council of Chief State School Officers, 2010). A key shift in the CCSS-M that was informed by research in mathematics education (National Mathematics Advisory Panel [NMAP], 2008; NRC, 2001) was the focus on a balanced approach to integrating ways of knowing mathematics. These ways of knowing, emphasized in the CCSS-M, include conceptual understanding, procedural fluency, and application through problem-solving. In *Adding It Up,* published by the NRC (2001), additional specification was provided to call out strategic competence, adaptive reasoning, and productive disposition as important dimensions of mathematical proficiency (these dimensions of mathematical proficiency are defined elsewhere in this volume). In addition to these dimensions of knowing, mathematical practices were embedded in the CCSS-M and in most states' content standards. Mathematical practices represent ways in which students engage with the mathematical content through solving meaningful real-world problems, reasoning abstractly and quantitatively, constructing viable arguments about mathematical conjectures, modeling, appropriately using mathematical tools, engaging with precision, discerning patterns and structures in mathematics, and searching for regularity when solving like problems.

I intentionally define and call out these curricular expectations for several reasons. First, the definition of mathematical proficiency just described represents an integrated understanding of what mathematical competency means and looks like, and by extension, how we can evaluate performance. Hunt and Tzur contend that mathematical proficiency and performance are separate constructs that lead to different interpretations of students' knowing and thinking in mathematics. However, in an aligned system, formative and summative assessment practices lead to performance that directly indicates and/or illustrates the depth and breadth of students' proficiency across these dimensions of mathematics. When assessments are not aligned to the curricular expectations, performance may not be indicative of mathematics proficiency. I recommend we focus on building aligned systems of instruction and assessment to facilitate teachers' decision-making and, ultimately, students' development of mathematical proficiency.

Second, as the guidepost for an aligned system, the curricular expectations should be at the forefront of all instructional opportunities and assessment practices. It is essential that all students—whether or not they are experiencing difficulties in learning mathematics—have opportunities to learn the depth and breadth of the curricular expectations and demonstrate their understanding. Zhang, Maher, and Wilkinson pointed out an over-emphasis on procedural fluency during instruction and on assessments for students experiencing difficulties in learning mathematics. I contend that if instruction does not emphasize the range of learning expectations specified in the content standards, then the curriculum is narrowed, and students are denied the opportunity to learn essential mathematics content that has been deemed important and necessary for them to learn in each grade. A growing number of interventions and intervention frameworks are available that effectively represent the depth and breadth of the curricular expectations (c.f., Fuchs et al., 2017; Powell et al., 2020; Zhang et al., 2014). An exception to alignment with grade-level content standards may exist for students with disabilities whose IEPs specify that they receive a modified curriculum and are assessed based on alternate content and performance standards. For these students, the IEP team determined that the most suitable instruction and assessment should focus on modified content standards.

Relatedly, for assessments and tasks that are intended to inform teachers' decision-making about students' development of mathematical proficiency across the range of curricular expectations, the content should be representative of the depth and breadth of the content standards. In other words, items or tasks should align with the content and the strand(s) of mathematical proficiency that is intended by the standard. For example, consider the Common Core State Standard in Mathematics (2010):

> 7.EE.4. Use variables to represent quantities in a real-world or mathematical problem, and construct simple equations and inequalities to solve problems by reasoning about the quantities.

This standard focuses on expressions and equations, in particular the use of variables. Students engage their conceptual understanding of variables to represent problem situations, procedural knowledge to solve problems, and strategic competence to construct equations and inequalities, and quantitative reasoning to consider the real-world outcomes. To adequately assess this standard, the range of proficiency should be elicited and will require multiple items. If, instead, this standard was only assessed with items that elicited students' procedural fluency, interpretations about students' level of understanding and reasoning related to this standard may not be valid.

To support some decisions, teachers may need data about specific aspects of students' mathematical proficiency that are inclusive but not exhaustive of the curricular expectations. For example, Zhang, Maher, and Wilkinson referenced the need to understand students' reasoning to make decisions about their abilities to develop arguments and justify their solution frameworks when solving problems. An assessment aligned with this purpose may not include items that assess other aspects of mathematical proficiency such as conceptual understanding, procedural fluency, or strategic competence. Similarly, as described by Zhang, Maher, and Wilkinson, CBMs in mathematics often focus on computations. Results from these assessments can provide teachers with valuable insights into students' procedural fluency and can guide instructional decisions relevant to the accuracy and efficiency in which students execute procedures, but not their conceptual understanding, strategic competence, or reasoning. In both of these examples, the intended construct of the tests is narrower than the range of proficiencies specified in the curricular expectations. Alone, results from these tests do not provide a comprehensive view of students' mathematical proficiency. However, because the decisions resulting from these assessments are specific to the assessed content (and, therefore, not the depth and breadth of the curricular expectations), these assessments may continue to have value for teachers. As I hope this discussion illustrates, it is important to keep an assessment's intended uses and interpretations in the forefront when examining alignment and evaluating validity.

Various item or task formats can adequately assess a range of students' knowledge, skills, and abilities. Although student interviews or individually administered tasks as described in the chapters by Hunt and Tzur and Zhang, Maher, and Wilkinson may provide direct evidence of student thinking, this item or task format may not be feasible within all situations due to the time needed for administration and interpretation. Selected response items (e.g., multiple choice, true–false, matching) may aid in efficiency of administration and scoring, while still assessing multiple dimensions of student proficiency. For example, well-designed multiple-choice items can assess students' conceptual understanding, reasoning, and other higher-order thinking skills (Downing, 2006; Haladyna, 2004; Schneider et al., 2013) and, when intentionally designed to do so, may provide insights into the students' thinking based on their selection of distractors (Briggs et al., 2006; Luecht, 2007).

Third and lastly, it is important to remind ourselves that we—public school teachers, administrators, curriculum developers, assessment designers, and education researchers—are facilitators of the learning process and should engage in activities that support student learning of the depth and breadth of the curricular expectations. We are responsible for providing each student with carefully constructed opportunities to meet these expectations. In most instances, a state board of education or a national ministry of education has decided what students are required to know and be able to do; we cannot and should not change these expectations. As such, our efforts to facilitate mathematical proficiency should be in service of the ratified content standards. A teacher may have concerns about the nature of their state or country's content standards, particularly as they represent ways of knowing of the dominant culture and themselves are socially, politically, and historically bound. These concerns should be communicated with the school and district leadership, and/or state or national legislators. However, while these content standards remain the legislated mandate, it is imperative that educators provide each and every student with opportunities to learn the breadth and depth of these curricular expectations.

Designing Assessments and Tasks to Promote Fairness, Accessibility, and Validity Another consideration when intentionally designing assessments and tasks is fairness and accessibility, because they directly impact the validity of the intended uses and interpretations of assessment and task results. Zhang, Maher, and Wilkinson emphasize the role accommodations play in supporting access to the assessed construct for students with disabilities. They also note that standardized accommodations may not live up to this goal because the variety of students' needs that may not be well matched with standardized accommodations. I would like to build on this discussion by recentering the purpose of accommodations within the perspectives of fairness and validity.

To support fair and valid decision-making, assessments must be free from bias, minimize the impact of sources of construct-irrelevant variance, and be based on statistical analyses (e.g., item calibrations, percentile and cutoff scores, validity evidence) that were obtained from a representative sample of students. Items, and the assembled test, should yield results that can fairly inform decision-making regardless of the examinee's race or ethnicity, gender, socioeconomic status, educational classification, or any other demographic variable. To verify comparability, sufficient evidence is needed that documents the absence of differential item functioning (DIF) and differential classification outcomes.

Regardless of the type of assessment, some students with disabilities will need testing accommodations to accurately demonstrate their knowledge, skills, and abilities. Zhang, Maher, and Wilkinson introduced the purpose of testing accommodations and provided an important distinction from test modifications. To further elaborate, testing accommodations are designed to mitigate the impact of students' personal characteristics that negatively interact with item design features and subsequently lead to construct-irrelevant variance (CIV) in students' scores. As an example of this interaction, consider a student who may have difficulty retaining information in short-term memory. This student may experience additional difficulty when solving multi-step word problems that require the student to use information temporarily stored in working memory to solve subsequent components of the problem. In these instances, the student may benefit from using a graphic organizer as an accommodation to record intermediary steps and solutions within the problem. However, when the same student is responding to items that elicit conceptual understanding or application of single-step procedures that do not require extensive use of working memory, the student may not encounter additional challenges that result in CIV. In those instances, the student may not benefit from the accommodation. Comparatively, a student who has difficulty decoding the text used in multi-step word problems may benefit from a reading-based accommodation (e.g., read aloud, simplified language) and may not benefit from using a graphic organizer.

The primary purpose of administering tests is to capture variability in student performance that can be attributable to variability in construct-relevant knowledge, skills, and abilities. However, the examples above illustrate how students' personal characteristics influence how they engage with test items in construct-irrelevant ways. Moreover, the examples also highlight how variations in item design features may support or hinder students' engagement depending upon their personal characteristics. It is because of this interaction that the evidence about the effectiveness of accommodations is inconclusive. I have long argued that accommodations should be assigned at the item level where these interactions impact students' ability to accurately demonstrate their construct-relevant knowledge, skills, and abilities (c.f., Ketterlin-Geller, 2008; Ketterlin-Geller, 2016). Accommodations can be effective only when they are applied at the juncture where CIV occurs.

As this discussion illustrates and is noted by Zhang, Maher, and Wilkinson, assigning test accommodations should not be conceptualized as a one-size-fits-all approach. IEP teams need to consider the student's personal characteristics and item and test design features to determine the probable sources of CIV. IEP teams can then assign allowable accommodations that may mitigate the negative interaction. Our previous research (c.f., Ketterlin-Geller et al., 2014) provides guidance on this process.

I want to explicitly note that accommodations are not intended to promote a deficit view of students with disabilities. Differences in students' cognitive processing, attention, language or linguistic processing, and physical characteristics are not deficiencies and may promote mathematical sense making in other ways (Healy & Powell, 2013). Insomuch as the test design features can support accurate elicitation of students' mathematical knowledge and skills, there may be little or no CIV introduced into students' scores. However, in instances where the test design features may hinder students' ability to demonstrate their mathematical understanding, accommodations will be needed to provide equitable opportunities for students with disabilities to demonstrate their mathematical sense making. Alternatively, tests can be designed to minimize the reliance on construct-irrelevant skills by applying the principles of universal design (Ketterlin-Geller, 2008; Ketterlin-Geller, 2016). For example, reading is often a leading cause of CIV on mathematics tests for many students with and without disabilities. Tests can support variability in students' reading skills by providing options to have the test items read aloud to anyone. Although it is not possible to mitigate all sources of CIV through intentional test design, universally designed tests may reduce the need for externally applied accommodations.

In summary, building on the theme introduced by Hunt and Tzur and Zhang, Maher, and Wilkinson, it is important to carefully consider the design of assessments and tasks when using data to guide instructional decisions for students who are experiencing difficulties in learning mathematics. Valid uses and interpretations of results are incumbent on teachers' access to data that are aligned with curricular expectations and instructional opportunities and that accurately reflect students' knowledge, skills, and abilities without bias or CIV.

7.3.3 Assessments and Tasks to Support Diagnostic Inferences

Quite often, mathematics classrooms do not comprise a homogeneous grouping of students. Instead, many of today's classrooms are characterized by considerable heterogeneity across a variety of factors, including students' background knowledge or experiences with formal and informal mathematics, prior conceptualizations of mathematical concepts, and facility with different mathematical representations. As a result, students within a typical mathematics class may benefit from different approaches to instruction, such as representing mathematical concepts in multiple ways, altered amounts of practice opportunities, and different levels of scaffolding. Students may also need varying levels of supplemental instructional support to reach their goals, with some benefiting from minimal additional support and others needing considerably more to meet the curricular expectations. Some schools use a system-level framework (e.g., multi-tiered system of support) to coordinate instructional delivery, while other schools use a more diffuse model. Within either approach, diagnostic data may be needed to efficiently and effectively design supplemental instructional support to address these students' learning needs.

Diagnostic assessment practices form an important part of an integrated and aligned system of assessments. As described in Chaps. 5 and 6 of this volume, teachers often seek diagnostic information to generate hypotheses about the nature of the instructional support from which a student who is experiencing difficulties in mathematics may benefit. These hypotheses can be based on information such as the student's understanding of specific mathematical concepts or skills, persistent errors, or naive conceptions the student makes while completing their work, the strategies the student uses to arrive at a solution and/or the efficiency with which the student executes the strategy, and/or the representations they use to demonstrate understanding. This information recognizes the student's prior knowledge and their ways of knowing and representing mathematics—and helps identify areas in which additional instruction could support future learning. With this information, the teacher can design a supplemental instructional plan that builds on the student's current understanding and facilitates the next steps in the student's learning.

There are various approaches to designing diagnostic assessments to inform these instructional hypotheses and, ultimately, guide the design of supplemental instructional opportunities. Zhang, Maher, and Wilkinson provide an overview of some published diagnostic assessments, as well as various approaches that can be applied without the use of standardized assessments, such as clinical interviews, error pattern and strategy analysis, and strategy analysis with a microgenetic approach. Hunt and Tzur describe an approach to gathering diagnostic information through interviews that scaffold tasks through the learning process. Each chapter documents the procedures of these approaches and offers examples in practice.

Another approach to gathering information that may be particularly useful for guiding the design of supplemental instructional opportunities for students who are experiencing difficulties in learning mathematics is the use of learning progression-based diagnostic assessments. In the remainder of this section, I provide an overview of learning progressions in mathematics, discuss ongoing development work to design diagnostic assessments based on learning progressions, and explain how these assessments may provide unique information that can accelerate the learning process.

Learning progressions (or learning trajectories) are theoretical models of learning that describe the development of sophistication in students' thinking within a discipline. While different disciplines tend to use one term over the other (e.g., science education tends toward learning progressions, while mathematics education often uses learning trajectories), in her exhaustive review, Confrey (2018) noted that there are minimal differences in the propositions underlying these terms. As such, I will use the term learning progressions with the intention of being inclusive of both learning progressions and learning trajectories.

Learning progressions illustrate a sequential progression of understanding within a discipline that leads from foundational knowledge to more advanced thinking (Bennett, 2015). Knowledge and skills are integrated together in a meaningful way to build deeper understanding. Inherently, learning progressions present an asset-based model of learning by framing students' development of understanding as continuing from their available conceptualizations through to more sophisticated and comprehensive understandings (Alonzo, 2018). Confrey (2018) likened this development to a climbing wall in which students deepen their understanding (move up the wall) by combining knowledge and skills (using each hand and foot hold) in a thoughtful, purposeful, and sometimes iterative way. As what happens with a climbing wall, the specific starting point and the pathway each child takes may look different, but the outcome is comparable (reaching the top of the wall, or completing a course).

When most learning progressions are described, they begin with a set of foundational skills that mark the entry point or lower boundary. Similarly, they have an upper boundary that identifies the targeted learning goals for that progression (Confrey, 2018; Corcoran et al., 2009). The lower and upper boundaries are connected by a series of intermediary phases of learning that are akin to the hand and foot holdings on the climbing wall mentioned earlier. These phases of learning make up the network of knowledge, skills, and processes that build in sophistication and complexity as they progress from the lower toward the upper bounds. Each student's pathway through this network may be different and will be guided by their prior experiences and exposure to the content, as well as their instructional opportunities.

The intermediary phases of learning can exist with varying levels of specificity from coarse-grained representations of concepts to fine-grained micro-conceptualizations of student thinking. The granularity in which the learning progression is specified depends on the learning outcomes. For example, the Common Core State Standards in Mathematics are based on the principles of learning progressions and are specified at a relatively coarse grain size. Conversely, the learning progression explicating the process by which students learn to divide fractions articulated in Ketterlin-Geller et al. (2013) is specified at a relatively fine-grained size.

The level of specificity has important implications for informing teaching and learning, notably the design of instruction and assessment. For the purposes of this commentary, I discuss the design of diagnostic assessments to inform supplemental instruction for students who are experiencing difficulties in learning mathematics.

Diagnostic assessments based on learning progressions provide teachers with information about the nature of students' learning, so as to directly inform instruction. Teachers are able to interpret student performance from a lens of students' knowing and understanding and move away from dichotomizing interpretations of performance as "they got it" or "they didn't get it" (Alonzo, 2018). For example, teachers can identify student understanding in relation to their foundational knowledge and skills (e.g., where they started on the climbing wall). This information can inform where instruction may intersect with students' prior conceptualizations. Relatedly, teachers can monitor students' movement through the intermediary phases of learning to determine the knowledge, skills, and processes students have developed and what instructional actions should come next to support learning. Also, teachers can examine fine-grained information within the intermediary phases of learning to better understand the students' conceptualizations, integration of prior learning, facility with mathematical representations, and other aspects of students' mathematical thinking. Understanding students' partial, emerging, or naive conceptions may help teachers identify prior knowledge that can be integrated with future learning to create a more complete representation of knowing in the domain (Confrey, 2018). This information can directly inform the design of instruction.

To illustrate the decisions teachers can make using learning progression-based classroom assessments, I briefly describe a learning progression that illustrates the knowledge, skills, and reasoning that students in kindergarten through grade 2 develop when reasoning about numeric relations. In the project *Measuring Early Mathematics Reasoning Skills* (NSF #1721100), we articulated a learning progression through an iterative and systematic process specified by Ketterlin-Geller et al. (2013). This process integrated evidence from (a) theoretical propositions underlying learning in the domains, (b) detailed reviews by nationally recognized mathematicians and mathematics educators, (c) in-depth analyses of students' thinking as elicited by cognitive interviews conducted with children in kindergarten through grade 2, and (d) input from teachers about their understanding of child development cultivated through extensive observations. The resulting learning progression for numeric relational reasoning has three targeted learning goals (relations, composition and decomposition, and properties of operations), each with three- or four-core concepts (see Kuehnert et al., 2020 for more information). The overall structure is displayed in Fig. 7.2.

Each core concept is further explicated into finer-grained subcomponents that describe the intermediary phases of learning, including the conceptualizations, reasoning, and strategic processes underlying learning.

To illustrate the way in which learning progressions can inform instructional decision-making, I will elaborate on the first core concept in relations, labeled *"Comparison of Unequal Numbers."* The intermediary phases of learning include the following:

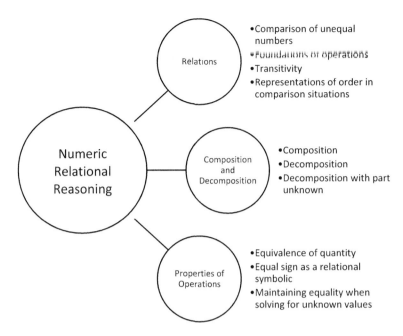

Fig. 7.2 Visual display of the learning progression for numeric relational reasoning

- Compare two quantities to find which has more or fewer items using one-to-one matching and counting strategies without a specific arrangement.
- Compare two quantities to find which has more or fewer items when given a specific arrangement.
- Compare two quantities grouped in tens and ones to find which has more or fewer items using one-to-one matching and counting strategies.
- Compare two quantities to find which has more or fewer items when given a specific arrangement with place value (e.g., place value blocks).
- Compare two numbers using the understanding of the number sequence to determine which is larger or smaller.
- Compare two numbers using written number lines to determine which is larger or smaller.
- Compare two numbers using open number lines to determine which is larger or smaller.
- Compare two numbers using symbolic notation.

These intermediary phases of learning occur iteratively as students build fluency and flexibility working with increasingly larger number ranges (e.g., from 1–5 to 1–19). Through each intermediary phase, students deepen their understanding and ability to apply number relations concepts. Moreover, they are introduced to the number line as an important tool for modeling mathematics. Ultimately, these intermediary phases culminate in a strong conceptual understanding of the number sequence, magnitude, place value, and the usefulness of communicating mathematical concepts through visual representations. These concepts lay the foundation for future work with operations.

Classroom assessments aligned to this learning progression can provide teachers with diagnostically useful information about students' conceptualizations, reasoning, and strategic processing at each intermediary phase as the student progresses to larger number ranges. For example, consider this sample item for the first subcomponent, *compare two quantities to find which has more or fewer items using one-to-one matching and counting strategies without a specific arrangement*. This item would be presented to a student who is working within the number range of 0–10 in an interview format with manipulatives (e.g., counters).

Stimulus	Prompt to elicit conceptualizations	Prompt to elicit reasoning	Prompt to extend students' thinking
Look at these white and black counters.	Without touching these counters, are there more black counters or white counters?	Show me how you got your answer. You may touch the counters.	Show me how you would make the groups so there are more white counters.

Because the purpose of administering diagnostic assessments is to gather data to generate hypotheses about student learning to inform the design of supplemental instruction, students' responses to these types of questions do not need to be scored in the traditional sense (e.g., scored for correctness, evaluated against a rubric). Regardless of whether the assessment includes constructed or selected response items, students' responses provide windows into their thinking. Teachers can interpret students' responses in relation to the learning progression to identify the students' prior conceptualizations. Moreover, teachers may get a glimpse into the strategies students use when solving problems and the varied ways of knowing and making sense of mathematics.

Using the item above as an example, a student may be able to count the number of counters by color and be able to state which group has more counters only when allowed to rearrange the counters so as to employ a one-to-one matching strategy. In this instance, the teacher gains insights into the student's understanding of relationships between quantities and the strategies they use to understand these relationships. The teacher may infer that the student is progressing in their understanding of the number sequence, has a grasp of cardinality, and has a strategy for comparing quantities using physical manipulatives. In turn, the teacher can use this information to understand

- what content to teach next and to what level of intensity;
- student's conceptualizations underlying their responses, which may illuminate varying levels of sophistication and efficiency in the strategies; and
- student's mental models of mathematics concepts.

As these interpretations inform the teacher's inferences about the student's thinking, they serve to guide instructional decisions to support future learning.

In sum, within an integrated and aligned system of assessments, diagnostic assessments serve an important role in promoting equitable outcomes in mathematics classrooms by providing teachers with valuable information to support inferences about student learning. These inferences may include an understanding of students' current conceptualizations, persistent errors, or naive conceptions, strategies the student uses and/or the efficiency with which the student executes the strategy, and the representations the student uses to demonstrate understanding. From this information, teachers can formulate hypotheses about the nature of the instructional support from which students may benefit. In the example described above, the teacher may facilitate the student's future learning by emphasizing the meaning of the number sequence, extending the range of strategic approaches to comparing quantities (including introducing visual and abstract representations of quantities), and expanding the number range in which the student is making comparisons—all of which align with the intermediary phases of the learning progression. As this example illustrates, because learning progressions are based on theoretical models of learning, they provide a unique lens through which students' knowing and thinking can be interpreted to directly guide instructional design decisions.

7.4   Conclusions

Students who are experiencing difficulties in learning mathematics may be supported in a variety of classroom settings within a school. It is essential that the values, norms, and beliefs that define a school's culture recognize the assets all students bring to the learning environment and take actions that signify these beliefs. Importantly, all students should view themselves as valued contributors to the learning environment who are capable of learning complex mathematical concepts and have unique and important perspectives that shape their learning.

When intentionally designed, assessments can help teachers align instruction with students' learning needs to improve and accelerate student learning. Designing assessments to provide meaningful, trustworthy, and reliable data to inform classroom decisions requires careful consideration of several factors, such as the depth of understanding elicited by the items, content representation at the item and test levels, and the procedures for administration, scoring, and interpretation. In particular, although there are various approaches to designing diagnostic assessments, aligning the content with learning progressions may help teachers make inferences about students' knowing, which can be translated into actionable steps for designing learning opportunities.

On a final note, I want to acknowledge and thank the editors for inviting me to contribute this commentary. We often point to differences in the ways in which special education and mathematics education researchers approach supporting students who are experiencing difficulties in learning mathematics (c.f., Woodward &

Tzur, 2017). In this commentary, I have tried to emphasize the commonalities both between the chapters and also among the broader community of educators who are committed to improving outcomes for students who are experiencing difficulties. I urge us—the community of educators, researchers, and other stakeholders—to build on our shared goals and vision of all students realizing positive outcomes in mathematics. The more we can work toward understanding and away from emphasizing difference, the better equipped our community will be to support all students, especially those who are experiencing difficulties in learning mathematics.

References

Alonzo, A. C. (2018). An argument for formative assessment with science learning progressions. *Applied Measurement in Education, 31*(2), 104–112.

Bennett, R. E. (2015). The changing nature of educational assessment. *Review of Research in Education, 39*, 370–407. https://doi.org/10.3102/0091732X14554179

Bishop, J. P. (2012). "She's always been the smart one. I've always been the dumb one": Identities in the mathematics classroom. *Journal for Research in Mathematics Education, 43*(1), 34–74. https://doi.org/10.5951/jresematheduc.43.1.0034

Bransford, J., Brown, L. A., & Cocking, R. R. (1999). *How people learn: Brain, mind, experience, and school*. National Academy Press.

Briggs, D. C., Alonzo, A. C., Schwab, C., & Wilson, M. (2006). Diagnostic assessment with ordered multiple-choice items. *Educational Assessment, 11*(1), 33–63.

Confrey, J. (2018). *Future of education and skills 2030: Curriculum analysis – A synthesis of research on learning trajectories/progressions in mathematics* (EDU/EDPC[2018]44/ANN3). Organisation for Economic Co-operation and Development.

Corcoran, T., Mogat, F. A., & Rosher, A. (2009). *Learning progressions in science: An evidence-based approach to reform* (CPRE Research Report #RR-63). Consortium for Policy Research in Education.

Downing, S. M. (2006). Selected-response item formats in test development. In S. M. Downing & T. M. Haladyna (Eds.), *Handbook of test development* (pp. 287–301). Lawrence Erlbaum.

Fuchs, L. S., Malone, A. S., Schumacher, R. F., Namkung, J., & Wang, A. (2017). Fraction intervention for students with mathematics difficulties: Lessons learned from five randomized controlled trails. *Journal of Learning Disabilities, 50*(6), 631–639.

Haladyna, T. M. (2004). *Developing and validating multiple-choice test items* (3rd ed.). Lawrence Erlbaum.

Healy, L., & Powell, A. B. (2013). Understanding and overcoming "disadvantage" in mathematics learning. In M. A. Clements, A. J. Bishop, C. Keitel, J. Kilpatrick, & F. K. S. Leung (Eds.), *Third international handbook of mathematics education* (pp. 69–100). Springer.

Jordan, N. C., Kaplan, D., Locuniak, M. N., & Ramineni, C. (2007). Predicting first-grade math achievement from developmental number sense trajectories. *Learning Disabilities Research and Practice, 22*(1), 36–46.

Jordan, N. C., Kaplan, D., Ramineni, C., & Locuniak, M. (2009). Early math matters: Kindergarten number competence and later mathematics outcomes. *Developmental Psychology, 45*(3), 850–867. https://doi.org/10.1037/a0014939

Ketterlin-Geller, L. R. (2008). Testing students with special needs: A model for understanding the interaction between assessment and student characteristics in a universally designed environment. *Educational Measurement: Issues and Practice, 27*(3), 3–16. https://doi.org/10.1111/j.1745-3992.2008.00124.x

Ketterlin-Geller, L. R. (2016). Understanding and improving accessibility for special populations. In A. Rupp & J. P. Leighton (Eds.), *Handbook of cognition and assessment* (pp. 198–225). Wiley-Blackwell.

Ketterlin-Geller, L. R., Crawford, L., & Huscroft-D'Angelo, J. N. (2014). Screening to assign accommodations: Using data to make decisions. *Learning Disabilities: A Multidisciplinary Journal, 20*(2), 61–74.

Ketterlin-Geller, L. R., Yovanoff, P., Jung, E., Liu, K., & Geller, J. (2013). Construct definition using cognitive-based evidence: A framework for practice. *Educational Assessment, 18*, 122–146. https://doi.org/10.1080/10627197.2013.790207

Kilgore, S. B., & Reynolds, K. J. (2011). *From Silos to Systems: Reframing schools for success. Thousand Oaks, CA: Corwin.*

Kuehnert, E. A., Geller, J., Perry, L., Hatfield, C., & Ketterlin-Geller, L. R. (2020). *Numerical relational reasoning (NRR): Learning progressions development* (Tech. Rep. No. 20-02). Southern Methodist University, Research in Mathematics Education. Available online at: https://blog.smu.edu/MMaRS/technical-reports/

Luecht, R. M. (2007). Using information from multiple-choice distractors to enhance cognitive-diagnostic score reporting. In J. P. Leighton & M. J. Gierl (Eds.), *Cognitive diagnostic assessment for education: Theory and applications* (pp. 319–340). Cambridge University Press.

National Governers Association & Council of Chief State School Officers. (2010). *Common Core state standards for mathematics.* Authors.

National Mathematics Advisory Panel. (2008). *Foundations for success: The final report of the National Mathematics Advisory Panel.* US Department of Education.

National Research Council. (2001). Adding it up: Helping children learn mathematics. In J. Kilpatrick, J. Swafford, & B. Findell (Eds.), *Mathematics learning study committee, center for education, division of behavioral and social sciences and education.* National Academy Press.

Powell, S. R., Berry, K. A., Fall, A.-M., Roberts, G., Fuchs, L. S., & Barnes, M. A. (2020). Alternative paths to improved word-problem performance: An advantage for embedding pre-algebraic reasoning instruction within word-problem intervention. *Journal of Educational Psychology, 113*(5), 898–910. https://doi.org/10.1037/edu0000513

Schneider, C. M., Huff, K. L., Egan, K. L., Gaines, M. L., & Ferrara, S. (2013). Relationships among item cognitive complexity, contextual demands, and item difficulty: Implications for achievement-level descriptors. *Educational Assessment, 18*, 99–121.

Tzur, R. (2013). Too often, these children are teaching-disabled, not learning-disabled. In *Proceedings of the 11th annual Hawaii international conference on education.* Author (DVD).

Woodward, J., & Tzur, R. (2017). Final commentary on the cross-disciplinary thematic special series: Special education and mathematics education. *Learning Disability Quarterly, 40*(3), 146–151.

Zhang, D., Xin, Y. P., Harris, K., & Ding, Y. (2014). Improving multiplication strategic development in children with math difficulties. *Learning Disability Quarterly, 37*(1), 15–30.

Part III

Chapter 8
Supporting Diverse Approaches to Meaningful Mathematics: From Obstacles to Opportunities

Carla Finesilver, Lulu Healy, and Ann Bauer

Abstract In this chapter, we explore how classroom cultures might be established that are more congruent with the diverse needs of all the learners who compose them and identify features that support the development of conceptual understanding of mathematics in learners currently marginalized by normative classroom expectations. We focus in particular on two interlinked issues, *time* and *tools*. First, we address how time-related pressures particularly affect students with learning differences, difficulties, or disabilities and suggest ways of thinking about the constructs of efficiency and economy that respect student difference. We then address these constructs in relation to tool use, looking at some ways by which tools (in a broad sense, including manipulatives, calculation and communication devices, memory aids, and representational strategies) may be particularly helpful for students who have not been able to access mathematics teaching in the same ways, or as successfully, as their more typically developing peers.

Keywords Inclusive education · Special educational needs · Disability · Mathematics education · Representation

8.1 Introduction

There are many approaches to working mathematically, and expressing and exploring mathematical ideas, but both students' and teachers' choices may be restricted by beliefs about the ways of working that are and are not valued or "legitimate" in school mathematics (Healy & Powell, 2013), or the purposes and statuses of different modes (Coles & Sinclair, 2018). Legitimated approaches and their associated tools are not equally congruent with the needs of all student groups, meaning that when certain practices or tools are undervalued or disallowed, opportunities for

C. Finesilver (✉) · L. Healy · A. Bauer
King's College London, School of Education, Communication & Society, London, UK
e-mail: carla.finesilver@kcl.ac.uk

© The Author(s), under exclusive license to Springer Nature Switzerland AG 2022
Y. P. Xin et al. (eds.), *Enabling Mathematics Learning of Struggling Students*,
Research in Mathematics Education,
https://doi.org/10.1007/978-3-030-95216-7_8

some students to fully participate in mathematical activities can become severely limited (Fernandes & Healy, 2020; Finesilver, 2017b). This affects not only their mathematical performances, but also their relationships with mathematics and perceptions of themselves as learners (Bauer, 2020).

In arguing for diverse approaches to pedagogy in order to develop mathematics education that is both meaningful and inclusive, we address some underlying issues and assumptions relating to "special education" in general and then focus on two issues in particular around which incongruences between learners' individual characteristics and educational environments have been identified, illustrated with three vignettes, one from each of our research. The first issue relates to *time*, in particular to the consequences of pedagogies which associate the quantity and speed of "work done" with mathematical prowess and being "slow" as evidencing lack of ability in mathematics. The second issue relates to *tools*, in particular the consequences of seeing certain tools and representations as temporary scaffolds, rather than as part and parcel of mathematical thinking.

In the following sections, we address some theoretical underpinnings regarding diversity, difference, disability, and "special educational needs" (hereafter SEN) and then focus on *time* and *tools* first separately and then interactionally. To illustrate the points, we draw on data from our different past and ongoing projects involving the *arithmetical-representational strategies* of learners who were/are considered to have "SEN and/or disabilities." These are as follows:

- Collaborative participatory action research projects conducted that collectively compose the ongoing research program, Towards an Inclusive Mathematics Education.[1] (Healy)
- Problem-solving interviews focusing on co-creation of arithmetical-representational strategies, conducted for a microgenetic study focusing on emerging and developing multiplicative thinking. (Finesilver)
- Interviews focusing on learners' and educators' relationships with mathematics learning and professional practice as a specialist teacher of young people with SEN. (Bauer)

Through these, plus our respective professional experience, we explore how classroom cultures might be established that are more congruent with the diverse needs of all the learners who compose them and identify some features that support the development of conceptual understanding of mathematics in learners currently marginalized by normative classroom expectations.

8.2 Diversity, Difference, Disability, and SEN

We begin this section by considering the case of a young blind learner, who we will call Vitória.[2] We have chosen this as an opening story, in part because the issues of time and tools permeate its telling in ways we will draw out in the sections and

---

[1] Publications associated with this program are available for download from http://www.matematicainclusiva.net.

[2] The data from which Vitória's story has been synthesized are available in Freire (2017).

Vignette 1: Number Activities in the Absence of Numerical Symbols

Vitória (10) lived in the Brazilian state of Sao Paulo and was invited to participate in a project intended to explore the understandings of multiplicative structures developed by learners who do not see with their eyes. By the fourth grade, it is expected that students will have encountered multiplicative structures of a variety of forms in their formal schooling, but this turned out not to be the case for Vitória. At the beginning of the project, she was still unfamiliar with the place value system, though she had memorized numbers up to thirty-two and a limited number of "sums" (1 + 1; 2 + 2). She seemed to have developed a sense of one-to-one correspondence and was able to count unit cubes, correctly separating the number requested by the researcher, although she had a tendency to lose count if more than 12 were requested. She very quickly became anxious to the point of crying if she was asked to engage with a problem whose answer she had not yet memorized. She was also unfamiliar with Braille at the beginning of the project, and her attendance at school was sporadic.

In her second meeting with the researcher, she was encouraged to work with some small wooden pieces (Fig. 8.1), which, along with a set of Dienes cubes, rods, and block, completely changed her relationship with mathematics.

These wooden pieces were small enough to fit comfortably in Vitória's hand and were easier for her to keep track of than the smaller unit cubes. The wood was smooth and comfortable to hold, with the corners filed so the edges were rounded. She very quickly developed ways of using them to engage with addition and subtraction, which she could apply not only to sums in which the number operations were made explicit but also to word problems, involving small quantities. Place value, and hence larger numbers (which she was very keen to make sense of), was introduced via the Dienes material. Vitória's confidence and self-esteem were almost palpable when she realized she could go on counting forever.

The wooden pieces also became fundamental in working on multiplication and division. She developed a strategy in which each piece could represent any number she chose. In the first instance, this seems to have been motivated by the researchers' attempt to introduce the concept of one to many, by suggesting she imagined the piece represented bags containing five sweets each. With two wooden pieces in front of her, following the researcher's suggestion, she counted to five while tapping on one of them, then placed her hand on second, and, without counting, moved it toward her and said quietly to herself "*five*." Later in the same session, to convince herself that four times five gave the same result as five times four she separated four pieces and counted in

(continued)

Vignette 1 (continued)

fives (rhythmically tapping on each piece in turn) to twenty. She then repeated
the process with five pieces and counted in fours to twenty. She seemed to
have no difficulty in treating the wooden piece as a kind of placeholder that
could represent any number. It seems that for Vitória, a tangible object had
become a way of representing a variable.

In her work on division problems, the pieces had a similarly important
role. Given the problem shown in Fig. 8.2, she chose to count up in threes,
selecting a wooden piece for each three, until she reached eighteen, and then
counting the number of pieces selected before announcing "each child will
receive six reais."

Fast forward almost four months, Vitória had begun to learn Braille, but
was still not confident in reading and especially writing numbers. Her encounters with multiplicative structures had widened, though rather than working
with these in conventional symbolic/written ways, she continued to make use
of the wooden pieces. For example, when asked to calculate 2/5 of 40, she
proceeded by dividing 40 by 5, selecting a wooden piece to represent each 5
as she counted and tapped "five, ten, fifteen, twenty, twenty-five, thirty, thirty-
five, forty," and then counting how many pieces she used. She hence determined that 1/5 is 8, and therefore, 2/5 would be 16. Interestingly, as she came
to this decision, she separated two wooden pieces, tapping again, as she says
"eight" then "sixteen."

Fig. 8.1 Wooden pieces
designed for Vitória

> Carminha won eighteen reais. She is going to divide the money
>
> equally between her three children. How much will each of
>
> Carminha's children receive?

Fig. 8.2 A division problem

further vignettes that follow, but also because it provides a framing context for the theoretical constructs that ground our research.

8.2.1 Cultural Norms and Pathologization of Difference

Many, if not most, education systems have exclusive origins, where access to formal schooling may have been limited to certain subgroups of the population (e.g., selection by wealth, gender, or ethnicity), and while their history has been one of slowly increasing inclusion and widening participation, children with disabilities and/or SEN (or variants of this terminology) are still predominantly a marginalized group (Mitchell, 2005; Mittler, 2002), seen as separate and categorically different from "normal" children, whose "integration" is thrust upon unwilling or dubious teachers and classmates (Avramidis & Norwich, 2002). Inclusion is then positioned as something additional to "normal" teaching and learning. A fundamental step in creating inclusive classrooms is to proceed instead from a new norm—that education systems are meant for the whole population and should be designed accordingly.

With this kind of statement, people often think first of physical school environments being designed to suit only a typical bodily configuration, while creating barriers for others. For example, Vitória's participation might be severely limited in an environment that assumes we all see with our eyes. Curricula too are frequently designed for an imagined typical student (e.g. "a normal 12-year-old"), who is expected to be in possession of a particular set of knowledge, capabilities, and skills, which they demonstrate in a particular way (with greater or lesser success). If an individual does not demonstrate this in the expected manner, they can become pathologized in ways that include being seen as lacking or having something wrong with them (McDermott, 1993; Finesilver, 2019); the lack or wrongness is not so commonly located in the education system. This misaligning of individuals and inflexible systems can be seen as creating *misfits* (Garland-Thomson, 2011) of those who do not perform successfully. Responses to this might be exclusion (excision of the "problem," passing it on elsewhere), attempts to "cure" (trying to change the pathologized individual to become or appear "normal," so they may perform within the expected range in a standardized system), or "accommodations" that allow those in certain medical or psychiatric categories to do certain things differently from their peers. In this way, even changes to any of the institutional education structures, and particularly the removal of obstacles within curriculum and assessment systems, are directed at a target population of special education, rather than to the education system as a whole.

8.2.2   Individual Accommodation or Structural Reorganization

Here, we might point to a tension between the ideologies associated with the movement for inclusive education and the policies and guidelines regarding "special education" provided for use in different education systems.

Norwich (2013) refers to this tension as the dilemma of difference. Inclusive education implies the reorganization of the structures, policies, and practices of mainstream schools, in order that they might respond to diversity in ways that value all learners equally, rather than labeling some as fundamentally more needy than others. This might suggest that an SEN label can only function to create divisions between those seen as "normal" and "others" deemed deficient or disordered (Ferri, 2012). Difference here is seen as stigmatizing, which might be used to imply that forms of specialization or differentiation that are congruent with particular body-minds[3] are viewed as problematic rather than potentially enabling (Norwich & Koutsouris, 2017). Focusing on the practices of particular student groups should not necessarily be associated with labeling them as problematic. It should instead be viewed as a way of enriching and expanding mathematics teaching possibilities.

In this vein, almost 100 years ago, Vygotsky (1924-31/1993) proposed that we look qualitatively at students' potential for learning and the conditions that support this development, rather than comparing their achievement to some imposed norm measured in conditions that do not match with the ways they experience and interpret the world. Although his work did not focus on all those students who would currently be considered as part of the population targeted by "special education," we believe it makes sense to extend his ideas this way. Using today's terminology, we might paraphrase Vygotsky's view thus: *If a child with SEN achieves the same level of development as a child without SEN, then they achieve this in another way, by another course, by other means. And, for the teacher, it is particularly important to know the uniqueness of the course, along which to lead the child. The key to originality transforms the barriers associated with the impediment into the potential for development.*[4]

There are a number of points we might highlight in relation to this view. Firstly, it is very different from positioning the learner with SEN as lacking or deficient in relation to others. There is none of the talk of catching-up, remediating, or curing

---

[3] We use this term following (Price, 2015, p.270), as "the imbrication (not just the combination) of the entities usually called 'body' and 'mind.'"

[4] The paraphrased original (in English translation) reads "If a blind or deaf child achieves the same level of development as a normal child, then the child with a defect achieves this *in another way, by another course, by other means*. And, for the pedagogue, it is particularly important to know the *uniqueness* of the course, along which he must lead the child. The key to originality transforms the minus of the handicap into the plus of compensation" (Vygotsky, 1924–31/1993, p. 34).

that characterizes many intervention programs for those experiencing difficulties with mathematics (Fuchs & Fuchs, 2006; Holmes & Dowker, 2013). In second place, it is the learning environment that is seen as presenting certain obstacles to learning, rather than their location solely within the learners' bodyminds. For example, for Vygotsky the solution for the inclusion of disabled learners in educational (or, more generally, cultural) activities lies in seeking ways to substitute the conventional means of participating (which they may lack or have limited access to), with alternate means that are more congruent with the bodyminds of the individuals in question. If we consider Vitória's story for example, her participation in and relationships with mathematics changed fundamentally when she was provided with tools to substitute the function of her eyes in mathematical activity and offer her different ways of seeing.

This signals the importance Vygotsky placed on the role of tools in shaping activity. He treated tools as defining components of human cultures, created at particular moments in their trajectories in response to the demands of a particular practice. Tools are created to enable us to transcend our bodily constraints, although they are not independent of them. As products of cultures, tools do not appear and become useful by magic: They take time to invent, build, test, and become familiar with; they may also affect the amount of time spent on any task for which they are invented or appropriated. Tools are hence fundamental in any attempt to reorganize educational structures and cultures, a point we will return to as this chapter develops.

Vignette 2: Emerging Conceptions of Multiplicative Structure through Student-Led Visuospatial Representation

This pair of examples focuses on single meaningful moments created by two girls, Tasha (12) and Wendy (13), who were in different classes at the same inner London secondary school. The researcher was informed that Tasha was diagnosed with ADHD and Wendy dyslexia. They both had very low prior performance in school mathematics and were identified by their teachers as struggling significantly (compared to peers) with acquiring numeracy skills and concepts. They could count and carry out addition of natural numbers fairly reliably, but avowed a particular dislike—hatred, even—for multiplication and division. Despite having put a great deal of time and effort over the years into attempting to commit multiplication facts to memory, this had been an almost complete failure; Tasha now exhibited a mixture of anger and distress when faced with multiplication or division, while Wendy had resigned herself to quiet hopelessness. In this context, the researcher aimed to work sensitively with various visuospatial representations to "nudge" students to better understanding of and relationship with multiplicative reasoning.

(continued)

Vignette 2 (continued)

Two overall methodological principles were key: encouragement of students' own representational ideas, needs, and preferences, and complete absence of time pressure on tasks. Here, the unrestricted time principle was taken a stage further: not stopping students when the expected end point had been reached, but allowing them to continue directing their own work with a given tool.

Tasha had been provided with a solid $3 \times 4 \times 5$ block of multilink cubes (Fig. 8.3), and set the task of working out how many of the unit cubes were present without taking it apart (effectively a calculation of volume, or three-dimensional multiplication, although not expressed in these terms, which would have been alienating to her). With minimal nudge prompts (see Finesilver, 2017a), she was able to perceive the spatial structure of the cuboid as four layers of 15 cubes and agreed to have a go at the addition (pictured) and, despite exclaiming this was "not easy!", carried it out successfully. When the block of cubes was reoriented with the 4×5 face uppermost, she took up her pen again and spontaneously carried out the corresponding addition for this too. This may surprise the reader, as it would seem to be a simple case of object permanence. An alternative suggestion is that this indicates a student whose prior experiences of school mathematics had not led her to believe that it was congruent with common sense and logic. She added three lots of 20 (corresponding to the three current layers) to get "60 again!" She then turned the block so that the 3×4 face was uppermost, commenting of the new layers "these are 12," and contemplated whether to carry out a third calculation, before deciding it was unnecessary, as she now had confidence that orientation would not change the result. This student-led task extension is notable in that it stemmed from a student's doubtfulness about something that would be unlikely to occur to a mainstream teacher of 12-year-olds as possibly contentious and her drive to collect the empirical evidence to "prove" the identities to her own satisfaction.

Wendy had been set the task of working out how 30 biscuits might be packaged in equal-sized packets (effectively finding factors, but within an imaginable scenario, and again in non-alienating language). She had recently been working with hand-drawn rectangular dot arrays and drew a 3×10 array. She then set about rhythmically ringing sets of different sized groups to produce an array-container blend[5] with all the different factorizations of 30 (see Fig. 8.4, which also includes a second task carried out in the same way). These were not attempts to *work out* the factors—she checked whether they would work before committing ink to paper—but an expressive external visualization of the results of her thinking about the arithmetical relationships

---

[5]Further details on container- and array-based configurations may be found in Finesilver (2014)

(continued)

Vignette 2 (continued)

between the numbers. Not only was this clearly a satisfying and absorbing activity for her at the time, but appeared to be meaningful to her grasp of multiplicative structure, as on later occasions she referred back to these images. This was not in order to look up any particular multiplication fact, but, I suggest, an instantiation and instant visual reminder of the relationship between multiplication and division, the knowledge that a given total may be divided up in different ways, and the commutative property of multiplication. She would certainly have been taught these principles in some manner at some point, but benefited from being allowed the time to work through them with her chosen tool.

Fig. 8.3 A cuboid block of multilink cubes and Tasha's two calculations

Fig. 8.4 Two array container blend configurations drawn by Wendy

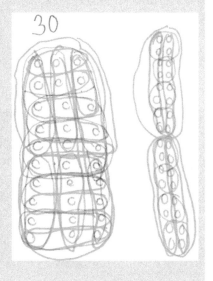

8.3   Time

The roles of *time* in mathematics education require critical consideration. Working "too slowly" is commonly considered a marker of low ability; i.e., a judgment of mathematical worth is made based on the quantity of "work" produced within an arbitrary time period independent of its quality (a judgment that, we note, is rarely applied to advanced mathematicians). However, a standard assessment accommodation offered for students with SEN is extra time. In other words, it is considered important for them to have enough time to demonstrate their knowledge and understanding effectively. While this acknowledgment is certainly welcome, it raises further questions. Is the restriction of working time important or not? In what contexts? Why? Does it indicate anything useful about their potential for more advanced mathematical development? What might be the consequences of removing some of the time constraints on mathematical activity?

Our second vignette is chosen to illustrate two outcomes of this last question. While the use of tools was central to the mathematical activity described, these instances of mathematical meaning-making were a direct result of having unrestricted time to engage in individual exploration of these tools' potential.

In class, students are all too aware of the prizing of speed/quantity over depth/quality. Struggling students see their peers call out answers to teachers' questions on demand, or produce pages of organized mathematical symbols, as if by magic, without visible thought going into the calculations, and so in response develop the imitative *maths-like behaviors* (Finesilver, 2017b) of ritual participation (Ben-Yehuda et al., 2005). In observing classrooms over the years, we have all seen instances of teachers, with the best of intentions, praising a performance of "quick answering" of questions—even when, for some struggling students, it involves the essentially unmathematical behavior of guessing at answers (or in written work, shuffling around partially remembered sequences of symbols). This unhelpful situation is compounded by, and further compounds, teacher beliefs that certain students are incapable of reasoning and should be restricted to low-level recall—or worse, that ritual participation is the most that can be expected of them (e.g., Watson, 2001). In contrast, here Tasha and Wendy were not pressured to produce quantities of symbolic busywork but engaged, by their own choice, in genuine arithmetical reasoning. As with Vitória, they gained satisfaction from working with their preferred tools to solidify their conceptual understanding of arithmetical content that had previously prompted negative emotions, refusal, and withdrawal. It can be seen that in environments where learners are not artificially pressured and consistently encouraged to use whichever forms and elements they find most congruent with the current state of their bodyminds, they can unlearn the bad habits of guessing, symbol shuffling, and other maths-like behaviors and instead engage thoughtfully with mathematics (Finesilver, 2014).

When students are allowed to take their time in this way, there are benefits to their understanding. However, depending on the tools and representational forms employed in "taking their time," another consequence might be a reduction in quantity of "expected classroom behavior and observable completion of clerical work" (Watson, 2006, p. 104). It might be argued that if students were seen to be doing fewer calculations than their peers, or in different ways, this could cause low self-esteem. However, we would argue that this effect is already happening in many classrooms and that the response to it should not be to fix the individual's pace of visible work production (at the expense of meaning), but to create an environment where quantity is not prized over quality, and where slow, thoughtful, in-depth working is valued, and taking time out from completing calculations to visualize patterns and principles is encouraged.

Inclusive education demands care and consciousness of how one uses the limited time that is available, but there may be external factors at work: A teacher may want students to understand the material conceptually, may understand that slow work can provide opportunities for deep understanding, and furthermore that this may be best demonstrated through potentially more time-consuming representational forms and know that the fact that a given individual did not demonstrate an objective in a given arbitrary timeframe does not mean the individual lacks the necessary knowledge or capability—yet be externally pressured to set timed tests, to perform economy at curriculum level (by "getting through" topics at pace) and efficiency at task level (by exercise books that evidence a high number of questions answered, while using only standardized tools and representation). This may well be false economy and false efficiency, in terms of long-term mathematical development, and so the challenges of creating more inclusive classroom practices will need support from the wider school community.

Vignette 3

Kevin was 11 and formally diagnosed with developmental coordination disorder (DCD), developmental language disorder (DLD), and visual perceptual difficulties. He was anxious about mathematics and perceived failure at every turn. One of Kevin's main difficulties was in keeping his place and negotiating visuospatial layouts when reading text, numbers, graphs, tables, and diagrams; this was exacerbated by untidy or crowded layouts, or distracting color choices on the whiteboard or in printed materials. Kevin had difficulties with concepts expressed through language and had been using Dienes blocks to gain a more secure understanding of place value up to thousands. He now expressed a desire to read and decode the meaning of larger numbers. Using blocks becomes unwieldy once tens of thousands have been reached, and the prepackaged place value cards that were available presented difficulties due to their color scheme, busy design, and the need to hold them in one's hand. Instead, the teacher and Kevin created tables (with columns grouped in threes to coordinate with the linguistic structure of larger numbers).

(continued)

Vignette 3 (continued)

This may seem straightforward: The important point to note is that Kevin was fully involved and had control of the colors and sizes (including of the decimal point) so they could be optimized for his perceptual processing (Fig. 8.5), thus removing as much of the additional cognitive effort that had been going into just keeping his place. The table pictured shows an extension into the decimal range that was subsequently made possible, and there are others that go into the millions and onward. These color blocks gave Kevin a method of "seeing" the decimal structure of the numbers in order to read, write, and speak them, understanding the relationship of each digit with those around it.

H of Th	T of Th	U of Th	H	T	U	●	Tenths

Fig. 8.5 A place value table created by Kevin and his teacher

8.4 Tools

Our third vignette is chosen for its focus on another classroom tool, which is familiar and commonly used in support work, but, for this very reason, deserves a closer look. It derives from practitioner 1:1 tuition and illustrates the way that tools need to be used with nuance, in an individually tailored and co-creative way.

The previous section highlighted how discourses that position "quick" as good and "slow" as bad can come to be associated with exclusionary—and false—concepts of efficiency and economy. In relation to tool use in mathematics education, we might point to another false binary construction: dependence/independence, which similarly disables certain groups of students. In discourses that position independence as desirable and associate dependence with weakness, then doing one's mathematics via certain tools is frequently judged as evidence of lack of mathematical ability. Making use of conventional mathematical tools, which are normalized for "typical" students, is not generally construed as "dependence" in this way, though the difference is cultural rather than theoretical.

Take, as an example, a stance recently promoted by the UK's National Centre for Excellence in the Teaching of Mathematics, which states that "Representations used in lessons expose the mathematical structure being taught, the aim being that students can do the maths without recourse to the representation" (NCETM, n.d.). The implication here is that the representation (or tool) is merely a scaffold that can be taken away once the mathematical structure is exposed. This negative positioning dependence on representation/tool may initially appear uncontroversial when considering pedagogy. However, from our point of view, this is not only potentially disabling, but it is also a little extraordinary. If one removes *all* the representations, it is not clear what mathematics will be left at all! It seems, then, that the term "representations" is being used in a particular way: In this context, it is not considered "dependence" to utilize symbolic representations of number, algebraic symbols, or geometrical figures, but only manipulable or nonstandard visual aids, because these are seen as "on the way" to formal representations rather than as valid alternatives. If this interpretation is correct, it adds another angle to the binary oppositions that construct SEN, positioning "concrete" alongside "slow" as an undesirable but hopefully temporary state, and "abstract" beside "fast" as the desired end state.

The positioning of the concrete as intellectually inferior to the abstract has roots in Piagetian descriptions of development. A view of mathematics learning as moving from the concrete to the abstract, for all learners, continues to dominate what happens in school mathematics today and is now embedded in many mathematics curricula and teaching approaches. Examples include the Bruner-inspired "Concrete, Pictorial, Abstract" (CPA) approach advocated by the Ministry of Education in Singapore since the 1980s (Leong et al., 2015) and the similarly named US-based intervention "Concrete, Representation, Abstract" (CRA), which "has been shown to be effective for remediating deficits in basic mathematics computation" (Flores, 2010, p. 195). Our first concern with such approaches is that, while they may lead to success for some students some of the time, if the assumption that all learners follow the same hierarchies all of the time, then those whose learning deviates from this curriculum-embedded path may again be pathologized. Secondly, there is the danger that the "concrete"/enactive and the pictorial/iconic are not seen as valid mathematical representations in their own right, but are instead demoted to the status of transitory aids eventually to be replaced by the conventional symbols, which are commonly equated with abstraction.

Vitória's story is not well captured by a view that mathematics learning proceeds from the concrete to the abstract (and perhaps especially not via "pictorial" representational modes). The wooden pieces she worked with certainly have a concrete, material quality, and yet, we would argue, served a role in shaping her conceptual understanding of numbers. That is, the pieces are neither simply concrete nor abstract but were used by Vitória to simultaneously enact processes of concretion (senses of particular numbers and operations) and abstraction (possibilities for generalizing, connecting, extending). Vitória's relationships with the wooden blocks support the view that doing mathematics involves traversing between the concrete and the abstract (see also Wilensky, 1991; Healy & Fernandes, 2011; Coles & Sinclair, 2019), rather than replacing one with the other. Her case, like all those we have presented here, also suggests that different representational modes enable different expressions of mathematical objects, properties, and relationships—and that concretion and abstraction feature in all. The students' interaction with the different tools also challenges any idea of a linear development between representation modes. We note that drawing (i.e., the middle "P(ictorial)" step of CPA) did not figure in the activities of Vitória; neither would "flat" inscription have been logical in Tasha's activity due to the three-dimensional nature of the multiplicative structure instantiated in the physical array of cubes. In contrast, Wendy chose drawing-based representations to communicate her perception of factors, using a model that could not seamlessly be expressed physically or symbolically. For Kevin, what might be described as a hybrid visual-symbolic concretization of the decimal system was helpful, perhaps not least because, as with all these students, he was an active participant in the representational strategy development.

Classifying mathematical representations as either concrete or abstract (or even on a concrete-abstract spectrum) is actually a rather problematic affair. Epistemologically speaking, a mathematical object can be conceived as a virtual, potential (abstract) entity manifested and interacted within some material form, through the use of a perceivable (concrete) tool—a representation of the object in question. So, in some sense at least, *all* representation tools are concrete. Doing and learning mathematics involve participating in activity with such tools. Irrespective of the representational form of the tool—that is, whether it is a tangible 3D object, a visual 2D figure, or a symbol—it is the *ways that it is used* in mathematical activity, which defines whether abstraction, concretion, or some synthesis between the two is played out. Rather than restricting teaching to a fixed progression of (supposedly increasingly sophisticated) representations, we suggest that more inclusive approaches to tool use involve employing and even inventing tools and representations that are congruent with the bodyminds of a wide diversity of learners.

8.5 Tensions/Discussion

Time and *tools* are deeply interconnected at multiple levels. We now look at these interconnections and some of the practical tensions with regard to *testing* and *teaching*. It may seem odd to address testing before teaching; we have done so to emphasize the bidirectional relationship, and in particular the phenomenon of "teaching to the test," which is unfortunately widespread, and we consider particularly detrimental to learners with SEN and/or disabilities who are perceived as lacking in comparison with their peers.

8.5.1 Time, Tools, and Testing

The ways in which it is decided whether a learner is in possession of the desired knowledge or skills make a significant difference to outcome: Assessment is not neutral. Most formal mathematics assessments include time limits of different kinds (per question, or for the whole test) and regulate tool use. When engaging in assessments that purport to test mathematical ability, it is worth questioning what, in fact, is being tested and what (if anything) it tells us about a learner's mathematical capacities and potential. One way of doing this is by looking critically at the allowances and restrictions (for both the "general" and "special" student populations), and their underlying assumptions and biases. The specific examples in this section are from the UK, but it is expected that the general issues raised will resonate with readers from various other education systems.

In some assessment systems, certain individuals with formal SEN diagnoses may be granted additional time. But is it not important for all students to have as much time they need? In which assessments (if any) should speed be the primary arbiter of success? Who is being advantaged and disadvantaged by time-restricted tests, and is this intended or indeed acknowledged? When artificial time restrictions are placed on tasks, this indicates that speed of working is prioritized; that quantity of product is more valued than quality of thought process. It also has implicit or explicit implications. For example, England's current "Multiplication Tables Check" assessment has a 6-second time limit on questions; this has the stated purpose to "demonstrate [students'] recall of multiplication tables, whilst limiting [their] ability to work out answers to the questions" (Standards and Testing Agency, 2018, p.8). That is, it aims to test learners' ability to rote memorize a set of statements of number relationships, regardless of whether they are understood, and to prevent them from making creative use of derived fact strategies to work out multiplicative relationships they do not immediately recall from ones they do (which one might argue is a more intrinsically mathematical skill to have). It is designed to reward students who are better at memorization of information and to punish those with memory difficulties (including many neurodiverse groups).

In assessments that purport to test mathematical reasoning and problem-solving abilities rather than simply information storage, tools also become a contested issue. For example, in another test, no "number apparatus, counters or number squares" can be used (Standards and Testing Agency, 2020, p. 13). We assume that the reasoning behind this is that access to these would be considered as giving "the pupil an unfair advantage" (ibid. p.10)—although any such unfairness might be addressed simply by allowing anyone who wished to use these tools to do so. The question should instead be whether individuals with certain bodymind configurations are currently unfairly disadvantaged. While there have been moves in this direction (e.g., specific accommodations in the case of certain officially diagnosed and documented impairments), the principles could be applied more generally.

A variety of other tools (e.g., calculators, laptops, vocabulary or formulae booklets, multiplication or logarithm tables) have been variously allowed or disallowed in formal assessments across different times and places; in each case, they affect which cognitive capacities are in fact being tested for, alongside or instead of mathematical ones. Furthermore, these decisions regarding restriction of time and tools are inextricably linked. Consider some tools that support memory: If a tables square is allowed, students with weaker memories will not have to work out the individual calculations that are required for solving more complex problems. If the area of a trapezium is included in a formula booklet, students will not be forced to work it out from principles. If the quadratic formula is included, they may compute roots without an understanding of factoring or differences of squares. All these tools support quicker answering—but in which cases is this desirable? We do not insist on any particular decision, only that the decision-making is thoughtful about the exact time and tools allowed, and what this means for individuals who, because of this, may or may not achieve the requisite number of correct answers for a passing grade.

We realize that individual teachers generally do not have control over the formal assessments their students are subjected to (e.g., national examinations) and, further, may be required to prepare students for these through practice under similar conditions, i.e., by restricting learners from taking time to reflect, or from making creative use of nonstandard tools and representations that help them make sense of mathematics. However, the nature of such testing and the resulting pedagogic practices should be assessed critically and challenged where possible.

8.5.2 Time, Tools, and Teaching

We have given examples of how allowing more thinking time and/or more diverse tools in the short term provides opportunities for engendering conceptual progression with long-term benefits to learning. These are not linear relationships: Depending on the tools and individuals involved, they may mean that the pace of visible "work done" needs to slow down (e.g., exploration in multiple representational forms), or they may mean it speeds up (e.g., using memory aids to allow a learner to focus on the mathematical reasoning/problem-solving aspects rather than basic arithmetical calculations—as is the case with their non-memory-impaired peers). However, the most commonly mentioned tension (in our experience) is that when a student uses nonstandard representational tools (e.g., involving physical objects, drawn or mixed-mode/media representations) on tasks, they may take longer to complete. This is essentially a false comparison, as it is usually comparing the time taken by a given individual to complete a calculation in a way that is meaningful to them not with the alternatives *available to that individual at that point* but with the time taken by a different, imagined, individual with different capacities, for whom standardized memory-based methods are a good fit. We have several times mentioned "false" efficiencies and economies: How, then, should they be construed for inclusive mathematics education? Efficiency is better thought of not as a comparison of calculation methods such that some are deemed absolutely more efficient than others, but in terms of strategic choices that best fit the context of both individual learner and task characteristics; similarly, economical use of time is better considered as a function of quality of mathematical activity achieved (e.g., engagement in arithmetical reasoning) rather than quantity of answers provided (Finesilver, 2017b).

A second tension is regarding the time taken for both students and teachers (and anyone else involved in education) to learn about the different types of tools that are available, to become familiar with their potential uses, and to work out which ones are most suitable for both individual needs or preferences, and different kinds of task. Popular criticisms of various educational tools often rest on the fact that (a) they do not work equally well for all students and (b) they cannot be put in front of students and immediately solve problems (Ball, 1992). We realize that the vignettes we have used are from 1.1 intervention situations, where this may be done in a more direct way. It is also reasonably straightforward in small groups, with the added benefit of collaborative work with peers, and the sharing of strategies. To achieve this with a full class of students, often a cultural change is necessary, including the deliberate reevaluation of the roles of time and tools (for all, and not just those students classified as "different"). Even then, teaching staff may well need additional time outside the classroom for familiarizing themselves with different tools prior to incorporating them in their planning. A lack of such time, and of appropriate training, is widespread (Blatchford et al., 2009; Webster and Blatchford, 2015), even when staff state a clear desire for it (Bauer, 2020).

We see another time/tool interaction in the view, held by some, that students with disabilities and/or SEN develop along a *delayed* version of the typical pathway: that is, that all students learn in the same manner but some take longer; that some require additional steps or supports along the way, in order that the same end goal can and should be reached by all. While this might turn out to be a useful plan for some students, for others typical interventions may never be appropriate. Rigid adherence to a pathway incongruent with a learner's sense of self is not only uneconomical, but it may also serve to repeatedly accentuate publicly and privately what the learner cannot do.

An alternative is to value and legitimize a wide range of different representational modes for mathematical activity, including tools to support functions such as memory and attention (as the representations in our vignettes may have done), as well as instantiating arithmetical relationships), selected *with* rather than *for* learners, so they feel empowered and have control and choice over the strategies they employ. Again, we find Vygotsky's position thought-provoking. He distinguished between what he called *physical* or *material tools* and *psychological* or *symbolic* tools (Vygotsky, 1981). Physical tools restructure physical activity, while psychological tools function in the development and shaping of psychological processes. In both cases, tools make it possible to engage in activity that would not be possible in their absence. Perhaps, though, the material/psychological distinction is unfortunate, as it could be interpreted as implying a hierarchy of tools rather than their different roles in interaction when participating in mathematical activity. Looking at the distinction from the point of view of participation highlights another problematic dilemma. Let us imagine two students. The first is a student whose vision has diagnosed and identified limitations. The function of her eyes might be magnified/transformed by the use of glasses, designed according to the specific nature of her visual limitations, a physical tool in that it restructures physical activity. With the exception of activities specifically aimed to emphasize tactile or auditory representations of mathematics, any suggestion that her glasses be removed would be absurd, and it would be extraordinary to forbid the use of glasses specifically for formal assessment. Our second student is one whose memory has been diagnosed and identified limitations. There are a range of tools that could be used to magnify/transform the functions of the memory in mathematical activity, which could be designed according to the specific nature of her memory limitations. Is it because these could be classified as "psychological" rather than "physical" tools, that the removal of these tools from mathematical activity is not seen as absurd, but instead as something desirable? The mathematical achievement of the first student is not judged on the basis of what she can do without the relevant tool, but the mathematical achievement of the second student is.

8.6 Summary

We have argued in this paper that time and tools are two interconnected factors fundamental in shaping the cultures of mathematics classrooms and the experiences of students with SEN and/or disabilities within them. Both time and tools constrain and afford different mathematical practices and come to be associated with different images of what counts as successful mathematics learning. The conscious decisions and implicit assumptions regarding them can create obstacles or opportunities. Where educational systems are tightly regulated, limiting time and tools in line with assumptions about how the "typical student" learns, and are replete with discourses that privilege certain ways of being over others, the result is classrooms, assessment regimes, and curricula that serve the processes of classification, marginalization, and exclusion rather than inclusion. We have argued that current conceptualizations of efficiency and economy collaborate to maintain a deficit view of nonconforming learners labeled as "SEN," and a pathologization of disability. Time and tools are being used in ways that are complicit in disabling learners. This does not need to be the case: They could instead be deployed to support and emancipate. For this to happen, we need to learn to see differently, to create classroom cultures that legitimize and validate "disabled embodiment and sensibility" (Overboe 1999, p. 22) as lived experiences that enrich pedagogic possibilities.

Difference is not deficiency; difference is diversity, inspiration, expansion, challenge, disruption, and renewal. Different ways of interacting with the physical world, and with others in it, impact differently on our mathematical experiences and understanding—and this is both valuable and desirable. We should seek to support such processes, not snuff them out by insisting that everyone does mathematics in the same ways, using the same tools, following the same paths. Vitória's ways of seeing, for example, though unconventional, breathed a mathematical life into seemingly rather uninspiring pieces of wood; Wendy brought an expressive joyfulness to dot arrays. Price (2015) talks of *crip politics* and ways of getting things done "by infusing the disruptive potential of disability into normative spaces and interactions" (p. 269). Dare we suggest that creating more inclusive mathematics involves this kind of disruption: A cripping of school mathematics?

References

Avramidis, E., & Norwich, B. (2002). Teachers attitudes towards integration - inclusion: A review of the literature. *European Journal of Special Needs Education, 17*, 129–147.

Ball, D. L. (1992). Magical hopes: Manipulatives and the reform of math education. *American Education, 16*, 14-I 8.

Bauer, A. (2020). *An exploration of feelings, memory and time in the mathematical education of pupils with special educational needs, including mathematical learning difficulties* [Unpublished doctoral dissertation]. King's College London.

Ben-Yehuda, M., Lavy, I., Linchevski, L., & Sfard, A. (2005). Doing wrong with words: What bars students' access to arithmetical discourses. *Journal for Research in Mathematics Education, 36*(3), 176–247.

Blatchford, P., Bassett, P., Brown, P., & Webster, R. (2009). The effect of support staff on pupil engagement and individual attention. *British Educational Research Journal, 35*(5), 661–686.

Coles, A., & Sinclair, N. (2018). Re-thinking 'normal' development in the early learning of number. *Journal of Numerical Cognition, 4*(1), 136–158. https://doi.org/10.5964/jnc.v4i1.101

Coles, A., & Sinclair, N. (2019). Re-thinking 'concrete to abstract' in mathematics education: Towards the use of symbolically structured environments. *Canadian Journal of Science, Mathematics, and Technology Education, 19*(4), 465–480. https://doi.org/10.1007/s42330-019-00068-4

Fernandes, S. H. A. A., & Healy, L. (2020). Mathematics education in inclusive, plurilingual and multicultural schools. In L. Leite (Ed.), *Science and mathematics education for 21st century citizens: Challenges and ways forwards*. Nova Science Publishers.

Finesilver, C. (2014). *Drawing division: Emerging and developing multiplicative structure in low-attaining students' representational strategies* [Unpublished doctoral dissertation]. University of London: Institute of Education. http://eprints.ioe.ac.uk/20816/

Finesilver, C. (2017a). Between counting and multiplication: Low-attaining students' spatial structuring, enumeration and errors in concretely-presented 3D array tasks. *Mathematical Thinking and Learning, 19*(2), 95–114. https://doi.org/10.1080/10986065.2017.1295418

Finesilver, C. (2017b). Low-attaining students' representational strategies: Tasks, time, efficiency, and economy. *Oxford Review of Education, 43*(4), 482–501. https://doi.org/10.1080/03054985.2017.1329720

Finesilver, C. (2019). Pre-service secondary mathematics teachers' attitudes and knowledge regarding inclusion and SEN/D. *Proceedings of the British Society for Research into Learning Mathematics, 38*(3) http://www.bsrlm.org.uk/wp-content/uploads/2019/02/BSRLM-CP-38-3-07.pdf

Flores, M. M. (2010). Using the concrete-representational-abstract sequence to teacher subtraction with regrouping to students at risk for failure. *Remedial and Special Education, 31*(3), 195–207.

Ferri, B. A. (2012). Undermining inclusion? A critical reading of response to intervention (RTI). *International Journal of Inclusive Education, 16*(8), 863–880. https://doi.org/10.1080/13603116.2010.538862

Freire, P. C. (2017). *Uma jornada dos números naturais aos racionais com uma aluna com deficiência visual* [A journey from natural to rational number with a visually impaired student]. [Unpublished doctoral dissertation]. Universidade Anhanguera de São Paulo, São Paulo, .

Fuchs, D., & Fuchs, L. S. (2006). Introduction to response to intervention: What, why, and how valid is it? *Reading Research Quarterly, 41*(1), 93–99. https://doi.org/10.1598/RRQ.41.1.4

Garland-Thomson, R. (2011). Misfits: a feminist materialist disability concept. *Hypatia, 26*(3), 591–609.

Healy, L., & Fernandes, S. H. A. A. (2011). The role of gestures in the mathematical practices of those who do not see with their eyes. *Educational Studies in Mathematics, 77*, 157–174.

Healy, L., & Powell, A. (2013). Understanding and overcoming "disadvantage" in learning mathematics. In M. A. Clements, A. Bishop, C. Keitel, J. Kilpatrick, & F. Leung (Eds.), *Third international handbook of mathematics education* (pp. 69–100). Springer.

Holmes, W., & Dowker, A. (2013). Catch Up Numeracy: A targeted intervention for children who are low-attaining in mathematics. *Research in Mathematics Education, 15*(3), 249–265.

Leong, Y. H., Ho, W. K., & Cheng, L. P. (2015). Concrete-Pictorial Abstract: Surveying its origins and charting its future. *Mathematical Education, 16*(1), 1–8.

McDermott, R. (1993). The acquisition of a child by a learning disability. In J. Lave & S. Chaiklin (Eds.), *Understanding practice: Perspectives on activity and context* (pp. 269–305). Cambridge University Press.

Mitchell, D. (2005). *Contextualizing inclusive education: Evaluating old and new international paradigms*. Routledge.

Mittler, P. (2002). Educating pupils with intellectual disabilities in England: Thirty years on. *International Journal of Disability, Development and Education, 49*(2), 145–160. https://doi.org/10.1080/103491220141730

National Centre for Excellence in the Teaching of Mathematics. (n.d.). *Five big ideas in teaching for mastery.* https://www.ncetm.org.uk/teaching-for-mastery/mastery-explained/five-big-ideas-in-teaching-for-mastery/

Norwich, B. (2013). *Addressing tensions and dilemmas in inclusive education: Living with uncertainty*. Routledge.

Norwich, B., & Koutsouris, G. (2017). Addressing dilemmas and tensions in inclusive education. In Oxford Research Encyclopedia of Education. https://doi.org/10.1093/acrefore/9780190264093.013.154

Overboe, J. (1999). Difference in itself: Validating disabled people's lived experience. *Body & Society, 5*(4), 17–29. https://doi.org/10.1177/1357034X99005004002

Price, M. (2015). The bodymind problem and the possibilities of pain. *Hypatia, 30*(1), 268–284.

Standards and Testing Agency. (2018). *KS2 Multiplication tables check: assessment framework.*; https://assets.publishing.service.gov.uk/government/uploads/system/uploads/attachment_data/file/755745/2018_MTC_assessment_framework_PDFA.pdf

Standards and Testing Agency. (2020). *Key stage 2: Access arrangements guidance.* https://assets.publishing.service.gov.uk/government/uploads/system/uploads/attachment_data/file/940830/2021_KS2_AA_V1.0.pdf

Vygotsky, L. S. (1924–31/1993). Introduction: The fundamentals of defectology. In R. W. Rieber & A. S. Carton (Eds.), *The collected works of L S Vygotsky. Volume 2, the fundamentals of defectology,* (pp. 29–51). Plenum Press.

Vygotsky, L. S. (1981). The instrumental method in psychology. In J. V. Wertsch (Ed.), *The concept of activity in Soviet psychology* (pp. 134–143). M.E. Sharpe.

Watson, A. (2001). Instances of mathematical thinking among low attaining students in an ordinary secondary classroom. *The Journal of Mathematical Behavior, 20*(4), 461–475. https://doi.org/10.1016/S0732-3123(02)00088-3

Watson, A. (2006). *Raising achievement in secondary mathematics*. Open UP.

Webster, R., & Blatchford, P. (2015). Worlds apart? The nature and quality of the educational experiences of pupils with a statement for special educational needs in mainstream primary schools. *British Educational Research Journal, 41*(2), 324–342.

Wilensky, U. (1991). Abstract meditations on the concrete and concrete: Implications for mathematics education. In I. Harel & S. Papert (Eds.), *Constructionism* (pp. 193–204). Ablex.

Chapter 9
Engaging Multilingual Learners with Disabilities in Mathematical Discourse

Erin Smith and R. Alex Smith

Abstract Multilingual learners (MLs) with mathematical difficulties (MD) are a diverse group of students who represent an increasing demographic within U.S. public schools (National Center for Educational Statistics: English Language Learners in Public Schools. Author, 2016) with dual learning goals of mathematics and language. To ensure MLs with MD can be academically successful, they must engage in mathematical discourse. However, for teachers to effectively engage MLs with MD in mathematical discourse, they should (1) employ general approaches to teaching MLs as well as students with MD, (2) foster learning environments conducive to participation, (3) strategically use discourse, (4) advance English language acquisition, and (5) enhance mathematics curriculum. In this chapter, we explore these five areas, drawing from research with MLs, students with mathematics difficulties/disability, and MLs with MD.

Keywords Multilingual learners · Mathematical difficulties · Mathematics teaching · Teaching practice · Discourse · Discussion · Curriculum · Learning disability · At-risk students · Language learners · Curriculum

Multilingual learners (MLs) are a diverse group of students who represent an increasing demographic within U.S. public schools (National Center for Educational Statistics [NCES], 2016). Even though increased attention has been paid to MLs within mathematics education research in general (de Araujo et al., 2018), as well as for students with mathematical difficulties/disability (Gersten et al., 2009), very little research has focused specifically on MLs with mathematical difficulties/disability (Orosco, 2014; Sanford et al., 2020). Consequently, in this chapter, we draw primarily from research focused on MLs and research on students with disabilities in the context of mathematics classrooms.

E. Smith (✉) · R. A. Smith
The University of Southern Mississippi, Hattiesburg, MS, USA
e-mail: erin.marie.smith@usm.edu

© The Author(s), under exclusive license to Springer Nature Switzerland AG 2022   177
Y. P. Xin et al. (eds.), *Enabling Mathematics Learning of Struggling Students*,
Research in Mathematics Education,
https://doi.org/10.1007/978-3-030-95216-7_9

This chapter is guided by a situated sociocultural perspective (Lave & Wenger, 1991; Moschkovich, 2002; Wenger, 1999), which posits that mathematical learning derives from discursive activity between teachers and students in a community of practice. Moreover, it asserts that students' diverse backgrounds, experiences, and linguistic resources are valuable assets that should be drawn on in mathematics instruction (Moschkovich, 2002). In this chapter, we pay specific attention to the way teachers' use of discourse creates opportunities for MLs with mathematical difficulties (MD) to engage in mathematical discourse. We begin with brief discussions about students who have difficulties in mathematics and MLs. Next, we define discourse and mathematical discourse. Finally, we describe research-based pedagogical practices to engage MLs with MD in mathematical discourse.

9.1 Students with Mathematical Difficulties

We use the term mathematical difficulty (MD) in this chapter to refer to students who have a mathematics learning disability or who are described as 'at-risk'. We combine these two student groups because they face significant challenges with mathematics achievement, experience limited access to standards-based mathematics instruction, and benefit from similar types of instruction and scaffolds (Lambert & Tan, 2017; Zhang & Xin, 2012). In combining these two student groups, students with MD represent approximately 12–15% of the U.S. student population (Geary, 2014).

Mathematics education emphasizes the importance of classroom discussion to foster students' conceptual understanding (NCTM, 2000), yet students with MD have historically been provided limited exposure to such classroom experiences (Lambert & Sugita, 2016; Lambert & Tan, 2017). One reason for this may be a perception by practitioners that students with MD cannot learn from or engage in discussion-based learning because of their disability (Lambert, 2015; Lambert & Tan, 2017). Although students with MD may face specific challenges participating in student-centered, communication-based instructional practices (Bottge et al., 2002), effective planning and scaffolding (such as discussed in this chapter) can support students' engagement and facilitate access to mathematical discussions (Hord et al., 2016; Hunt & Tzur, 2017; Lambert & Sugita, 2016; Woodward & Tzur, 2017).

9.2 Multilingual Learners (MLs)

We use the term ML to refer to students who are in the process of acquiring English as an additional language. We also use this term to refer to the research of scholars who use the term English language learner (ELL) or English learner (EL). We use the term ML as a way to de-privilege English, call attention to the resources students possess instead of what they lack, and challenge historic deficit positions of students via labeling. In the United States, MLs represent a linguistically and culturally diverse student population, whose language competencies can vary from limited to nearly proficient. In the United States, MLs also represent about 10% of the K-12 student population and are the fastest growing sub-population of students (Kena et al., 2016). Since the population of MLs is growing, it is anticipated that the number of MLs with MD is increasing as well. This necessitates the need for teachers to implement practices that show promise for MLs with MD.

9.3 Discourse and Mathematical Discourse

Students learn mathematics by communicating their thinking and negotiating meaning by participating in discourse-rich classrooms (Cobb et al., 1993; NCTM, 2000; Vygotsky, 1978). As students engage in discourse, they refine their ability to question, critique, and explain mathematics—practices that can lead to "doing mathematics" (Stein & Smith, 1998, p. 270)—and advance their language acquisition (Lightbrown & Spada, 2013; Moschkovich, 2002; NCTM, 2014). In this section, we briefly describe what we mean by discourse and mathematical discourse and why engaging in mathematical discourse is important for MLs with MD in particular.

9.3.1 Discourse

"Discourse" captures the ways that we communicate with each other, including oral and written interactions, as well as gestures and body language (Gee, 2011). The ways we communicate with each other through our discourse vary by context. For example, think about the ways your discourse varies across these situations:

- You ask your boss for a salary raise
- You describe what an array is to your students
- You coo to a baby
- You respond to a student's question

In each situation, you may have thought about the words, phrases, or gestures you might use and the way you position yourself (e.g., content expert, caregiver, subordinate). These different ways we communicate and position ourselves and others can be thought of as *discourse* (Fairclough, 2010; Gee, 2011).

When people use discourse, they relay messages to others about what is important and culturally valued. For instance, *what* students are asked to communicate about, *who* students are asked to communicate with (e.g., teacher, peer, city council), and the overall *purpose* of communicating relays messages to students about the value, importance, and nature of classroom discourse. For example, students in a mathematics class who are only tasked with providing numerical solutions or explaining steps in a procedure may view mathematical discourse in a limited way (e.g., used to state answers and steps). This can be opposed to students in another classroom who are frequently asked to use discourse in a range of ways (e.g., justify, argue, define, etc.). Or, if students are only asked to communicate with the teacher, they may not understand how mathematics can be used to engage with multiple audiences, such as city council members or government, community organizations, parents/guardians, and the general public. Consequently, as teachers we are better when we consider the who, what, and why of our discourse and what messages we relay as a result.

9.3.2 Mathematical Discourse

Mathematical discourse (e.g., discourse characteristic of mathematics classrooms) has been defined in a range of ways, from narrow to expansive. In our work, we take an expansive view of mathematical discourse and use Moschkovich's (2003) definition of "talking and acting in the ways that mathematically competent people talk and act when talking about mathematics" (p. 326). This broad definition not only captures specialized ways of communicating mathematically (e.g., use of specialized syntax or vocabularies), but also informal ways, such as using everyday or colloquial language, which is important for MLs since they may draw on other language resources to supplement or augment their mathematical discourse as they acquire English. This expansive definition also allows for variability in what "counts" as mathematical discourse, because through classroom interactions mathematical discourse can be (re)defined (Moschkovich, 2003).

9.3.3 The Importance of Mathematical Discourse for MLs with MD

Communicating mathematical ideas clearly, coherently, and effectively to teachers, peers, and others has been a clear goal of mathematics instruction for over 30 years (NCTM, 1989). Consequently, the acquisition of mathematical discourse is critical to any students' academic success, particularly when the use of mathematical discourse is often equated with mathematical competence (Schleppegrell, 2007). Yet, for MLs with MD to engage in mathematical discourse, it is essential teachers make engagement in mathematical discourse a priority (Harper & De Jong, 2004; Kayi-Aydar, 2014; Khisty, 1995; Lambert & Sugita, 2016). When this occurs, teachers create opportunities for every student to develop fluency and competency in mathematical discourse (Celedón-Pattichis & Turner, 2012; Khisty & Chval, 2002). Although research is limited specifically for MLs with MD, research does indicate that MLs (de Araujo et al., 2018; Moschkovich, 2007; Pinnow & Chval, 2015;

Turner et al., 2013) and students with MD (Lambert & Sugita, 2016; Xin et al., 2016; Xin et al., 2020) can be fully engaged in mathematical discourse with appropriate supports and scaffolds in place.

9.4   Pedagogical Practices for Engaging MLs with MD in Mathematical Discourse

Teachers that effectively engage MLs with MD in mathematical discourse (1) employ general approaches to teaching MLs as well as students with MD, (2) foster learning environments conducive to discourse participation, (3) strategically use discourse, (4) advance English language acquisition, and (5) enhance mathematics curriculum. In the remainder of this chapter, we explore each of these pedagogical practices situated in the context of general education mathematics classrooms.

9.4.1   General Approaches to Mathematics Instruction for MLs

To support teachers in addressing MLs dual-language goals of mathematics and English (in the case of the United States), World-Class Instructional Design and Assessment (WIDA) created descriptors by grade and language mode of what teachers can expect MLs to be able to do at different stages of their language acquisition process (see https://wida.wisc.edu/teach/can-do/descriptors). In addition, WIDA developed the English Language Development (ELD) Standards (2012) that represent the "social, instructional, and academic language that students need to engage with peers, educators, and the curriculum in schools" (p. 4). These Standards span multiple content areas and include a Standard specific to mathematics ("ELD Standard 3").

Within mathematics education, scholars have identified six broad recommendations for teachers of MLs (from Moschkovich, 2013, pp. 52–54):

- MLs should engage in the Standards for Mathematical Practice (National Governors Association Center for Best Practices & Council of Chief State School Officers, 2010).
- MLs should engage in mathematics tasks that are cognitively demanding, designed to foster students' conceptual understanding of mathematics, and incorporate and connect across multiple mathematical representations.
- MLs should engage in different kinds of mathematical content and reasoning with different purposes (e.g., justifying, explaining, describing).
- MLs should participate in a range of mathematical activities with different structures (e.g., partner, small group, whole class, formal/informal, student-led discussions).

- MLs' first language should be used as a resource for engaging in mathematical reasoning, sense making, and communication (e.g., encouraging students to make connections between their first language and English). MLs should also communicate with a range of audiences (e.g., peers, teachers, community members, parents) across different modes (e.g., writing, speaking).
- MLs should be prepared to engage with and understand typical kinds of text in mathematics, including a range of problem types (e.g., word, assessment, application, modeling) and formats (e.g., peer strategies, assessment).

These general recommendations provide the foundation for the instructional strategies we share in the remainder of this chapter.

9.4.2   Fostering an Equitable Learning Environment

There are multiple aspects of a learning environment, both physical and metaphorical. Although the physical environment may often be overlooked, it is a critical first step in promoting and supporting effective and accessible discourse-based instruction (Center for Applied Special Technology [CAST], 2016). For instance, consider if: students can quickly and easily find their materials, transition from whole class to small group to pair work quickly, and if the teacher can easily move between groups to listen to student discussion and/or redirect students when needed. As teachers plan their classroom spaces, it may be beneficial to consult the website for Universal Design for Learning by CAST.

Norms   A critical component of developing a classroom community is the establishment of norms (i.e., social expectations) and their ongoing review and revision across the school year as students' competencies in social interactions around mathematics develop. In an effort to give voice to students in the creation and ownership of classroom norms, we suggest teachers co-construct norms *with* students (Foote & Lambert, 2011). To do this, teachers can initiate discussions of what kinds of norms students want in their mathematics classroom and use these as the foundation for the class's norms. For instance, teachers may prompt discussion of norms by asking students about their prior experiences with and/or interactions about mathematics. These conversations allow many students to participate (since they likely had past experiences in mathematics classrooms), can illuminate your students' prior experiences, and can create opportunities for students to identify what are productive ways to interact to mitigate unproductive emotions in mathematics (e.g., fear, anxiety). (See the section, Eliciting Student Thinking, for additional strategies to foster MLs with MDs' participation in these discussions.) In these discussions, students may identify norms like listening while others speak, respectfully disagreeing with a peer's idea, and encouraging peers to participate in the conversation or discussion—each of which prepares *every* student to engage equitably and productively in classroom discourse. In such conversations, teachers should point out how different kinds of activities call upon different norms (e.g., whole class versus partner discussion). As teachers engage in these conversations, they would:

- Ensure classroom norms are explicit and can easily be referenced throughout lessons (e.g., chart or poster; Lambert et al., 2020).
- Ask students to clarify what are examples and non-examples of the norms, so students develop a deep understanding of what behaviors do or do not constitute a given norm.
- Positively state norms to draw attention to productive behaviors.
- Post norms in age-appropriate language with accompanying images (Lambert et al., 2020).
- Try to limit the norms to five at any given time. However, norms may need to be re-evaluated and revised throughout the year.
- Ensure the norms reflect learning mathematics as a process and mistakes are learning opportunities.

Although students are more likely to adhere to co-constructed norms, teachers should still plan to spend substantial time modeling (i.e., teacher modeling and student modeling), reinforcing, and reteaching norms, particularly at the start of the school year.

Students may occasionally act in ways that contradict the norms they co-constructed. Situations like these provide opportunities to revisit the norms and clarify any ambiguities that may be present. At this time, it may also be useful to role-play different scenarios of ways students engage with one another while calling explicit attention to what are productive (and unproductive) behaviors. Use of visuals and permanent reminders, like a poster, should be referred to often. For example, the teacher may refer to the poster and remind students of specific norms (e.g., always provide an explanation for your answer) prior to transitioning into pair work or when redirecting unproductive behavior during pair work. The teacher could also ask the class to evaluate how their behavior aligned with the norms after they have worked collaboratively in a pair or small group (e.g., draw attention to pairs that always provided explanations).

Another way to support students to productively engage with one another and act in ways that align with the classroom norms is to provide a resource students can consult while doing mathematics. For example, students could keep a notecard at their desk with questions and compliments they could ask of each other as they are doing mathematics. These statements should be general enough that they can be used across a range of mathematical activities over the course of the year (e.g., "What did you do first?" or "Your drawing really helped me understand your thinking."). Some students may also benefit from tracking the frequency in which they ask questions or provide compliments stated on the card, which could be tied to individual goal-setting activities and reinforcement. It is important to note that when using such a resource, teachers spend instructional time explicitly teaching students how to use the resource (modeling and practice), encourage students to use it during activities, and use reinforcement when students do use it. By providing physical resources that students can consult, students are better prepared to engage in respectful mathematical discussions.

Routines In any learning environment, a goal should be to maximize time and learning opportunities. To do this, teachers should use consistent routines within which a variety of activities can be embedded (Lambert & Sugita, 2016). By routines we mean structures or activities that occur on a regular basis at consistent times, such as opening each lesson with a quick math fluency warm-up in pairs, ending each lesson with an independent "exit-ticket" activity, or using the launch/ explore/discuss format. Consistent routines in which expectations regarding movement, grouping, getting materials, getting started, and responsibilities are made explicit increase the amount of time students can engage in mathematical activities because they limit instructional time spent on learning new routines or having to resume sporadically employed routines that may be forgotten. Ultimately, when teachers use consistent daily routines, students become efficient in transitioning between activities because they know what is expected of them during this time. Moreover, when routines provide students with opportunities to enact classroom norms (e.g., during pair or small group work), students can become more adept at working productively together while doing mathematics. As students develop competencies with any classroom routine, new routines may be introduced systematically, and norms revised accordingly. We caution teachers from drastically changing classroom routines as this could pose academic, social, and emotional challenges for many students.

Grouping Students Heterogeneous groups can benefit students' mathematical learning (Opitz et al., 2018), including working in pairs and small groups. In general, research indicates that pair work fosters the greatest engagement for students with MD (Lambert et al., 2020), allows teachers to formatively assess students' thinking, and increases opportunities for every student to engage in mathematical discourse. Therefore, teachers should prioritize pair work while keeping in mind the structures needed to encourage productive engagement, such as holding students accountable for their peers' thinking (e.g., asking students to report on their peers' thinking) and providing clear roles and responsibilities for each student during pair work. In addition, teachers should ensure that students take on equitable roles as they work in dyads so that one student is not always positioned as in need of help. Similar routines can also be incorporated in small group activities (e.g., 3–5 students) to maximize participation for each group member.

Although whole-class discussions are a critical component of a discourse-rich classroom, this format often poses two instructional challenges. First, whole-class discussion is frequently dominated by a small number of students (e.g., 2–3) who are often viewed as mathematically competent by the teacher, which can position peers (perhaps 30 others) as periphery participants and place increased demands on student attention, language comprehension, and metacognition (Baxter et al., 2002; Lambert et al., 2020). Second, research indicates, in general, teachers are less likely to purposefully engage MLs and students with MD in whole-class discussions and when they do it is often to respond to questions of low cognitive demand (e.g., provide yes/no answers, complete simple arithmetic; Baxter et al., 2002; Lambert et al., 2020). In response to these challenges, scholars have identified some instructional strategies to engage MLs with MD in mathematics discourse; however, successful implementation depends on teachers' beliefs about students' capabilities. One strategy is to split the class into two heterogeneous groups that engage in whole-group discussions at different times (Lambert et al., 2020). In using this approach, the number of students engaged in the whole-class discussions at any time is half the class, thus increasing the opportunities that each student can engage in the discussion at any point. Alternatively, teachers can embed brief partner activities, such as Think-Pair-Show-Share (see Hunt et al., 2018), and call on one or two individuals to share their own and/or their peer's thinking with the class. This type of activity, when interspersed with relative frequency, holds each student accountable to listening to their peers and increases every student's engagement in mathematical discourse even during whole-class discussions. Teachers may also embed other interactive activities, such as sharing a compliment, critiquing a peer's thinking, or offering an alternative strategy amidst the whole-class discussion. Lastly, another way to address these challenges is for teachers to continually track and reflect on which students they call on and what kinds of elicitations they use (Herbel-Eisenmann & Shah, 2019). Teachers may find the tracking chart provided in Herbel-Eisenmann and Shah's (2019) article a valuable tool in this process.

9.4.3 Teachers' Strategic Use of Discourse

The teacher's position in the classroom is one of a more experienced other (Lave & Wenger, 1991; Vygotsky, 1978). In this position, teachers are responsible for the establishment and maintenance of a classroom community that provides each student opportunities to participate and learn mathematics (Khisty & Chval, 2002; Moschkovich, 2002). However, if teachers are to effectively build and sustain a classroom community, it is essential they understand that the things they say and do in the classroom can influence opportunities to learn mathematics and English (Gibbons, 1992, 2003; Khisty & Chval, 2002; Lightbrown & Spada, 2013; Pinnow & Chval, 2014; Schleppegrell, 2007), reify (dis)ability (i.e., position the individual with MD as unable or incompetent; Heyd-Metzuyanim, 2013), and influence the development of students' mathematical identities (Turner et al., 2013; Yamakawa et al., 2009). Consequently, teachers should use their discourse in strategic ways. In this section, we discuss ways teachers can use their discourse to elicit student thinking and revoice student contributions.

Eliciting Student Thinking Eliciting student thinking is a critical component of effective mathematics instruction (NCTM, 2014). When teachers elicit student thinking, opportunities arise to create connections to students' prior knowledge and experiences, clarify student thinking, engage students in discussion, and advance lessons, among others (de Araujo et al., 2018; Hufferd-Ackles et al., 2004; NCTM, 2014; Wood et al., 1991). Moreover, elicitations that encourage extensive use of language can also be used to foster second language acquisition (Gibbons, 1992, 2008). Regardless of these benefits, prior research (e.g., Iddings, 2005; Planas & Gorgorió, 2004; Weiss et al., 2003) has found that elicitations requiring extensive use of language are often not employed in classrooms, particularly with MLs. We anticipate that this is also the case for MLs with MD. One reason for this may be teachers' perceptions that MLs are unable to participate in conversations due to their language development, but research has found MLs can participate when developing language (Gibbons, 2015; Moschkovich, 2002; Setati, 2005; Yoon, 2008).

Like in any discourse, teacher elicitations relay messages to students while simultaneously positioning them as mathematical thinkers (Turner et al., 2013; Yoon, 2008). To illustrate this, consider a teacher who, during a whole-class discussion, asks a student: "What do you think about Mariella's strategy?" With this question, the student is positioned as someone who is capable of thinking mathematically, has a mathematical idea worth hearing and sharing, and can contribute to the discussion at hand. Moreover, such a question can solidify community membership, foster students' mathematical identities, and reinforce the conception that, in this classroom, students are co-constructors of mathematics. Alternatively, consider a situation where a teacher identified some students to share their problem-solving strategies and accompanying representations during a whole-class discussion at the conclusion of a lesson. To introduce the first student who shares, the teacher states, "I asked Ahmad to present his work to us today because he represented his strategy in multiple ways and explained it well to his partner." This statement alone calls attention to Ahmad's mathematical ideas in front of his peers before he even begins to speak and reinforces the norms of this classroom to explain one's thinking and reason mathematical (Yackel & Cobb, 1996). It also emphasizes which aspects of Ahmad's mathematical thinking are notable, what peers should attend to in his presentation, and why he can explain his ideas and teach his peers. Moreover, by scanning and projecting Ahmad's work, he can use it as a visual referent to support his explanation—a useful instructional support for MLs (Chval et al., 2009). In both of these examples, the teacher's elicitations productively position the student they are directed at, which is important because peers reinforce positions circulated by the teacher (Turner et al., 2013; Yoon, 2008). This means that *how* teachers position MLs with MD via elicitations will likely be appropriated by peers, while also fostering their own self-worth as capable mathematical thinkers. Consequently, it becomes vital that teachers use elicitations with MLs with MD in ways that support their mathematical learning and position them as competent thinkers.

Although teachers can elicit student thinking in a range of ways, such as having students present their problem-solving strategies, explain their interpretation of a peer's thinking or representation, or ask questions of peers, much research has focused on how teachers use questions to elicit student thinking. Teachers' questions are important because they influence the learning opportunities that are provided to students (Hunt & Tzur, 2017). When teachers employ questions that merely direct (or funnel) students toward a specific strategy or answer, student thinking seems constrained and the focus is often on relaying information to the student (Boaler, 2003; Wood, 1998). Alternatively, teachers can use questions that focus student thinking, like "Chintan, can you explain why you did ____?" and "Zaynia, why do you think ____?" (Herbal-Eisenmann & Breyfogle, 2005; Sherin, 2002). These questions are characteristically different from funneling questions in that they do not have a predetermined direction decided by the teacher and often result in information being *elicited from* the student rather than *relayed to* the student (Wood, 1998). Teachers can further shape the discussion by asking students to compare and contrast mathematical ideas (e.g., "How are ____'s and ____'s the similar? How are they different?"; Sherin, 2002). Such techniques ensure student thinking remains the focus of the conversation while enabling students to retain agency over their learning (Sherin, 2002; Webel, 2010). Of course, this requires teachers to be highly attentive to student thinking.

Revoicing Revoicing is the practice of "re-uttering (oral or written) of a student's contribution by another participant in the discussion" (O'Connor & Michaels, 1996, p. 90). Most commonly, when teachers revoice, they restate, recast, and expand on a student's contribution. Revoicing can serve multiple functions, such as preserving a student's idea as the focus of discussion, indicating student competencies, providing linguistic models, and amplifying speech (Enyedy et al., 2008; Forman et al., 1998; Moschkovich, 1999). Revoicing can also be used to amplify and give oral language to students' thinking, particularly when teachers revoice non-verbal forms of communication (e.g., gestures, drawings). However, not all revoicing is useful to every student. Sometimes revoicing can restrict MLs' access to ideas if they are made only in the language of instruction (Enyedy et al., 2008). Moreover, teachers' revoicing can be problematic if it positions students in unproductive ways, such as suggesting an initial utterance by a ML with MD is unclear and requires teacher clarification.

Prior research (Enyedy et al., 2008) has found that some types of teacher revoicing—*revoicing to position*—can facilitate MLs participation in mathematical discourse. Revoicing to position "explicitly attributes authorship to the students, can be seen as an epistemic device that shares the intellectual authority with the students, and helps establish their role as one of contributing to the construction of knowledge" (p. 137). Revoicing to position has been found to serve multiple purposes, such as positioning students as mathematically competent or in relation to an idea, to a task, or to others (Enyedy et al., 2008; Turner et al., 2013). For example, a teacher could use revoicing to position MLs with MD as possessing valid ideas or ideas related to the task, acting in ways that align with classroom norms (e.g., listening intently, acting respectfully), or as agentive problem solvers (e.g., such as possessing mathematical justifications or ideas worth investigating further; Turner et al., 2013). Revoicing to position is a useful discursive move for teachers of MLs with MD because it can ensure students' ideas remain at the center of the conversation (Turner et al., 2013).

9.4.4 Scaffolding English Language Acquisition for MLs with MD

Broadly speaking, in mathematics instruction teachers should focus on the mathematical meaning and underlying reasoning of what students say and write; they should not get hung up on students' language accuracy (Moschkovich, 2010, 2013). Although some teachers may see this recommendation as counterintuitive for MLs (i.e., language should be corrected/addressed when encountered), prior research (e.g., Moschkovich, 2010) has found that this attention does not benefit students since teachers focus attention on language instead of students' mathematical ideas. That said, it is important that teachers of MLs with MD *do* plan for and implement language instruction if MLs with MD are to grow in their mathematics and language competencies.

Mathematical Language There is no doubt that mathematics has a specialized language, or register, that is unique to the discipline. This specialized language includes terms or phrases such as Pythagorean Theorem, hypotenuse, and algebra. In contrast to this specialized language, there are other terms or phrases used in mathematics that typically have other everyday uses, like table, leg, foot, and decompose. Recognizing the distinctions in everyday and specialized mathematical language as well as using precise mathematical language consistently is important for teachers and students, particularly MLs with MD. When teachers fail to recognize the language resources and understandings that students bring into the classroom, or use informal terms in lieu of specialized mathematical language (e.g., *diamond* instead of *rhombus*), MLs with MD can be faced with unnecessary obstacles to their mathematics understanding and language acquisition.

Oftentimes, teachers may find themselves reducing the linguistic demands of mathematical tasks or their language as a way to increase access to mathematics. For example, a teacher may simplify language or select tasks devoid of language, so the language does not obscure the mathematics. Yet, this seemingly intuitive avoidance ultimately restricts access to both mathematics and language, creates an overabundance of tasks devoid of context (by removing the language of the context) and/or highly proceduralized, and does little to foster the acquisition of mathematical discourse (e.g., eliminating opportunities to explain and justify mathematical thinking; de Araujo, 2017; Gibbons, 1992). Consequently, teachers of MLs with MD should "avoid reductionist approaches" (National Academies of Sciences, Engineering, And Medicine, 2018, p. 126) and, instead, craft their instruction to facilitate students' language acquisition by drawing from students' existing linguistic resources to build meaning for language *through* language (e.g., utilizing students' first language, everyday language, and/or cognates to promote conceptual understanding and bridge between informal and formal language; de Araujo et al., 2018; Moschkovich, 2002; Razfar, 2012; Schleppegrell, 2007).

Building Understanding for Terms and Phrases Teachers must explicitly teach mathematics language if students are to acquire it (de Jong & Harper, 2005). Yet, this does not mean language should be taught solely in isolation and devoid of context, since learning single terms or phrases does not mean students can effectively engage in mathematical discourse (Gibbons, 1992; Gottlieb & Ernst-Slavit, 2013; Moschkovich, 2002, 2010, 2013). Rather, vocabulary words and phrases should be taught in a variety of ways and students should have multiple opportunities to use the words and phrases in their oral and written discourse (Lei et al., 2020).

Prior to engaging in a unit/series of lessons, teachers should identify key terms and phrases that are essential to the target standards and conceptual understanding (Sanford et al., 2020). As recommended by Sanford et al. (2020), teachers should employ an explicit instruction model for teaching vocabulary, including (a) introducing the word, (b) providing a student-friendly definition that connects to prior learning (Tzur, 2000), (c) providing both examples and non-examples across multiple modalities (e.g., manipulatives, images), and (d) encouraging students to use the word/phrase in oral and written language (e.g., have students discuss the word and connect it to prior knowledge in pairs after it is first introduced and defined, have students write the word and its definition as well as use it within a sentence and keep that in a word book).

Finally, MLs with MD should be provided extensive opportunities to use the word/phrase in both oral discussions and in writing, especially within the context of problem-solving (Baker et al., 2014; Sanford et al., 2020). One way to provide context for vocabulary and problem-solving is through shared story reading (Whitney et al., 2017). Shared story reading is characterized by teacher modeling of thinking and question-asking while reading out loud and frequently eliciting students' thinking (see Courtade et al., 2013). Stories can provide a context for mathematical learning and many present rich mathematical problems (Whitney et al., 2017).

Scaffolds and Resources to Encourage Mathematical Discourse Beyond classroom norms and consistent routines/activities that encourage engagement in discourse, MLs with MD may benefit from additional language scaffolds and resources. Two such scaffolds and resources are the use of sentence frames and a "word-wall" or anchor chart.

Sentence frames are scaffolds that provide guidance or structure to students' productive use of language (i.e., oral and written), so they can engage authentically in classroom interactions (Zwiers, 2020). In addition, sentence frames can be used to encourage MLs with MD to use specific vocabulary or phrases in context as they develop their language competencies (Sanford et al., 2020). Some examples of sentence frames are as follows: "First I _____, then I _____.", "I agree/disagree because _____.", "Next time, I will _____", "_____ and _____ are the same/different because _____.", or "_____'s idea is _____ because _____." To make this resource accessible to students, teachers can post sentence frames for all to see or, as mentioned earlier, provide frames on an individual note card or graphic organizer. We recommend using sentence frames in conjunction with other instructional scaffolds designed to foster students' mathematical discourse because the provision of sentence frames alone (or any other single language scaffold) is not enough to ensure productive and authentic interactions (Turner et al., 2019; Zwiers, 2020).

A *"word-wall"* or *anchor chart* is a visual reminder of language (like the aforementioned word book) that students can reference as they engage in a mathematical activity. Word walls/anchor charts have historically been used in literacy instruction in the elementary grades but can be used in any discipline and grade level. The word wall/anchor chart is a visual display of key terms or phrases and can include student-friendly definitions, examples, translations, and/or cognates. Word walls/anchor charts should be student-centered artifacts and contain what students identify they need. For instance, teachers may ask where a word wall/anchor chart should be located in the classroom and what words/phrases should be contained on it. In contrast to a word wall/anchor chart that is designed for the entire class, a word book is an individualized resource for students. A word book is very similar to a word wall/anchor chart, yet students have more control over what is contained in it and where it is located. Students may add in words/phrases they think are interesting or are having difficulties with. Students can also include student-generated model sentences that use the word/phrase, cognates, and/or definitions of vocabulary words in their native language.

Like any scaffold or resource, it is essential that teachers spend time modeling, providing feedback, and encouraging use of word wall, anchor chart, and word book across various activities. In addition, it is important to highlight these scaffolds and resources are dynamic, not static or permanent. As MLs with MD develop their competencies in mathematics and English, these scaffolds and resources may be revised or modified. This will ensure the scaffolds and resources are timely and appropriate for what students need at the time of instruction.

Facilitating Students' Mathematical Writing Students benefit from writing across content areas (Graham et al., 2020), yet writing in mathematics classrooms is often overlooked (Powell et al., 2017). Students use writing in the mathematics classroom to "consolidate their thinking because it requires them to reflect on their work and clarify their thoughts" (NCTM, 2000, p. 61; Peregoy & Boyle, 1993). It is important to provide MLs with MD multiple opportunities to write, receive feedback, and revise their writing. Moreover, ask students to write to a range of audiences and utilize a range of genres in mathematics—beyond just writing explanations of procedures to their teacher (Moschkovich, 2015; Powell et al., 2017).

Teachers may refer to Graham et al. (2012) for guidance on teaching writing to students in the elementary grades and Graham et al. (2016) for high school students. Across both guides, one of the most effective instructional strategies for student writing is *Self-Regulated Strategy Development* (SRSD). SRSD combines instructional strategies (e.g., a strategy for students to use to organize their writing) with self-regulation supports (e.g., self-monitoring) by employing six common steps (see Graham et al., 2012; Graham et al., 2016). Although we only found one study examining the effectiveness of SRSD specifically in mathematical writing (Hughes et al.,

2019), results were promising and SRSD has been shown to be effective across genres, content areas, grades, and student populations in other studies.

The increasing availability of quality speech-to-text software may help some students overcome challenges associated with writing and spelling, though functionality may be limited for some MLs, depending upon the individual's accent. Prior research indicates the use of speech to text can increase the quantity of students' writing, but not necessarily quality of their writing (De La Paz & Graham, 1997). However, employing planning strategies, such as SRSD, in conjunction with speech-to-text can improve both the quantity and quality of student writing (Graham & Harris, 2013). As with any new technology, teachers need to plan for time for students to learn and practice the technology. It is also important to keep in mind that mathematical writing has special characteristics (e.g., symbolic notation, representations) that may stretch the functionality of many speech-to-text applications.

9.4.5   Enhancing Mathematics Curriculum for MLs with MD

Mathematics curriculum is not one-size-fits-all (Gutstein, 2003). It is important that teachers adapt their curriculum to meet the needs of and increase access for every student. Thus, it is not enough to rely on the curriculum itself to supply adequate modifications and adaptations to instruction, because oftentimes these recommendations do not align with mathematics education research (de Araujo & Smith, 2022). In this section, we outline some ways teachers can enhance curriculum specifically related to language, context, and visuals.

Adapting the Language of Mathematical Tasks   Prior research (Abedi & Lord, 2001) has identified ways teachers can adapt mathematical tasks to facilitate ML engagement and understanding as they develop their language competencies. These recommendations (adapted from Abedi & Lord, 2001, and Herbel-Eisenmann, 2007) include the following:

- Use active voice, not passive voice (e.g., instead of "Four pizzas were shared among Mariam, Ahmad, and Horatio" say "Mariam, Ahmad, and Horatio shared four pizzas"),
- Refrain from using long nominal phrases (e.g., instead of "the graph of the class's favorite fruits" use "the graph above"),
- Replace conditional clauses with two separate sentences (e.g., instead of "Margo has five boxes of oranges with each box having 15 oranges in it" say "Margo has five boxes of oranges. Each box has 15 oranges"),
- Use straightforward question phrases instead of complicated or obscure ones (e.g., replace "How much cake is leftover after Jonas ate?" with "What fraction of the cake is left?"),
- Use present tense instead of complex verb forms (e.g., change "Zaynab was watching TV for 3 h" to "Zaynab watches TV for 3 h."), and

- Instead of inanimate nouns or abstract objects, use pronouns or animate nouns (e.g., instead of "17 cars sold" say "Ms. Jones sold 17 cars").

Selecting and Using Contexts to Build Understanding and Meaning The context of mathematical tasks can either promote or constrain student reasoning and understanding. For example, if a student is posed a problem in the context of cricket, yet they have never heard of the game cricket nor understand its key terminology, the context of the problem restricts access to mathematics. In contrast, if the student was an avid watcher and player of cricket, they may be highly engaged and motivated. This example illustrates that some students may be familiar with a context while others may not be. Consequently, teachers need to recognize what contexts students may be familiar with and build meaning for those they are not (e.g., create a shared experience using books, images, or videos during a task launch; Jackson et al., 2012; Whitney et al., 2017).

Prior to selecting contexts, it is essential teachers first consider what mathematics students are learning through the task. All too often, contexts are used in mathematics that do not reflect actual experiences (e.g., counting animal legs to determine the total number of animals). One way to ensure mathematics is authentically embedded within contexts is to select contexts that reflect cultural practices (Wager, 2012). However, to be able to do this, teachers will need to spend time learning about and getting to know their students—a task that takes time. As teachers select contexts, it is important to consider whose experiences are being reflected. Frequently, contexts embedded in curricula tend to reflect the dominant culture's ideology and experiences. In mathematics education, scholars use the language of "mirrors" and "windows" to call attention to this (Gutiérrez, 2008, 2012; Rezvi et al., 2020). In this case, "mirror" is used to signify contexts that a student has experience with (i.e., their experiences are reflected back at them) and "window" is used to signify contexts that provide an opening into others' experiences. It is important for teachers to critically analyze which students are experiencing windows and mirrors with which contexts. It should not be the case that some students always have mirrors and other students always have windows. There should be a balance of mirrors and windows for every student, including MLs with MD.

Contexts can also provide rich opportunities to examine social inequities at local, state, national, and international levels through mathematics. For instance, students could examine the amount, distribution, and impact of local or state tax dollars; access to grocery stores with fresh produce, banks (as opposed to predatory banks and lenders), and recycling centers; and crime rates and policing. They could also explore issues at a national (e.g., charter school funding) or international level (e.g., population distributions, gross national products, cost and impact of cheap labor). For more information on teaching mathematics for social justice, see Gutstein and Peterson (2013).

Incorporating Visuals Teachers can use visuals (e.g., pictures, images, videos, drawings) to build meaning for language, contexts, and mathematics. However, *how* visuals are used is a critical component of instruction. Oftentimes, we have been struck by the inclusion of images on a mathematics handout or in lessons that seem irrelevant or are distracting. In such cases, the visuals distract from productive participation in discourse instead of facilitating it. When selecting and incorporating visuals, it is important to consider how the use of visuals will add to and support students' engagement, thinking, and communication. For instance, a teacher may want to use visuals to discuss the multiple meanings of a specific term to highlight which meaning is relevant. Alternatively, a teacher may want to use visuals to build meaning for a specific term that may be unfamiliar or typically poses challenges (e.g., row versus column). Visuals can also be used to build meaning for contexts with which students may be unfamiliar. For example, if teaching a lesson built around a factory context, it is likely some students may not know what a factory does or looks like inside. In this case, a video of a factory can be used to support students' understanding of the context. It is important to remember that including visuals in mathematics instruction needs not to take much time or financial investment. A quick Internet search can result in a multitude of images and videos that can be used. Alternatively, teachers can use the camera on their smartphones to take pictures or videos or ask students to take/share photos, which can be reused the following year.

9.5   Conclusion

In this chapter, we presented instructional strategies teachers can embed in their practice to engage MLs with MD in mathematical discourse—a critical component in fostering students' mathematical learning and language acquisition. We recommend that teachers making transformations to their practice do so incrementally. By incorporating a few strategies at a time, teachers can refine their pedagogy strategically over time rather than implementing many strategies haphazardly. We hope teachers of MLs with MD will begin to implement the strategies discussed in this chapter as they strive for equity by increasing access for every student.

References

Abedi, J., & Lord, C. (2001). The language factor in mathematics tests. *Applied Measurement in Education, 14*(3), 219–234.

Baker, S., Lesaux, N., Jayanthi, M., Dimino, J., Proctor, C. P., Morris, J., Gersten, R., Haymond, K., Kieffer, M. J., Linan-Thompson, S., & Newman-Gonchar, R. (2014). *Teaching academic content and literacy to English learners in elementary and middle school (NCEE 2014–4012)*. National Center for Education Evaluation and Regional Assistance (NCEE), Institute of Education Sciences, U.S. Department of Education.

Baxter, J., Woodward, J., Voorhies, J., & Wong, J. (2002). We talk about it, but do they get it? *Learning Disabilities Research & Practice, 17*(3), 173–185.

Boaler, J. (2003). *Studying and capturing the complexity of practice—The case of the 'dance of agency'*. Paper presented at the 27th International Group for the Psychology of Mathematics Education Conference Held Jointly with the 25th PME-NA Conference (Honolulu, HI, Jul 13-18, 2003)

Bottge, B. A., Heinrichs, M., Mehta, Z. D., & Hung, Y. H. (2002). Weighing the benefits of anchored math instruction for students with disabilities in general education classes. *The Journal of Special Education, 35*(4), 186–200.

Center for Applied Special Technology. (2016). *Top 5 UDL tips for learning environments*. Retrieved June 8, 2020 from http://castprofessionallearning.org/project/top-5-udl-tips-for-learning-environments/

Celedón-Pattichis, S., & Turner, E. E. (2012). "Explícame tu respuesta": Supporting the development of mathematical discourse in emergent bilingual kindergarten students. *Bilingual Research Journal, 35*(2), 197–216.

Chval, K. B., Chavez, O., Pomerenke, S., & Reams, K. (2009). Enhancing mathematics lessons to support all students. In D. Y. White & J. S. Silva (Eds.), *Mathematics for every student: Responding to diversity PK-5* (pp. 43–52). National Council of Teachers of Mathematics.

Cobb, P., Wood, T., & Yackel, E. (1993). Discourse, mathematical thinking, and classroom practice. In E. A. Forman, N. Minick, & C. A. Stone (Eds.), *Contexts for learning: Sociocultural dynamics in children's development* (pp. 91–119). Oxford University Press.

Courtade, G. R., Lingo, A. S., Karp, K. S., & Whitney, T. (2013). Shared story reading: Teaching mathematics to students with moderate and severe disabilities. *Teaching Exceptional Children, 45*(3), 34–44.

de Araujo, Z. (2017). Connections between secondary mathematics teachers' beliefs and their selection of tasks for English language learners. *Curriculum Inquiry, 47*, 363–389.

de Araujo, Z., Roberts, S. A., Willey, C., & Zahner, W. (2018). English learners in K–12 mathematics education: A review of the literature. *Review of Educational Research, 8*, 879–919.

De Araujo, Z., & Smith, E. (2022). Examining ELLs' learning needs through the lens of algebra curriculum materials. *Educational Studies in Mathematics., 109*(1), 65–87.

De Jong, E. J., & Harper, C. A. (2005). Preparing mainstream teachers for English-language learners: Is being a good teacher good enough? *Teacher Education Quarterly, 32*(2), 101–124.

De La Paz, S., & Graham, S. (1997). Effects of dictation and advanced planning instruction on the composing of students with writing and learning problems. *Journal of Educational Psychology, 89*(2), 203.

Enyedy, N., Rubel, L., Castellón, V., Mukhopadhyay, S., Esmonde, I., & Secada, W. (2008). Revoicing in a multilingual classroom. *Mathematical Thinking and Learning, 10*(2), 134–162.

Fairclough, N. (2010). *Critical discourse analysis: The critical study of language* (2nd ed.). Pearson Education.

Foote, M. Q., & Lambert, R. (2011). I have a solution to share: Learning through equitable participation in a mathematics classroom. *Canadian Journal of Science, Mathematics and Technology Education, 11*, 247–260.

Forman, E. A., Larreamendy-Joerns, J., Stein, M. K., & Brown, C. A. (1998). "You're going to want to find out which and prove it": Collective argumentation in a mathematics classroom. *Learning and Instruction, 8*, 527–548.

Geary, D. (2014). Learning disabilities in mathematics: Recent advances. In H. Swanson, K. Harris, & S. Graham (Eds.), *Handbook of learning disabilities* (pp. 239–255). Guilford Press.

Gee, J. P. (2011). *How to do discourse analysis: A toolkit*. Routledge.

Gersten, R., Chard, D. J., Jayanthi, M., Baker, S. K., Morphy, P., & Flojo, J. (2009). Mathematics instruction for students with learning disabilities: A meta-analysis of instructional components. *Review of Educational Research, 79*(3), 1202–1242.

Gibbons, P. (1992). Supporting bilingual students for success. *Australian Journal of Language and Literacy, 15*(3), 225–236.

Gibbons, P. (2003). Mediating language learning: Teacher interactions with ESL students in a content-based classroom. *TESOL Quarterly, 37*, 247–273.

Gibbons, P. (2008). "It was taught good and I learned a lot": Intellectual practices and ESL learners in the middle years. *Australian Journal of Language & Literacy, 31*(2), 155–173.

Gibbons, P. (2015). Scaffolding language and learning. In *Scaffolding language, scaffolding learning. Teaching English language learners in the mainstream classroom* (2nd ed., pp. 1–22). Heinemann.

Gottlieb, M., & Ernst-Slavit, G. (2013). *Academic language in diverse classrooms: Mathematics, grades 3–5.* Corwin.

Graham, S., Bollinger, A., Booth Olson, C., D'Aoust, C., MacArthur, C., McCutchen, D., & Olinghouse, N. (2012). *Teaching elementary school students to be effective writers: A practice guide (NCEE 2012–4058).* National Center for Education Evaluation and Regional Assistance, Institute of Education Sciences, U.S. Department of Education.

Graham, S., Bruch, J., Fitzgerald, J., Friedrich, L., Furgeson, J., Greene, K., Kim, J., Lyskawa, J., Olson, C. B., & Smither Wulsin, C. (2016). *Teaching secondary students to write effectively (NCEE 2017–4002).* National Center for Education Evaluation and Regional Assistance (NCEE), Institute of Education Sciences, U.S. Department of Education.

Graham, S., & Harris, K. R. (2013). Common core state standards, writing, and students with LD: Recommendations. *Learning Disabilities Research & Practice, 28*(1), 28–37.

Graham, S., Kiuhara, S. A., & MacKay, M. (2020). The effects of writing on learning in science, social studies, and mathematics: A meta-analysis. *Review of Educational Research, 90,* 179–226.

Gutiérrez, R. (2008). A "gap-gazing" fetish in mathematics education? Problematizing research on the achievement gap. *Journal for Research in Mathematics Education, 39,* 357–364.

Gutiérrez, R. (2012). Context matters: How should we conceptualize equity in mathematics education? In B. Herbel-Eisenmann, J. Choppin, D. Wagner, & D. Pimm (Eds.), *Equity in discourse for mathematics education: Theories, practices, and policies* (pp. 17–33). Springer.

Gutstein, E. (2003). Teaching and learning mathematics for social justice in an urban, Latino school. *Journal for Research in Mathematics Education, 34*(1), 37–73.

Gutstein, E., & Peterson, B. (2013). *Rethinking mathematics: Teaching social justice by the numbers* (2nd ed.). Rethinking Schools.

Harper, C., & De Jong, E. (2004). Misconceptions about teaching English-language learners. *Journal of Adolescent & Adult Literacy, 48,* 152–162.

Herbal-Eisenmann, B. A., & Breyfogle, M. L. (2005). Questioning Our "Patterns" of Questioning. *Mathematics Teaching in the Middle School, 10*(9), 484–489. http://www.jstor.org.lynx.lib. usm.edu/stable/41182145

Herbel-Eisenmann, B. (2007). From intended curriculum to written curriculum: Examining the "voice" of a mathematics textbook. *Journal for Research in Mathematics Education, 38,* 344–369.

Herbel-Eisenmann, B., & Shah, N. (2019). Detecting and reducing bias in questioning patterns. *Mathematics Teaching in the Middle School, 24*(5), 282–289.

Heyd-Metzuyanim, E. (2013). The co-construction of learning difficulties in mathematics—teacher–student interactions and their role in the development of a disabled mathematical identity. *Educational Studies in Mathematics, 83,* 341–368.

Hord, C., Tzur, R., Xin, Y. P., Si, L., Kenney, R. H., & Woodward, J. (2016). Overcoming a 4th grader's challenges with working-memory via constructivist-based pedagogy and strategic scaffolds: Tia's solutions to challenging multiplicative tasks. *Journal of Mathematical Behavior, 44,* 13–33.

Hufferd-Ackles, K., Fuson, K. C., & Sherin, M. G. (2004). Describing levels and components of a math-talk learning community. *Journal for Research in Mathematics Education, 35,* 81–116.

Hughes, E. M., Lee, J. Y., Cook, M. J., & Riccomini, P. J. (2019). Exploratory study of a self-regulation mathematical writing strategy: Proof-of-concept. *Learning Disabilities–A Contemporary Journal, 17*(2), 185–203.

Hunt, J. H., MacDonald, B., Lambert, R., Sugita, T., & Silva, J. (2018). Think-pair-show-share to increase classroom discourse. *Teaching Children Mathematics, 25*(2), 78–84.

Hunt, J. H., & Tzur, R. (2017). Using a constructivist-based pedagogy for students labeled as learning-disabled: Ana and Lia's advance to an anticipatory stage of a unit fraction scheme. *Journal of Mathematical Behavior, 48*, 62–76.

Iddings, A. C. D. (2005). Linguistic access and participation: English Language Learners in an English-dominant community of practice. *Bilingual Research Journal, 29*(1), 165–183.

Jackson, K. J., Shahan, E. C., Gibbons, L. K., & Cobb, P. A. (2012). Launching complex tasks. *Mathematics Teaching in the Middle School, 18*(1), 24–29.

Kayi-Aydar, H. (2014). Social positioning, participation, and second language learning: Talkative students in an academic ESL classroom. *TESOL Quarterly, 48*(4), 686–714.

Kena, G., Hussar, W., McFarland, J., de Brey, C., Musu-Gillette, L., Wang, X., Zhang, J., Rathbun, A., Wilkinson-Flicker, S., Diliberti, M., Barmer, A., Bullock Mann, F., & Dunlop Velez, E. (2016). *The condition of education 2016 (NCES 2016–144)*. U.S. Department of Education, National Center for Education Statistics.

Khisty, L. L. (1995). Making inequality: Issues of language and meanings in mathematics teaching with Hispanic students. In W. G. Secada, E. Fennema, & L. B. Adajian (Eds.), *New directions for equity in mathematics education* (pp. 279–297). Cambridge University Press.

Khisty, L. L., & Chval, K. B. (2002). Pedagogic discourse and equity in mathematics: When teachers' talk matters. *Mathematics Education Research Journal, 14*, 4–18.

Lambert, R. (2015). Constructing and resisting disability in mathematics classrooms: A case study exploring the impact of different pedagogies. *Educational Studies in Mathematics, 89*, 1–18.

Lambert, R., & Sugita, T. (2016). Increasing engagement of students with learning disabilities in mathematical problem-solving and discussion. *Support for Learning, 31*(4), 347–366.

Lambert, R., Sugita, T., Yeh, C., Hunt, J. H., & Brophy, S. (2020). Documenting increased participation of a student with autism in the standards for mathematical practice. *Journal of Educational Psychology, 112*(3), 494.

Lambert, R., & Tan, P. (2017). Conceptualizations of students with and without disabilities as mathematical problem solvers in educational research: A critical review. *Education Sciences, 7*(2), 51.

Lave, J., & Wenger, E. (1991). *Situated learning: Legitimate peripheral participation*. Cambridge University Press.

Lei, Q., Xin, Y. P., Morita-Mullaney, T., & Tzur, R. (2020). Instructional scaffolds in mathematics instruction for English learners with learning disabilities: An exploratory case study. *Learning Disabilities: A Contemporary Journal, 18*(1), 123–144.

Lightbrown, P. M., & Spada, N. (2013). *How languages are learned* (4th ed.). Oxford University Press.

Moschkovich, J. (1999). Supporting the participation of English Language learners in mathematical discussions. *For the Learning of Mathematics, 19*(1), 11–19.

Moschkovich, J. (2002). A situated and sociocultural perspective on bilingual mathematics learners. *Mathematical Thinking and Learning, 4*, 189–212.

Moschkovich, J. (2003). *What counts as mathematical discourse?* Paper presented at the 27th International Group for the Psychology of Mathematics Education Conference Held Jointly with the 25th PME-NA Conference (Honolulu, HI, Jul 13-18, 2003)

Moschkovich, J. (2007). Using two languages when learning mathematics. *Educational Studies in Mathematics, 64*, 121–144.

Moschkovich, J. N. (2010). Language(s) and learning mathematics: Resources, challenges, and issues for research. In J. N. Moschkovich (Ed.), *Language and mathematics education: Multiple perspectives and directions for research* (pp. 1–28). Information Age.

Moschkovich, J. (2013). Principles and guidelines for equitable mathematics teaching practices and materials for English language learners. *Journal of Urban Mathematics Education, 6*(1), 45–57.

Moschkovich, J. N. (2015). Scaffolding student participation in mathematical practices. *ZDM, 47*(7), 1067–1078.

National Academies of Sciences Engineering and Medicine. (2018). *English learners in STEM subjects: Transforming classrooms, schools, and lives.* The National Academies Press.

National Center for Educational Statistics. (2016). *English language learners in public schools.* Author.

National Council of Teachers of Mathematics. (1989). *Curriculum and evaluation standards for school mathematics.* Author.

National Council of Teachers of Mathematics. (2000). *Principles and standards for school mathematics* (Vol. 1). Author.

National Council of Teachers of Mathematics. (2014). *Principles to action: Ensuring mathematical success for all.* Author.

National Governors Association Center for Best Practices & Council of Chief State School Officers. (2010). *Common core state standards for mathematics.* Author.

O'Connor, M. C., & Michaels, S. (1996). Shifting participant frameworks: Orchestrating thinking practices in group discussion. In D. Hicks (Ed.), *Discourse, learning and schooling* (pp. 63–103). Cambridge University Press.

Opitz, E. M., Grob, U., Wittich, C., Häsel-Weide, U., & Nührenbörger, M. (2018). Fostering the computation competence of low achievers through cooperative learning in inclusive classrooms: A longitudinal study. *Learning Disabilities: A Contemporary Journal, 16*(1), 19–35.

Orosco, M. J. (2014). Word problem strategy for Latino English language learners at risk for math disabilities. *Learning Disability Quarterly, 37*(1), 45–53.

Peregoy, S. F., & Boyle, O. F. (1993). *Reading, writing, and learning in ESL* (pp. 119–131). Longman.

Pinnow, R. J., & Chval, K. B. (2014). Positioning ELLs to develop academic, communicative, and social competencies in mathematics. In M. Civil & E. Turner (Eds.), *Common core state standards in mathematics for English language learners: Grades K-8* (pp. 21–34). TESOL International Association.

Pinnow, R. J., & Chval, K. B. (2015). "How much You wanna bet?" Examining the role of positioning in the development of L2 learner interactional competencies in the content classroom. *Linguistics and Education, 30*, 1–11.

Planas, N., & Gorgorió, N. (2004). Are different students expected to learn norms differently in the mathematics classroom? *Mathematics Education Research Journal, 16*(1), 19–40.

Powell, S. R., Hebert, M. A., Cohen, J. A., Casa, T. M., & Firmender, J. M. (2017). A synthesis of mathematics writing: Assessments, interventions, and surveys. *Journal of Writing Research, 8*(3), 493–526.

Razfar, A. (2012). ¡Vamos a Jugar Counters! Learning mathematics through funds of knowledge, play, and the third space. *Bilingual Research Journal, 35*(1), 53–75.

Rezvi, S., Han, A., & Larnell, G. V. (2020). Mathematical mirrors, windows, and sliding glass doors: Young adult texts as sites for identifying with mathematics. *Journal of Adolescent & Adult Literacy, 63*, 589–592.

Sanford, A. K., Pinkney, C. J., Brown, J. E., Elliott, C. G., Rotert, E. N., & Sennott, S. C. (2020). Culturally and linguistically responsive mathematics instruction for English learners in multitiered support systems: PLUSS enhancements. *Learning Disability Quarterly, 43*(2), 101–114.

Schleppegrell, M. J. (2007). The linguistic challenges of mathematics teaching and learning: A research review. *Reading & Writing Quarterly, 23*(2), 139–159.

Setati, M. (2005). Teaching mathematics in a primary multilingual classroom. *Journal for Research in Mathematics Education, 36*, 447–466.

Sherin, M. G. (2002). A balancing act: Developing a discourse community in a mathematics classroom. *Journal of Mathematics Teacher Education, 5*, 205–233.

Stein, M. K., & Smith, M. S. (1998). Mathematical tasks as a framework for reflection: From research to practice. *Mathematics Teaching in the Middle School, 3*, 268–275.

Turner, E., Dominguez, H., Maldonado, L., & Empson, S. (2013). English learners' participation in mathematical discussion: Shifting positionings and dynamic identities. *Journal for Research in Mathematics Education, 44*, 199–234.

Turner, E., Roth McDuffie, A., Sugimoto, A., Aguirre, J., Bartell, T. G., Drake, C., ... Witters, A. (2019). A study of early career teachers' practices related to language and language diversity during mathematics instruction. *Mathematical Thinking and Learning, 21*(1), 1–27.

Tzur, R. (2000). An integrated research on children's construction of meaningful, symbolic, partitioning-related conceptions, and the teacher's role in fostering that learning. *Journal of Mathematical Behavior, 18*(2), 123–147.

Vygotsky, L. S. (1978). *Mind in society: The development of higher psychological processes.* Cambridge University Press.

Wager, A. A. (2012). Incorporating out-of-school mathematics: From cultural context to embedded practice. *Journal of Mathematics Teacher Education, 15*(1), 9–23.

Webel, C. (2010). Shifting mathematical authority from teacher to community. *The Mathematics Teacher, 104*(4), 315–318.

Weiss, I. R., Pasley, J. D., Smith, P. S., Banilower, E. R., & Heck, D. J. (2003). *Looking inside the classroom: A study of K-12 mathematics and science education in the United States.* Horizon Research.

Wenger, E. (1999). *Communities of practice: Learning, meaning, and identity.* Cambridge University Press.

Whitney, T., Lingo, A. S., Cooper, J., & Karp, K. (2017). Effects of shared story reading in mathematics for students with academic difficulty and challenging behaviors. *Remedial and Special Education, 38*, 284–296.

Wood, T. (1998). Alternative patterns of communication in mathematics classes: Funneling or focusing. In H. Steinbring, M. G. Bartolini Bussi, & A. Sierpinska (Eds.), *Language and communication in the mathematics classroom* (pp. 167–178). National Council of Teachers of Mathematics.

Wood, T., Cobb, P., & Yackel, E. (1991). Change in teaching mathematics: A case study. *American Educational Research Journal, 28*(3), 587–616.

Woodward, J., & Tzur, R. (2017). Final commentary to the cross-disciplinary thematic special series: Special education and mathematics education. *Learning Disability Quarterly, 40*(30), 146–151.

World Class Instructional Design and Assessment. (2012). *Amplification of the English language development standards kindergarten–grade 12.* Board of Regents of the University of Wisconsin System.

Xin, Y. P., Chiu, M. M., Tzur, R., Ma, X., Park, J. Y., & Yang, X. (2020). Linking teacher–learner discourse with mathematical reasoning of students with learning disabilities: An exploratory study. *Learning Disability Quarterly, 43*(1), 43–56.

Xin, Y. P., Liu, J., Jones, S., Tzur, R., & SI, L. (2016). A preliminary discourse analysis of constructivist-oriented math instruction for a student with learning disabilities. *The Journal of Educational Research, 109*(4), 436–447. https://doi.org/10.1080/00220671.2014.979910

Yackel, E., & Cobb, P. (1996). Sociomathematical norms, argumentation, and autonomy in mathematics. *Journal for Research in Mathematics Education, 27*, 458–477.

Yamakawa, Y., Forman, E., & Ansell, E. (2009). Role of positioning: The role of positioning in constructing an identity in a third grade mathematics classroom. In *Investigating classroom interaction: Methodologies in action* (pp. 179–201). Sense Publishers.

Yoon, B. (2008). Uninvited guests: The influence of teachers' roles and pedagogies on the positioning of English language learners in the regular classroom. *American Educational Research Journal, 45*, 495–522.

Zhang, D., & Xin, Y. P. (2012). A follow-up meta-analysis for word-problem-solving interventions for students with mathematics difficulties. *The Journal of Educational Research, 105*(5), 303–318.

Zwiers, J. (2020). *The communication effect: How to enhance learning by building ideas and bridging information gaps.* Corwin.

Chapter 10
Equitable Co-teaching Practices in Mathematics

Michelle Stephan and Lisa Dieker

Abstract STEM jobs are growing at twice the rate of other professions. Due to sophisticated technological advances, current mathematics jobs in particular require *less* calculational skills and *deeper* conceptual understanding. Mathematics education research has shown that traditional direct instruction has not prepared students to think critically and advocates more exploratory, inquiry teaching methods. Given that direct or explicit instruction has been the primary approach used with students with disabilities, we explore the feasibility of using an inquiry approach in inclusion settings. In this chapter, we present the results of 7 years of inquiry mathematics co-teaching research and present several co-instructing and co-planning practices that are associated with inquiry mathematics.

Keywords Co-teaching practices · Inquiry mathematics · Co-planning practices

Are students ready for the fourth Industrial Revolution? According to the U.S. Chamber of Commerce, the workforce is undergoing a fundamental change in the way humans work, live, and interact with one another, fueled in large part by technological innovations. Experts predict jobs of the future will require both STEM *content* and numerous *soft skills* such as communication and collaboration (U.S. Chamber of Commerce, 2017). In particular, the Partnership for twenty-first Century Skills and the US Chamber of Commerce argue K-12 education must prioritize teaching that fosters students' critical thinking, communication, collaboration, and problem-solving skills (Partnership for twenty-first Century Skills, 2015). To compete globally in university and positions in science, technology, engineering, and mathematics (STEM), students need deep understanding in mathematics (Adams et al., 2014; Pasko et al., 2013; Rissanen, 2014).

M. Stephan (✉)
University of North Carolina at Charlotte, Charlotte, NC, USA
e-mail: michelle.stephan@uncc.edu

L. Dieker
University of Central Florida, Orlando, FL, USA

A review study of a series of twenty-first Century Skills projects by Voogt and Pareja (2010) suggests a broader list of competencies such as critical thinking and problem-solving, collaboration across networks, agility and adaptability, initiative and entrepreneurialism, effective communication, accessing and analyzing information, and curiosity and imagination. In terms of mathematics specifically, Gravemeijer et al. (2017) ask *"What mathematics education may prepare students for the society of the future?"*. They argue current mathematics teaching does not prepare students for jobs in the digital age and challenge the education community to revise both the content and practices that are taught in mathematics classrooms around the world. We extend this question to ask "What mathematics education may prepare *students with disabilities for the society of the future?"*. If critical thinking, problem-solving, and communication are the key skills needed for jobs of the future, to what extent are we preparing students with special needs in these areas? Does explicit instruction provide the opportunities that students with disabilities need in order to develop these skills? We argue that despite a clear place and time for students to receive explicit instruction, if this is the primary means of learning for students with disabilities, they will not develop the critical thinking about mathematics required to compete and collaborate with their peers for future jobs.

10.1   Teaching for Intellectual Autonomy

Mathematics education researchers contend that one of the most important contributions education can make in individuals' lives is to their development of intellectual autonomy (e.g., Kamii, 1982; Piaget, 1948/1973; Yackel & Cobb, 1996). Kamii (1982) explained that intellectual autonomy is the ability to think for oneself and make decisions independently of the promise of rewards or punishments. Rather than viewing mathematics as a set of rules and facts to be memorized, students who have autonomy believe they are responsible for creating meaningful mathematical solutions to problems. In contrast, intellectual heteronomy is the belief someone else is the authority of one's mathematical thinking, and the goal is to understand and reproduce his work. Heteronomy can be seen when students say, "but the calculator told me I was wrong," "can you [the teacher] just tell me how to do it," and "is this the right way?" Each of these comments implies the student relies on someone or something else to determine the validity of their thinking, lacking autonomy.

Our central premise in this chapter is that fostering intellectual autonomy in all students is the most effective way to prepare them for the future workforce, as well as global and local citizenship. Students who follow and/or apply rules without meaning are at the mercy of others and are typically unable to solve problems with creativity and imagination. Additionally, with technological advances, developing critical thinking skills is more important for future employment and citizenry than performing calculations. We argue implicit (inquiry) instruction offers the best chance for developing intellectual autonomy since students are encouraged to create their own personally meaningful strategies during problem-solving (León et al.,

2015) rather than rely on an outside authority to tell them how to do it. We also argue that explicit instruction, as it is defined in special education, is set up to teach students for heteronomy. This is an equity issue. Why should students with disabilities receive instruction that does not prepare them for current and future employment in a world where machines do the calculating and humans do the critical thinking?

We use the term *explicit* rather than *direct instruction* purposely. The term *direct instruction* denotes a traditional lecture-style format of teaching by telling. The term explicit instruction integrates more inquiry-type teaching *structures* such as manipulatives, multiple representations, mathematical model construction, and partner work. The use of explicit instruction is an evidence-based practice from the What Works Clearing House (U.S. Department of Education, n.d.) in literacy and mathematics. It is combined with Self-Regulated Strategy Development (SRSD) for instance, which begins with teacher direction, but ends with students applying their own strategy through planning and organizing their own ideas (also an evidence-based strategy per the What Works Clearing House). Explicit instruction is an attempt to build a bridge between the worlds of mathematics education and special education research by utilizing some of the *structures of inquiry*. We maintain, however, that despite using some inquiry features, explicit instruction still promotes heteronomy rather than the autonomy that is critical in inquiry.

To begin our argument, it may be useful to distinguish between explicit instruction as rooted in a certain epistemological tradition and explicit instruction as a set of teaching moves. From an epistemological perspective, explicit instruction is deeply rooted in behaviorism and teaches for heteronomy by having a teacher or computer serve as the mathematical authority. When teaching for heteronomy, the teacher shows the steps for problem-solving to the student so she can mimic them for success. If the student uses the teacher's steps enough times, she is considered to have mastered the procedure. This is in direct contrast to teaching for autonomy in which the teacher gives the student a problem and allows the student to create her own solution. Inquiry instruction, on the other hand, has its origins in constructivism and positions the student as the *doer of mathematics* rather than the *receiver of mathematical rules*. This dissonance makes it impossible to "blend" the two approaches. Explicit approaches still teach for heteronomy despite using some of the features of an inquiry approach (e.g., Hudson et al., 2006; Scheuermann et al., 2009). By the same argument, the inquiry approach teaches for autonomy even though utilizing some explicit moves. We will explore the latter case in this chapter by showing how co-teachers in inquiry settings use some explicit teaching moves while still teaching for autonomy. For now, a simple example may be useful. One inquiry teaching *strategy* that is used within an explicit SRSD *program* involves using manipulatives. In the explicit approach, a teacher introduces a manipulative and then direct models for all children the way to use the materials, which is a form of teaching by telling. Direct modeling would be described as *explicit teaching* because the mathematics is becoming more explicitly available to students with the manipulative (and the teacher's help). Once the students can replicate the modeled strategy, they can then go on to solve problems on their own. However, the opportunity for students to develop autonomy is low because she has relied on the modeling of the teacher and learned the steps that work for *someone else* rather than developing her own strategy. Direct modeling with concrete materials is an example of implementing an inquiry *strategy* but teaching for heteronomy, the opposite of the inquiry epistemology.

Articles that describe an inquiry approach in co-taught classrooms are rare (exception Akyuz & Stephan, 2020) and empirical results even scarcer. What does teaching for autonomy, the hallmark of the inquiry approach, look like in co-taught classrooms then? This question is one we have attempted to answer over the last 20 years of co-teaching ourselves as well as in our working with several co-teaching pairs who use an inquiry approach in their inclusive mathematics classrooms. Our work with teachers who use the inquiry approach to foster intellectual autonomy in co-taught mathematics classrooms is presented below in the form of several co-teaching practices for inquiry mathematics. We use the term *co-teaching* to refer to the activities of co-planning, co-instructing, and co-assessing. In the section that follows, we zoom in on two general practices of co-teaching (planning and instruction) in order to examine the role the mathematics and special education teachers play in an inquiry mathematics classroom.

10.2 Co-planning Practices in Inquiry Classrooms

10.2.1 Planning in Inquiry Mathematics Classrooms

From the mathematics education sector, co-planning involves lesson imaging (Stephan et al., 2015), a process through which teachers work together to imagine how students will engage with the materials planned. The first part of lesson imaging involves *determining the content standards* in the class and then *choosing/adapting a high cognitive demand task or tasks* for students to solve. The task(s) must be accessible to all students (often referred to as Universal Design for Learning; Gordon et al., 2016), regardless of ability, but demanding enough that a solution strategy is not known immediately to students. The teachers then imagine how they will *launch the task* to students without compromising the demand. Next, the teachers solve the task themselves in multiple ways to *anticipate the numerous correct and incorrect solution paths* students might create. In this way, teachers are better prepared to build on students' mathematics during the whole-class discussion that follows students' problem-solving and are less likely to be blind-sided by an "out of the box" strategy. Building on students' strategies is similar to adaptive student pedagogy described in Tzur et al. (2013) and Xin et al. (2019). Then, teachers imagine how the *flow of the whole-class discussion* might unfold, provided they get the solutions they anticipated. The teachers weigh the pros and cons of which strategies to share, in what order, and what questions to ask to provide the best opportunity for students to abstract the mathematics (Smith & Stein, 2011; Stephan et al., 2015).

As the discussion on lesson imaging above indicates, planning for inquiry mathematics instruction requires a different set of skills and knowledge than does a lecture format. Although deep knowledge of mathematics content is critical for both formats, inquiry instruction requires sophisticated knowledge of the *mathematics of students*. In order to anticipate the diverse ways students might solve a problem, the teacher must be familiar with the mathematical domain as well as how students construct knowledge of it in different ways. How does lesson imaging change, if at all, when considering students with Individualized Education Programs (IEPs) and 504 plans in the classroom? What knowledge does each co-teacher bring to the partnership during co-planning?

10.2.2 Co-planning in Inclusive, Inquiry Mathematics Classrooms

The previous discussion illuminates some of the requisite knowledge, skills, and resources that are necessary for co-teachers to plan for inclusive, inquiry classrooms: *Choice of mathematics instructional resources, Knowledge of students' mathematics, and Selection of instructional format.* Special education preparation programs generally do not require content specialization; hence, special educators with deep mathematics content knowledge may be rare. Conversely, an equal barrier to success of co-teaching can be the lack of preparation on the part of mathematics teacher to plan for and accommodate the mathematical, learning, and behavioral needs of students with disabilities (Leko & Brownell, 2009). Allowing time for general and special education teachers to co-plan with one another supports growth in both teachers (Dieker & Berg, 2002; Eisenman et al., 2011; LaShorage & Thomas-Brown, 2015).

10.2.2.1 Co-planning Practice 1: Choosing Appropriate Mathematics Instructional Resources

Designing instructional materials for mathematics has emerged differently in the mathematics and special education fields. Special education designers suggest a concrete semi-concrete abstract (CSA) or concrete representational abstract (CRA) approach to task design/choice (Gersten et al., 2009; Hughes et al., 2018; Maccini & Hughes, 2000; Witzel et al., 2003). In this approach, the teacher should sequence tasks by beginning with concrete (C), manipulative-based activities. Students would manipulate the concrete materials to make sense of the mathematical quantities represented. Then, as students become more knowledgeable, teachers pose tasks that push them to represent (R) their concrete activity with semi-concrete (S) drawings or other inscriptions. As the students apply what they have learned in the concrete situations to drawings, their level of abstract reasoning grows. Finally, teachers should pose activities that are more abstract (A) using more conventional, symbolic notation.

Mathematics education researchers similarly advocate designing instructional materials that sequence tasks, building from students' informal to more formal strategies (Cobb et al., 1997; Gravemeijer, 1994). The instructional design approach written about the most in mathematics is the Dutch approach called realistic mathematics education (RME) (Freudenthal, 1973). RME consists of three main heuristics: (1) Students' initial work with a concept should be grounded in experientially real contexts; (2) problem-solving should be arranged such that students create models of their informal activity and move toward models for abstract reasoning; and (3) instructional materials should be designed to encourage students' reinvention of key mathematical concepts. Similar to CRA/CSA, problems are posed informally at first (in context), sometimes with manipulatives. Students invent mathematical strategies to solve dilemmas that characters in the story context encounter. Next, students solve problems by drawing pictures of their activity and then move toward abstract symbolic manipulations.

As an example of the type of mathematical materials inquiry co-teachers might use, consider the instructional sequence on ratios and rates (https://www.nc2ml. org/6-8-teachers/6-2/). We have used this particular series of problems in co-taught classrooms ourselves as well as observed other co-teachers implement it in their own classrooms.

In choosing such materials, co-teachers consider characteristics such as the following:

- Does the unit include appropriate mathematical standards? In what ways will we need to supplement?
- Are the initial problems grounded in a context (experientially real) allowing for all students, but particularly those with disabilities, to enter into problem-solving?
- Is engagement possible for all students *throughout* the unit?
- Are the problems sequenced to help students build from concrete/informal to abstract/formal reasoning?
- Are the problems high in cognitive demand without showing students a method for solving? Do they build students' intellectual autonomy?

When asking these questions of the ratio unit, co-teachers have routinely concluded that it meets the criteria for inquiry mathematics co-teaching. The special educators typically notice the starting point of the aliens is relatable and experientially real enough for students with disabilities to want to engage (see Fig. 10.1). They also spot the pictorial support at the beginning intended to help students create a strong multiplicative link between number of aliens and number of food bars that will feed them. Both special and general education teachers note subsequent pages of pictures support those who are not ready to inscribe their thinking in ratio tables. The problem-solving throughout the unit moves from reasoning with pictures of aliens/food bars to ratio tables and finally to conventional proportions. The general

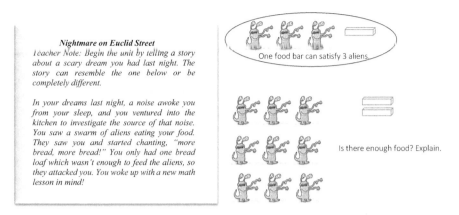

Fig. 10.1 Experientially real context for the ratio unit and first problem

education teacher notes the particular mathematics content standards infused throughout and, depending on the background knowledge of the students, can decide which problems may not be needed. Finally, both teachers notice the pages are comprised of open-ended problems sequenced in a very particular order, with no solution strategies modeled on the pages, thus preserving student autonomy.

Although both co-teachers can contribute in ways that draw on their expertise, typically the role of the general education teacher is to ensure (1) the inclusion of mathematics standards in the materials, (2) the problems build from informal to formal and conventional reasoning, (3) the appropriate representations of the problem are included as scaffolds for student thinking, and (4) the materials begin at a place that is accessible to students (UDL). For her part, the special education co-teacher verifies (1) the use of appropriate instructional scaffolds for students with particular disabilities on the IEPs and 504 s, (2) supports of areas of difficulty (UDL) that may not be supported without additional tools (e.g., text to speech, braille, calculator, vocabulary instruction), (3) behavioral supports (e.g., Class Dojo points, behavioral coaching, positive redirection) for individual students to ensure positive peer interactions, and (4) accessibility of the context and problems to students with disabilities.

10.2.2.2 Co-planning Practice 2: Using the Mathematics of Students to Plan for Instruction

A deep knowledge of mathematics at one time was the main and perhaps the only requirement to be a "teacher." Hill et al. (2008) have shown that teachers who produce strong outcomes also have knowledge of *students' mathematical thinking.* What work can co-teachers do to understand the mathematics of their students without waiting until the students show them during instruction?

Interview Students

During the second year of co-teaching mathematics together, Stephan and her co-teacher realized they knew very little about how middle school students reasoned about ratios naturally. How were they going to anticipate student thinking with ratios if they did not know what type of reasoning was possible? One way to gather this knowledge is to interview students in the classrooms (Stephan et al., 2014). Stephan and her co-teacher gathered a list of open-ended ratio problems from some research articles (see two examples in Fig. 10.2) and conducted interviews with five students. The five students represented a wide range of abilities as indicated by their scores on annual assessments, goals on IEPs, and students' work in class up to that point.

As examples of the type of reasoning students used *prior to* ratio instruction, we present the work of Bradley, Nuria, and Arthur. Bradley had an IEP for both dyscalculia (mathematics disability) and emotional behavioral disordered (often presenting disengagement or anger if unsuccessful in front of peers). Nuria had scored just at the proficient level on her previous years' state test. Arthur had not been diagnosed formally with a disability but was scoring below average consistently on state tests, thus having, at minimum, a mathematics difficulty (Lewis, 2014). Bradley solved the balloon problem by attempting to keep an organized list of ratios (see Fig. 10.3a). He began with 3 balloons for $2, 6 for $4 and then began to lose track when he started to increment by twos (8 balloons) and then found himself with 15 and $3. What Stephan and her co-teacher learned from Bradley is that he could establish the link between 3 balloons and $2 and maintain that for the next iteration (add 3 balloons or double the 3, so add $2 or double the 2). Arthur, for his part, did not establish a link between the 3 balloons and the $2 and began to operate with numbers he found in the word problem, without clear meaning for his actions (Fig. 10.3b). Finally, Nuria drew a picture of 3 balloons with $2 underneath it and continued with that picture until she had drawn 24 balloons (in sets of 3) for a total of $16 (Fig. 10.3c).

From these short interviews, we learned some students would have difficulty setting up the link between both units in a ratio (e.g., Arthur) so our instruction would need to focus heavily on the relationship between 3 balloons and $2, for example, and how they change together. We knew that some students, like Bradley, would be able to link the two units together and begin to iterate them without breaking the 3–2 link, but worried about Bradley's frustration level when he was not instantly

Ellen, Jim, and Steve bought three helium-filled balloons and paid $2.00 for all three. They decided to go back to the store and get enough balloons for everyone in their class. How much did they have to pay for 24 balloons?

Lisa and Rachel drove equally fast along a country road. It took Lisa 6 minutes to drive 4 miles. How long did it take Rachel to drive six miles?

Fig. 10.2 Balloon problem (**a**) and driving problem (**b**)

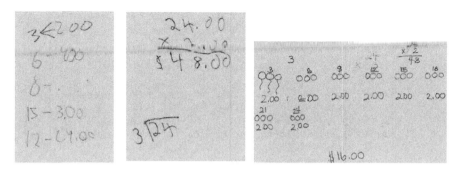

Fig. 10.3 (a–c) Mathematics of students before ratios were taught

successful. We knew we would need to support students like Bradley with some visual representations to help him keep track of his iterating. Finally, we learned from Nuria, some students would be able to link the 3 balloons and $2 together and iterate them multiple times, keeping track of both quantities simultaneously. Nuria's picture inspired us to pick problems grounded in pictures as a starting point for students.

Much can be learned from conducting interviews with a handful of diverse students prior to starting a unit. With the three students' mathematical reasoning in mind, we were better able to predict some of their strategies during lesson imaging. In addition, these students gave us ideas about how we could best support them during instruction (organizing tool, pictures, etc.). When interviews cannot be conducted, sometimes samples of student solutions can be found in research articles (e.g., for ratio unit, Battista & Borrow, 1995).

Draw on Professional Knowledge

With some new ideas about the mathematics of students, co-teachers engage in lesson imaging by discussing how they will launch the unit and first problem. Some co-teachers have launched the ratio unit with the Nightmare on Euclid Street scenario while others have shown a 3- to 5-minute video clip of the Gremlins™ or Men in Black™ movies to help students understand that the "fate of the world" is at stake if they do not feed the aliens. Both the general and special education teacher can contribute equally in imaging the launch. In particular, the special educator probably has more knowledge of the ways in which students may struggle with reading the story or making sense of the situation and can offer literacy and behavioral supports.

Once the launch is imaged, the co-teachers solve the first page of alien problems from both their own points of view and imagining what their students might do. For example, when imaging students' mathematics on problem 1 on the second page of the ratio instructional sequence (see Fig. 10.4), one special education teacher suggested some students might draw pictures of 3 aliens and one food bar, like Nuria

One food bar can satisfy 3 aliens.

Solve each of the problems below. Draw pictures if necessary or show your work.
1. Will 12 food bars be enough to feed 36 aliens? Explain.
2. Will 24 food bars be enough to feed 72 aliens? Explain.
3. Will 6 food bars be enough to feed 18 aliens? Explain
4. Will 8 food bars be enough to feed 20 aliens? Explain
5. How many food bars are needed to feed 39 aliens? Explain.

Fig. 10.4 Second page of the ratios instructional sequence

with the balloons, until they had drawn 12 food bars, then count the number of aliens. The general education teacher suggested that students may start drawing aliens and food bars but then stop and make a list of numbers that resemble a table format: 1 for 3, 2 for 6, 3 for 9, etc. Another general education teacher noted the different types of division problems students might write such as $36 \div 12$ or $36 \div 3$.

Having anticipated the mathematics of students, the co-teachers then imagine the flow of the whole-class conversation that occurs after students solve the five problems. The special educator notes some of the IEPs require modifications to the quantity of problems a student will solve. With that in mind, the two teachers discuss which of the five problems is essential for all students to solve to have a productive whole-class discussion. Agreeing that, say 1, 2, and 4 are the most critical for engagement in the discussion, the teachers now decide which strategies they would like to have presented in class and in which order. Drawing on her knowledge of supporting students with disabilities, the special educator typically argues the picture should be the first presented, if someone uses that technique. The general educator agrees and suggests that the table-esque list should be second because it is a more efficient, abstract model of the picture. If a student presents that strategy just after the picture, students are more likely to want to use the list instead because it is quicker than drawing a picture. The co-teachers also agree that the table/list would be good to have presented because the teacher can simply put lines through the list to introduce the semi-formal ratio table. Finally, the co-teachers agree that the number sentences may not be the easiest to explain in meaningful ways and are the most abstract, so those should be presented last.

Looking over IEPs and 504 s, the co-teachers make sure they have provided the appropriate accommodations to students and determine whether they think all students will be able to do at least one of the strategies they imagined. Since some students may be assigned fewer problems than others, the teachers decide which of the five is essential for all students. Those problems are the ones assigned to students who need the accommodation of less problems.

10.2.2.3   Co-planning Practice 3: Selection of Instructional Format

Once the overall mathematical concept and students' potential reasoning and patterns are discussed, the next step in the co-teaching process is to determine how to ensure parity and co-presence in the lesson (Dieker et al., 2014). Friend et al. (1993) describe five approaches co-teachers can use to support students and each other in an inclusive environment: One teach-one assist, station co-teaching, parallel co-teaching, alternative co-teaching; and team-co-teaching. These five types of co-teaching structures are typically found in practice with a sixth type often mentioned, but we believe should only be used within a lesson for approximately five minutes no more than every two weeks. This sixth type is called one teach and one observe. This type can be very helpful in doing an environmental scan (student's desk compared to height, noise near a student, grouping, lighting, being able to hear throughout the room, or any other variable one might want to observe), or to monitor and take data on specific IEP goals (e.g., monitoring on-task behavior, observing a student's test taking skills). We recommend the observation role be switched among the math and special education teacher with each taking 5 minutes during the week to observe and make notes of issues found during the observation to discuss in future planning sessions.

10.3   Co-instructing in Inquiry Mathematics Classrooms

Instructional materials that contain carefully sequenced problems are critical for inquiry co-teaching. However, having the materials in hand does not necessarily mean that they will be implemented in an inquiry manner. A teacher might take the first alien problems and "model" how to do the first one, with the genuine intent of helping students be successful. Yet, explicit modeling at the beginning of exploration steals the cognitive demand from students because the mathematical reasoning is done by the teacher. Once high-demand tasks are chosen, a lesson is carefully imaged, and a type of co-teaching is selected, what are the instructional strategies that the co-teachers use to ensure students to develop intellectual autonomy? This section builds off the study conducted by Akyuz (2010) which documented the practices of an inquiry mathematics teacher.

10.3.1   Co-instructing Practice #1—Creating and Sustaining Social Norms

Social norms for inquiry classrooms are critical for supporting students' intellectual autonomy and refer to the accepted ways students and teachers will communicate with each other during classroom discussions. Social norms for inquiry mathematics classrooms include encouraging students to (a) explain and justify their reasoning, (b) indicate respectful agreement and disagreement, (c) ask questions when clarification is needed, and (d) attempt to understand the reasoning of others (Cobb et al., 1992).

Both co-teachers play a role in creating and sustaining these social norms for their co-instruction. Building social norms should begin on the first day of class and be renegotiated as the need arises later in the school year. As an example, consider an episode that occurred on the fifth day of class in one of our co-teaching partners' classroom. One of the co-teachers placed a problem on the board (see Fig. 10.5), launched it and then allowed approximately 17 minutes for students to work independently and then in groups.

As the small group exploration ended, the co-teachers met to decide which students would put their strategies on the board. The general education teacher began the whole-class discussion as follows:

> GT: We're going to come back together and we're going to start sharing some strategies. Now, here is where some of you are gonna get some clarity and you're gonna get some ideas. This was hard for quite a few of you. And we have lots of different strategies and we have lots of different answers. And that doesn't bother me, but we have to figure out what thinking is the most accurate and is gonna tell us whether or not the people at both tables get the same amount of pizza. Now for a lot of you, this is gonna be the first time coming up and talking about your math to your classmates. So, everybody at some point is gonna do this. SO, it's important that *we're respectful* and *we listen* and if *we see something we like, we add that strategy to your notebook*. So, Camden, I'm gonna let you go first. Can you go up there and kind of *explain how you were thinking*? We should all look at Camden's work and *try to make sense of it* as he talks to us and *be ready to ask any questions* if we have some.

In the excerpt above, the general education teacher announces her expectations that students listen to Camden as he explains his thinking. Stating this explicitly lets students know they are responsible for making sense of each other's thinking. The

The dining room at a camp has two sizes of table. A large table seats ten people, and a small table seats eight people. When the campers come for dinner one night, there are four pizzas on each large table and three pizza on each small table.

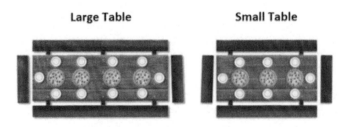

Fig. 10.5 Pizza problem given on the fifth day of class

teacher also states that students are to ask questions if they do not understand. The co-teacher reiterates these expectations both during whole-class discussion as well as during small group interactions. It is important that the co-teachers discuss these expectations prior to classroom instruction, so they can support these norms during their interactions with students. In other words, if the expectation set by the general education teacher is that students will explain their method, and the co-teacher walks over to a small group and shows a student with special needs how to do the problem, there would be a conflict in expectations. Therefore, both co-teachers need to be in tune with the expectations, so there is no confusion among students and all are held to the same expectations.

Another important aspect of creating and sustaining social norms is teaching students that, not only are mistakes acceptable, they are important for everyone's learning. Rather than trying to instruct students not to make mistakes, co-teachers actually ask students to present incorrect strategies in whole-class discussion and frame them as assets. Encouraging all students to take responsibility for their learning is also part of establishing norms. When students say, "I don't know how to do this. Can you come show me how to do it?" teachers should be prepared to let them struggle. We have heard co-teachers say, "I have faith in you! I can't wait to see what you come up with!" or "Talk with your partners. I bet you'll figure something out together!" If those suggestions do not work, the teachers can provide more supportive interaction.

10.3.2   Co-instructing Practice #2—Developing Small Groups as Communities of Learners

Both co-teachers play a role in helping students learn how to work in small group communities. Students with disabilities are likely accustomed to having a teacher provide explicit instruction (or, as we like to say, students with disabilities have been overdosed with help). In contrast, the inquiry approach encourages students to build on their own knowledge to develop a personally meaningful solution first, before explicit instruction is used. One of the first supports can and should be working with other peers in a well-constructed and supportive group. Collaboration with peers is a critical skill for the workplace and is essential for all students. To that end, all students need to learn how to contribute to the successful functioning of a group. The co-teachers should encourage students to explain their solutions to others in their group if the person does not have a way forward. Alternatively, if all students have a strategy, they should share it with their group mates with the expectation they will attempt to understand a strategy different than their own. We note this autonomy as a thinker is especially critical if a paraprofessional is in the room (they too need to be trained in the social norms) or even more important if a student has a one-on-one aide. Often times, these paraprofessionals have great intentions but may try to help too much due to a lack of background in mathematics, inquiry-based instruction, or from a belief their job is to provide the student with errorless learning.

An example from one of our co-teaching partners occurred half way through the school year. The students were working in small groups to solve the following problem (see Fig. 10.6). A co-teacher walked up to a group of three girls and proceeded to ask each one to explain.

Ms. Clemson:	Did you guys get the 100th one yet?
Roberta:	Yeah.
Ms. Clemson:	You did? Tell me what you did [squats down beside the student].
Roberta:	Cause you multiply the path number by 2. And then you add one.
Ms. Clemson:	Why do you do that?
Roberta:	Because the number of the path, the number of the path, there's always that right here [outlines one of the wings of the bird pattern] and then there's always one extra right here [points to the bottom of the V pattern].
Ms. Clemson:	Oh! [writing her strategy and name on her iPad]. Did you think about it the same way that she did?
Carly:	No, I did it a completely different way.
Ms. Clemson:	Ok, talk to me about what you did.
Carly:	Well, it depends on the path number. You would go up one. So, like if the path number was 5, you would go up to 6. Plus 5 cause it'd be like…you wouldn't double count it [the bottom dot of the V].
Ms. Clemson:	Oh! [writing her strategy and name on her iPad]. Could you show
me that on one of the pictures if I asked you to?	
Ms. Clemson:	[to the third member of the group] Now what about you?
Donna:	I didn't really get it.
Ms. Clemson:	Well, can you talk to your groupmates?
Donna:	Yeah.

The teacher, in this example, most notably asked students to explain their reasoning, asked higher order questions, and suggested students' reason with the drawing

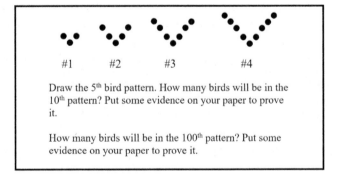

Draw the 5th bird pattern. How many birds will be in the 10th pattern? Put some evidence on your paper to prove it.

How many birds will be in the 100th pattern? Put some evidence on your paper to prove it.

Fig. 10.6 Pattern problem in small groups

to aid in their explanation. When she asked the third member about her thinking, the student curtly replied, "She did not get it." Rather than spend time giving her explicit instruction, the teacher asked her to discuss her difficulties with her groupmates who had two reasonable solution methods. Such a move reinforces the small group norms of members working together and asking questions when they are struggling. The special educator came back to that student later in the class to check on her progress and found she was successful in creating her own solution from dialog with her peers.

What do co-teachers do when they encounter a student who does not have a strategy, despite asking her group mates for help? First, the teacher might explore the student's understanding of the context of the problem. If we revisit the camping/pizza problem in Fig. 10.5, she might ask the student to identify the characters in the camping story. She might ask what challenge the campers are facing and ask the student to pick out parts of the story she does not understand. It might become known, for example, that the three campers at each end of the table have to share one pizza while the two campers facing each other in the middle get to share a pizza and that is not fair. Thus, the student's inability to engage with the problem may be due to a misinterpretation of the context, not a mathematical difficulty. The teacher can clarify the context and set the student on her way. If the issue is not contextual, but rather mathematical, the teacher can ask the student to draw a picture to represent pizzas being shared and encourage the student to partition them. Rather than explicitly instructing the student on a strategy the student should use, the teacher is suggesting that the student draw a picture and begin to solve the problem. Once a student's difficulty is uncovered, then the special educator can decide how to proceed from there, which may involve consulting the math teacher about other ways to move the student forward without explicit instruction as the first option.

Finally, the goal of teachers during small group time is NOT to fix students mistakes. In fact, there are times in which it is completely appropriate to allow students to go down a path that does not end in a correct answer. Sharing those solutions can help students unpack why the strategy did not work, allowing for a deeper understanding of mathematics. If the teachers corrected all of the mistakes during small group time, those gems would no longer be there for students' learning.

10.3.3 Co-instruction Practice #3—Facilitating Genuine Mathematical Discourse

A third instructional practice related to co-teaching in inquiry classrooms involves facilitating mathematical discussions. There is no simple recipe for engineering productive mathematical talk in whole-class discussions, yet we have a few tips from observing multiple teachers and co-teaching pairs.

10.3.3.1 Use Solutions Effectively to Engineer the Teacher's Summary

With the launch of the mathematical problem(s) students then move into small groups or independent time, while both co-teachers collect data as they monitor student engagement. The main goal during monitoring is to write down the different strategies students created. Then, just prior to the whole-class discussion, co-teachers can have a quick sidebar to compare notes. During this sidebar, the co-teachers decide which students will present their thinking and in what order. For example, during the small group time for the bird V pattern task (i.e., find the number of birds in the tenth pattern), one pair of co-teachers collected the following four strategies (Fig. 10.7):

As the two teachers brought their notes together, they decided to have all four strategies presented because the solutions contributed to the mathematical goal of identifying linear structures from spatial patterns and writing linear equations. Roberta's structuring involved seeing two "arms"/pattern #s and a leader bird at the bottom. Carly's spatial arrangement blended the leader bird with one of the arms to get pattern number (left arm) and pattern number plus 1 (right arm). Tyshawn presented a completely different structure in Wi-Fi picture in which he saw P number of sets containing two dots plus a leader bird. Finally, Nikki's strategy did not capitalize on the picture at all, and she reasoned if the fifth pattern yielded 11 dots, then the tenth pattern (double the five pattern) should contain twice as many dots. However, it is worth considering Nikki's strategy because she has double-counted the leader bird which can lead to a conversation about the structure of a linear function (a rate of change plus a constant), with the leader bird being the constant that does not change (get doubled). Roberta, Carly, and Tyshawn each present a way of structuring the picture eventually leading to different equations:

$$\text{Roberta's}: t = 2p + 1$$

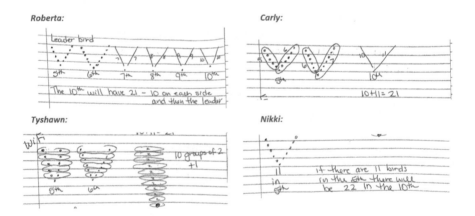

Fig. 10.7 Students' strategies for the number of birds in the tenth pattern

$$\text{Carly's}: t = p + (p+1)$$

$$\text{Tyshawn's}: t = p(2) + 1$$

Having had the discussion of the mathematical merit of these solutions, the co-teachers decided to have students present in this order: Roberta, Tyshawn, Nikki, and Carly. They felt that Roberta's leader bird method was the most popular and easily could be seen by students who may not have found a way. Tyshawn would be next because he, too, saw a leader bird yet he organized the dots on the arms as pairs of two, resembling a Wi-Fi symbol. The difference is two sets of the pattern number (Roberta) versus P sets of two (Tyshawn). Nikki's solution would be third because students who now understand the leader bird method may have a way to explain why her strategy provides one extra dot (extra leader bird that you do not repeat). Finally, Carly's example would be last because, it is a bit different, in that one of the arms absorbs the leader bird. The co-teachers would leave all four strategies on the board at the same time, thus giving students a chance to use each picture to interpret the others. Importantly, there is not one right way to sequence the solution methods during a whole-class discussion; however, the order may be more or less supportive of students accomplishing the mathematical agenda for the day.

10.3.3.2   Make Critical Discourse Moves During the Whole-Class Session

The previous tip involved actions that occur outside of the whole-class, public discussion space. Once students start sharing their reasoning with peers in a whole-class setting, the co-teachers can take specific actions to support equitable access to the mathematics in the discourse: (a) introduce mathematical vocabulary after students have invented an idea, (b) restate students' explanations in clearer or more advanced language, (c) ask questions that promote higher mathematical thinking, and d) allow students to use technology such as Flipgrid or text to speech to present their thinking if there is anxiety about sharing ideas in front of the class. Throughout the session, the co-teachers should take into account any anxiety, behavior, academic needs, reading skills, or thinking skills that may need to be considered to ensure successful interaction in both the small group and presentations to the larger group. Allowing students to "opt out" cannot be a regular option if autonomy is the goal, but allowing all students "options" as to how they want to share their idea should be at the core of any co-teaching relationship.

A common strategy in special education is to pre-teach mathematical vocabulary. However, co-teachers in inquiry mathematics introduce the more conventional language *after* students have already used the concept in their problem-solving. In this way, co-teachers build on students' invented solution processes as well as informal language to introduce the more conventional terms. For example, the leader bird in the bird V pattern problem is actually the "y-intercept" in a linear equation. However, introducing that term here would be counterproductive because the technical term does not make sense in this context. Students might use the informal term until they begin to graph the relationship between the number of birds and the pattern number on a Cartesian plane. Then, introducing the conventional term makes more sense because x and y variables appear on the graph.

A second way co-teachers can support mathematical discourse is in re-stating a students' strategy in clearer or more advanced language. This technique is especially productive in an inclusion class if there are students who have a language disability. Rather than discourage students with language difficulties from speaking publicly, a different approach is to ask them to present and then restate their words more clearly, if they have difficulty explaining their thinking. We have seen co-teachers restate the student's solution and then check with them to make sure they restated it correctly. In this way, the language disability is not seen as a deficit, and students have the ability to participate equitably in the mathematics discourse. Of course, students without a language disability also can have difficulty getting across complicated ideas, so the teachers must also restate when appropriate for any student.

A final strategy we have learned from co-teachers is to ask higher order questions that help bring the important mathematics to the fore. In the bird V pattern discussion, for example, some of the higher order questions we have heard teachers ask during or after explanations are *"What does the +1 stand for in your picture?"* *"When you say 2p, where is that in the picture?"* *"Whose strategy is represented by 2p and whose is p(2) and why?"* *"Why do you get one extra in Nikki's strategy?"* *"Will all patterns have a +1?"*. These questions use students' strategies yet ask them to think beyond just their pattern. The teachers use pictures as a way to help students connect the abstract symbols ($2p$) to something meaningful (two arms of p dots). They ask about how other unseen patterns might be similar to or different from the concrete one.

10.4   Conclusion

Although we cannot predict the future, we are confident the future will require not stronger procedural knowledge of mathematics (as a computer has that skill covered), but a deeper understanding of the processes, inputs, and outputs of computers as a twenty-first-century learner. We also are confident deeper understanding in STEM content aligns with future higher pay, but only for those workers who can engage in productive teamwork, discourse, and problem-solving. Our work suggests that students with and without disabilities can gain deep understanding of mathematics to become autonomous learners. Yet, without access to highly skilled mathematics teachers working alongside highly skilled learning and behavioral specialists promoting intellectual autonomy, students with disabilities will continue to be unemployed or underemployed compared with their peers.

We believe access to high-quality jobs can best occur with two teachers standing side-by-side in planning, instructing, and assessing a heterogeneous group of students who learn not through teacher talk, but through student talk. We see evidence from the literature the power of parity between co-teachers while still acknowledging, embracing, empowering, and ensuring the unique skillset of each teacher. Co-teachers who work together to blend their expertise in inquiry settings can provide the best learning outcome for all students. We see this occurring through a planning structure of teachers imaging their students' mathematical reasoning, while teaching with rich tasks and questioning of student thinking and while facilitating student-led discussions that are critiqued as a mathematical community. We see these teachers working in harmony to let students make mistakes, encourage them to persevere in problem-solving, and work as teams with peers—all twenty-first Century Skills.

For us, mathematics education should provide students access to high-paying jobs, support students to live, and communicate positively in a global economy and become productive citizens of the world. Developing students as intellectually autonomous beings who possess twenty-first-century competencies is crucial to this endeavor. We have presented a set of co-teaching practices that can be utilized by inquiry mathematics co-teachers that provide all, but most importantly students with disabilities, the opportunity to receive instruction preparing them for the future. Students with disabilities should have access to the same high-quality mathematics instruction as their peers; anything less is inequitable.

References

Adams, A. E., Miller, B. G., Saul, M., & Pegg, J. (2014). Supporting elementary pre-service teachers to teach STEM through place-based teaching and learning experiences. *Electronic Journal of Science Education, 18*(4), 1–22.

Akyuz, D. (2010). *Supporting a standards-based teaching and learning environment: A case study of an expert middle school mathematics teacher*. (Doctoral Dissertation). University of Central Florida.

Akyuz, D., & Stephan, M. (2020). Co-teaching practices that build autonomy for students with learning disabilities in mathematics. *International Journal of Mathematical Education in Science and Technology*, 1–28. https://doi.org/10.1080/0020739X.2020.1795286

Battista, M. T., & Borrow, C. V. A. (1995). *A proposed constructive itinerary from iterating composite units to ratio and proportion concepts*. Paper presented at the annual meeting of the North American Chapter of the International Group for the Psychology of Mathematics Education. Columbus, OH.

Cobb, P., Gravemeijer, K., Yackel, E., McClain, K., & Whitenack, J. (1997). Mathematizing and symbolizing: The emergence of chains of signification in one first-grade classroom. In D. Kirshner & J. A. Whitson (Eds.), *Situated cognition: Social, semiotic, and psychological perspectives* (pp. 151–233). Erlbaum.

Cobb, P., Wood, T., Yackel, E., & McNeal, B. (1992). Characteristics of classroom mathematics traditions: an international analysis. *American Educational Research Journal, 29*, 573–604.

Dieker, L. A., & Berg, C. A. (2002). Can secondary math, science and special educators really work together? *Teacher Education and Special Education, 25*, 92–99.

Dieker, L. A., Rodriguez, J. A., Lignugaris/Kraft, B., Hynes, M. C., & Hughes, C. E. (2014). The potential of simulated environments in teacher education: Current and future possibilities. *Teacher Education and Special Education, 37*(1), 21–33.

Eisenman, L. T., Pleet, A. M., Wandry, D., & McGinley, V. (2011). Voices of special education teachers in an inclusive high school: Redefining possibilities. *Remedial and Special Education, 32*(2), 91–104.

Freudenthal, H. (1973). *Mathematics as an educational task*. Reidel.

Friend, M., Reising, M., & Cook, L. (1993). Co-teaching: An overview of the past, a glimpse at the present, and considerations for the future. *Preventing School Failure, 34*(4), 6–10.

Gersten, R., Chard, D., Jayanthi, M., Baker, S., Morphy, P., & Flojo, J. (2009). Mathematics instruction for students with learning disabilities: A meta-analysis of instructional components. *Review of Educational Research, 79*, 1202–1242.

Gordon, D., Meyer, A., & Rose, D. H. (2016). *Universal design for learning: Theory and practice*. CAST Professional Publishing.

Gravemeijer, K. (1994). *Developing realistic mathematics education*. Utrecht, the Netherlands.

Gravemeijer, K., Stephan, M., Julie, C., Fou-Lai, L., & Ohtani, M. (2017). What mathematics education may prepare students for the society of the future? *International Journal of Science and Mathematics, 15*(1), 105–123.

Hill, H. C., Blunk, M. L., Charalambous, C. Y., Lewis, J. M., Phelps, G. C., Sleep, L., & Ball, D. L. (2008). Mathematical knowledge for teaching and the mathematical quality of instruction: An exploratory study. *Cognition and instruction, 26*(4), 430–511.

Hudson, P., Miller, S., & Butler, F. (2006). Adapting and merging explicit instruction within reform based mathematics classrooms. *American Secondary Education, 35*(1), 19–32.

Hughes, E. M., Riccomini, P. J., & Witzel, B. (2018). Using concrete-representational-abstract sequence to teach fractions to middle school students with mathematics difficulties. *Journal of Evidence-Based Practices for Schools, 16*, 171–190.

Kamii, C. (1982). *Number in preschool and kindergarten*. National Association for the Education of Young Children.

LaShorage, S., & Thomas-Brown, K. (2015). Enhancing teacher competency through co-teaching and embedded professional development. *Journal of Education and Training Studies, 3*(3), 117–125.

Leko, M., & Brownell, M. (2009). Crafting quality professional development for special educators: What school leaders should know. *Teaching Exceptional Children, 42*(1), 64–70.

León, J., Núñez, J. L., & Liew, J. (2015). Self-determination and STEM education: Effects of autonomy, motivation, and self-regulated learning on high school math achievement. *Learning and Individual Differences, 43*, 156–163. https://doi.org/10.1016/j.lindif.2015.08.017

Lewis, K. E. (2014). Difference not deficit: Reconceptualizing mathematical learning disabilities. *Journal for Research in Mathematics Education, 45*(3), 351–396.

Maccini, P., & Hughes, C. (2000). Effects of a problem solving strategy on the introductory algebra performance of secondary students with learning disabilities. *Learning Disabilities Research & Practice, 15*(1), 10–21.

Partnership for 21st century skills. (2015). *P21 Framework definitions*. Retrieved from http://www.nea.org/home/34888.htm. 7 Jan 2020.

Pasko, A., Adzhiev, V., Malikova, E., & Pilyugin, V. (2013). Advancing creative visual thinking with constructive function-based modelling. *Journal of Information Technology Education: Innovations in Practice, 12*, 59–71.

Piaget, J. (1948/1973). *To understand is to invent*. Grossman.

Rissanen, A. J. (2014). Active and peer learning in STEM education strategy. *Science Education International, 25*(1), 1–7.

Scheuermann, A., Deshler, D., & Schumaker, J. (2009). The effects of the explicit inquiry routine on the performance of students with learning disabilities on one-variable equations. *Learning Disability Quarterly, 32*(2), 103–120. https://doi.org/10.2307/27740360

Smith, M., & Stein, M. (2011). *5 Practices for Orchestrating Mathematics Discussions*. National Council of Teachers of Mathematics.

Stephan, M., McManus, G., & Dehlinger, R. (2014). Using research to inform formative assessing technique. In K. Karp (Ed.), *NCTM Annual Perspectives in Mathematics Education: Using Research to Improve Instruction* (pp. 229–238). Reston, VA.

Stephan, M., Pugalee, D., Cline, J., & Cline, C. (2015). *Lesson imaging in math and science: Anticipating student ideas and questions for deeper STEM learning*. ASCD.

Tzur, R., Johnson, H., McClintock, E., Kenney, R., Xin, Y., Si, L., Woordword, J., & Jin, X. (2013). Distinguishing schemes and tasks in children's development of multiplication reasoning. *Philippine Journal of Nursing, 7*(3), 85–101.

U.S. Chamber of Commerce. (2017). *Bridging the soft skills gap*. Paper retrieved from https://www.uschamberfoundation.org/reports/soft-skills-gap. 7 Jan 2020.

U.S. Department of Education, Institute of Education Sciences, National Center for Education Evaluation and Regional Assistance, What Works Clearinghouse.

Voogt, J., & Pareja, R. N. (2010). *21st century skills*. Universiteit Twente.

Witzel, B., Mercer, C. D., & Miller, M. D. (2003). Teaching algebra to students with learning difficulties: An investigation of an explicit instruction model. *Learning Disabilities Research and Practice, 18*, 121–131.

Xin, Y. P., Chiu, M. M., Tzur, R., Ma, X., Park, J. Y., & Yang, X. (2019). Linking teacher–learner discourse with mathematical reasoning of students with learning disabilities: an exploratory study. *Learning Disability Quarterly*, 1–14. https://doi.org/10.1177/0731948719858707

Yackel, E., & Cobb, P. (1996). Sociomath norms, argumentation, and autonomy in mathematics. *Journal for Research in Mathematics Education, 27*, 458–477.

Chapter 11
Commentary on Section III

Lisa A. Dieker and Amanda Lannan

Abstract The authors provide a summary of the three chapters focused on students with a range of neurodiversity. The authors provide their perspectives on each chapter through the lens as a parent, sibling, and person with a disability. The chapter is grounded in the cross-cutting themes presented by the authors and anchored in discussion on equity and access. The theme of how Universal Design for Learning could provide an anchor for equity and access is explored.

Keywords Universal Design for Learning · Disability · Accessibility · Discourse · Co-teaching · Equity · Access

Unlike the standards for mathematics, the field of science in the United States has approached standards differently. The Next Generation Science Standards (NGSS) includes the cross-cutting standards (National Research Council, 2015), which are considered core to transcend all areas of science. In these three chapters, we see a similar theme to NGSS cross-cutting standards. The authors collectively provide the field with cross-cutting considerations for the field of mathematics related to providing equity and access for students with a range of neurodiversity, including those with disabilities. The authors present ideas not nested in simply the content of mathematics, but in the art of using various approaches and practices of teaching mathematical knowledge to students who in the past have been forgotten, denied, or challenged to learn this content. Therefore, our commentary provides a review of the interwoven concepts to support these learners.

L. A. Dieker (✉)
University of Central Florida, College of Community Innovation and Education, Orlando, FL, USA
e-mail: lisa.dieker@ucf.edu

A. Lannan
University of Kentucky, Lexington, KY, USA

© The Author(s), under exclusive license to Springer Nature Switzerland AG 2022   221
Y. P. Xin et al. (eds.), *Enabling Mathematics Learning of Struggling Students*,
Research in Mathematics Education,
https://doi.org/10.1007/978-3-030-95216-7_11

The title of this book, *Enabling Mathematics Learning of Struggling Students: International Perspectives*, makes a clear statement that parallels the current events in our world, from the fight for equity of treatment of all humans, to the lack of equity and access for students of poverty during this pandemic. A gap already existed for students who are at the margins – whether it be poverty, second language, or disability – but what is made clear in these chapters is how this population can be renewed through effective practices, universal approaches to access, and by building bridges between teachers who serve students at the margins while maintaining high expectations and measuring for high outcomes.

As the commentators for these three chapters, we feel it essential to provide a summary of our personal and professional backgrounds to show the lens through which we view this work. We follow our personal positionality with a reflective summary of each chapter and conclude with a synthesis of our thoughts across chapters.

11.1   Personal Positionality of the Commentary Authors

I, Lisa Dieker, am the parent of a young adult who, in third grade, was diagnosed with dysgraphia, dyslexia, and Tourette syndrome. His deficits could have made him poor in mathematics, but he has had the opposite journey. His positive outcomes have been due to great teachers. My son, despite having a college degree, cannot fluidly name his multiplication facts; however, his conceptual underpinnings of mathematics are strong. As a result, he "truly" understands mathematical concepts.

My personal perspective on effective practices in the field of special education has been further shaped by the 40+ years of being both a sibling and sister-in-law of people with disabilities. I understand how equity and access changed their lives, especially during an era when people with disabilities had fewer civil rights than women or people of color in the United States. My personal experience also lies in the last 35 years of my career, as a teacher and professor. My knowledge has advanced, over the past 16 years, by leading the Ph.D. program in special education. I have had the privilege of chairing 29 dissertations and serving on 15 Ph.D. committees, with most aligned in the STEM area. My personal and professional knowledge has been further enhanced by having the privilege to direct a doctoral program that has graduated 30 Ph.D. students with disclosed disabilities.

As the second author, I, Amanda Lannan, present my commentary from the positionality of an individual with a disability. At 2-years-old, I was diagnosed with Leber's congenital amaurosis (LCA), a condition which affects the retina, causing severe vision loss in children. At the time I entered kindergarten, parental rights were limited and the common belief was children with disabilities could not learn. Luckily for me, and millions of other students with disabilities, in 1975, the Education for All Handicapped Children Act was passed, which guaranteed equal access to public education.

Despite my vision loss, I had an insatiable curiosity and a fierce determination to be independent. By the end of kindergarten, I could no longer see the huge writing, copied by my teacher, in thick magic marker, which led to the most important educational decision made on my behalf; I would learn braille. When I began high school, I discovered just how hard my parents worked to shelter me from society's doubts. Attending public school for my academic classes meant I had to advocate for the opportunity to do what was "expected" of general education students. During college and my early career, I began to see the varied societal perceptions of people with disabilities, not just blindness, but of anyone who did not fit neatly into the typical "box." Each organization promised an "equal opportunity", but the struggle to access that promise was real. Despite the struggles, I have completed my doctorate during a pandemic after teaching in the public-school system for more than 14 years. Both as a student with a disability, and as an educator, I have experienced, firsthand, and as so beautifully stated by these authors collectively, how teacher expectations play a critical role in student achievement.

The Education for All Handicapped Children Act was passed just 2 years prior to my start in school, and the Americans with Disabilities Act (ADA) was signed into law just before I graduated high school. Even though the U.S. Brown v. BOE passed in 1954 stating, "separate was not equal", which began the process of desegregating schools; it was not until 1975 that people with disabilities were legally allowed to walk through the door of a school. Today, although students with disabilities are allowed in schools, they still continue to struggle in mathematics due to a lack of access to strong and effective mathematics teachers. Teachers who work with students with disabilities often lack rich backgrounds in mathematical practices and content knowledge, as well as equitable approaches to give students with disabilities the chance to access advanced mathematics classrooms.

These three chapters align with our expertise in disability, and our commentary expands upon this work with the concept of Universal Design for Learning (UDL; Rose & Meyer, 2008). The concept behind UDL emerged from the field of architecture in making buildings accessible. The proponents of UDL (Burgstahler, 2015; Rose & Meyer, 2008) encourage the same approach to curriculum, such as mathematics. The UDL approach focused on "Equity and access" for all students is critical and is an underlying theme in these chapters.

The majority of students with disabilities are now in general education mathematics classrooms, but simply placing students in inclusive settings does not ensure equitable opportunities to participate in high levels of mathematical discourse. These authors remind us that all too often, students with disabilities are viewed through a deficit perspective, such as having gaps in mathematics knowledge, sending the message that they will not be successful in mathematics-related areas. However, these authors provide examples, hope, optimism, and evidence that, when given the right opportunities students in special education can be viewed as capable mathematics doers and thinkers. Unfortunately, the misconceptions surrounding disabilities and mathematics put students receiving special education services at a serious disadvantage due to a lack of access and potential teacher disposition. Students with disabilities have very limited power to make decisions regarding their education, such as placement in advanced mathematics courses. The current state of discriminatory practices reduces not only students with disabilities opportunities to access equitable academic instruction, but also significantly limits their access to postsecondary choices and career paths.

The legal parameters giving students with disabilities the right to a free, appropriate public education have been established in many countries like the United States, while research findings have documented the positive impact co-teaching and inclusive practices have on students with disabilities. As decades of mathematics education reform has demonstrated, the value of critical thinking and problem solving is essential to a high level of performance in mathematics. These authors remind us that now is the time to follow through high expectations for all by providing students with equitable opportunities to engage, learn, and contribute to the mathematical discourse taking place in inclusive settings. We strongly believe that the most prudent action we can take, regarding the education of students with disabilities, is to anticipate success. Just by raising the expectations, we will in turn improve confidence and increase achievement.

11.1.1 Current Status Within and Across Chapters

We begin with a summary of each chapter and then provide our synthesis through our own lens of the cross-cutting themes across chapters. Clearly, in all three chapters, a theme of students having the "right to mathematics," emerged. The importance of this right is summarized for each chapter.

11.2 Commentary on Each Chapter

11.2.1 Supporting Diverse Approaches to Meaningful Mathematics

In the chapter *Supporting Diverse Approaches to Meaningful Mathematics* Finesilver, Healy, and Bauer provide a rich thick description of the literature and practices needed for a positive classroom culture. These authors ground their discussion in two important considerations: time and tools. They provide readers with examples from the extensive work of each author to show how the intersection of time and tools creates a culture to support diverse learners in not just accessing mathematics, but as noted in their title "meaningful mathematics".

The chapter uses a case study approach to ground the reader in their individual and collective work. The lens of how international policy on inclusion intersects with the need for time and tools to ensure students are successful in these settings across the globe. Evoking the work of Vygotsky, aligned with looking at the child's potential, these authors argue students should be seen for what they can contribute and not their labels.

We strongly agree with Finesilver, Healy, and Bauer with the critical point that scaffolds are meant to be removed. We agree. If a student is in a wheelchair, that is a permanent tool, but tools to help build concrete or representational understanding should be removed as soon as possible for all students. This need to remove support over time is especially true for a population we tend to try and make dependent upon adult learning instead of, what is emphasized so clearly across all three chapters, engaging students in discourse with their peers to ensure *their* understanding, independent of adults!

These authors provide rich examples from their work with students with disabilities in their country and how access and equity are at the core of success. We believe a great summary of their work lies in a quote from their chapter: "a new norm…the education systems are meant for the whole population and should be designed accordingly". Once again, this quote aligns with what we see as the overarching theme of UDL across these chapters. Creating a culture norm of access for all creates a pathway of success for all. These authors provide a powerful argument that, when obstacles are removed for specific individuals, all benefit. We also agree with the author's premise that labels divide. When we instruct our methods students, we often emphasize that labels have value but really only on "jelly jars," where the jar tells you the difference between grape and blackberry but labels on humans really do not inform you of differences; typically they only permeate stereotypes. Yet, the juxtaposition is in our current model. Without labels, the proposal of co-teaching by Stephan and Dieker cannot occur. The time is right, aligned with Finesilver and colleagues thinking, to create cultures with learning specialists working in collaboration with teachers to help everyone learn.

We could not agree more with the authors' call for a reorganization of the structures to redefine an inclusive educational system. The current policies and practices of mainstream schools reflect the medical model of disability, which views disability as resulting from an individual person's physical or mental limitations, further perpetuating the marginalization of students with disabilities, whereas the social model reminds us to think creatively about accessibility and ways of accommodating to account for learner variability. By adopting teaching and learning practices aligned with the principles of universal design for learning, a culture of diversity will emerge.

The grounding of these authors' work in Vygotsky further amplifies the alignment to UDL. Vygotsky, as they noted, placed an emphasis on meeting students where they are at and providing tools to help them build their own knowledge and understanding. This theme permeates across all three chapters, with these authors providing clear examples of the danger of expectations of timed assessments that often result in failure instead of success. The power is in students' understanding, which they emphasize so eloquently, whereas the process and the product of a timed assessment (especially paper pencil) may not be a true reflection of learning. As their multiple examples show, with a UDL approach, through access and equity, students can come to their own understanding of more complex ideas than when taught isolated facts or rote memorization skills. The emphasis on how repeated failure in mathematics compounds motivation and anxiety, which we know impedes learning, further promotes the need for access and scaffolds for all.

The authors explain how time and tools are essential tools for special education students. Using vignettes, they address the common misconceptions surrounding certain accommodations, such as additional time or use of manipulatives, needed to fully participate in the same mathematical activities as other students in the class. These are allowable IEP accommodations, which have been chosen to level the playing field for students with disabilities. Frequently though, giving a student who has no means of writing the option to demonstrate their conceptual understanding of the lesson with blocks is somehow misinterpreted as unfair to students who are able to convey their thoughts through typical means. Should a student who is not yet proficient in braille and must rely on someone else to read the problems aloud be expected to complete as many problems? Consider a gifted third grade student, born without hands and feet; would the dictation tool be taken away, simply because she has an unfair advantage because the software can check her spelling? The resounding answer given by the authors of this chapter, and by special education professionals everywhere across the country, is no. The experiences of students with disabilities are often disregarded by the professionals responsible for their education. Despite wanting the same understanding of mathematics, students should be empowered to advocate for their needs and to achieve the same conceptual outcome without having to struggle with assignments that may take more time to produce the answer than to understand the concept.

Accessibility is not beneficial without ensuring usability. Students need to have the training and support to build competences. Even with a strong UDL approach to instruction, students need support from teachers who understand how to adapt instruction to meet their needs. Ultimately, UDL principles and practices do not exist if all learners are expected to engage, represent, and express themselves identically.

The concept of what counts as successful mathematics learning is evolving, the technology solutions are ever changing, and the pressure to prepare students with twenty-first century skills is increasing. Throughout this chapter, the authors show how problem solving was successfully used to provide scaffolding and support to foster positive, inclusive general mathematics experiences for all students.

The chapter concludes with a clear theme of focusing on care and concern in the culture of mathematical practices that provides tools and time for student success. The authors argue that access to more time may not be the answer. Instead teachers need to be better equipped to understand a student's individual needs in the context of time combined with personalized and individualized tools aligned with UDL principles for success. Grounding classroom mathematical practices and approaches in the mindset that all students learning the concepts presented in mathematics classrooms must come from equal production of problems is missing the point of understanding mathematics concepts. If student A can finish 20 problems, yet fails to understand anything but procedural knowledge, and student B with a disability completes two problems but can talk deeply about the mathematical concept –the outcome for student B is what is needed and should be expected for all learners.

11.2.1.1 Engaging Multilingual Learners' with Disabilities in Mathematical Discourse

In their chapter, Smith and Smith provide a strong summary of ways for teachers and researchers to consider how to engage with what they call multilingual learners in mathematics. They discuss in-depth ideas from both the mathematics and the special education literature for students with mathematics difficulties (MD), be it a disability or a struggle in general. They combine students with learning disabilities and those at-risk in mathematics. Although we see their chapter filled with best practice from both areas, we want to remind readers that those with a true disability in the area of mathematics may need an alternative pathway to learn math, much like a student who cannot walk and who needs a wheelchair. As educators, we need to ensure students with neurological disabilities are not expected to overcome their disability through effort. Instead, these students may need additional tools to ensure their success. A student who is at-risk, even one with a labeled disability, may need more intervention; as time alone may not be enough. We often see those students pulled out of class, away from the rich discourse of their general education peers, for fluency practice or for other services they might need (e.g., speech, physical therapy) leaving gaps in their foundational knowledge. Therefore, we want to remind readers to consider what needs a modification (changing what is taught) versus an accommodation (support to learn the same concept but in a unique way). Many students need accommodation in mathematics using the concept of a "wheelchair for their brain" to ensure they do not need modifications in the future.

We also celebrate the authors' approach to those who are learners from other language backgrounds, focusing on the strength of a second language instead of a deficit model. In our work with Title 1 schools, we see many of the great practices provided by English Language (EL) teachers as equally powerful and beneficial for students with MD and vice versa. Hence, the need to see the strengths in the students and those professionals in schools. This need is even greater when, as Smith and Smith note, a student is dually ML and MD. These numbers are rising as we better use learning sciences to understand the differences between language development and learning disabilities.

Throughout their chapter, Smith and Smith remind the field that discourse is essential for understanding what students do and do not know; yet, equally important, is the preparation of their teachers so these students have a future in advanced mathematics and employment. As so well stated by Smith and Smith, teachers need to consider the "who, what, and why of our discourse" to ensure the message relayed to these learners is aimed at an asset and not a deficit model. Yet, they so adeptly remind us that without appropriate "supports and scaffolds" (e.g., UDL). discourse is a moot point for students with MD and EL.

Smith and Smith then further remind readers of the critical importance of structure within discourse, which seems like an oxymoron. They skillfully share the importance of norms, which we often see in positive behavioral support schools, where the culture is set for positive discussion across the curriculum. Yet, in a curriculum where math anxiety is prevalent and where problems might arise as part of discourse, setting norms is critical. It is especially critical for two populations of disability, not directly discussed across these three chapters, which are students with emotional and behavioral disorders and those students on the autism spectrum disorder. The need for clarity and norms around discourse, as well as UDL approaches to discourse, are especially important for this population (Thomas et al., 2015).

The authors conclude the chapter by providing numerous ideas to support students with MD and ML in the environment including peer supports, revoicing, self-regulated writing strategies, scaffolding, building knowledge of terms, using visual cues, using context to build meaning, and aligning mathematical discourse with written language. These ideas all fall nicely under the UDL principles by providing multiple ways to represent ideas, and at the same time, provide support to the students who are MD and ML, while supporting and improving instruction for all learners in the environment. The authors conclude with a key point of their work and of the work across the three chapters: the ideas provided are moot to consider if access is not the first step for students who are MD and ML. We could not agree more!

11.2.1.2 Diversity in Mathematics

The final chapter by Stephan and Dieker provides an approach that contributes to struggling learners' success through the power of two teachers working together in the classroom. Their examples and approach are framed in the relationship between general and special education teachers, but could be between two content teachers or even a mathematics teacher and a multilingual teacher. These authors suggest that new approaches are needed to prepare students for future employment. They ground their work in Piaget and the need for students to develop "intellectual autonomy", specifically in mathematical thinking for future life, college, and career success. They provide a rich discussion grounded in differences between approaches in mathematics education and special education, specifically discussing the differences in inquiry versus direct instruction.

The authors open with the current status of the field for STEM education, advocating the need for high-quality instruction and access for students with disabilities. They advocate for the need for both STEM content and "soft skill" instruction to be a part of mathematics, especially for students with disabilities. They then frame the chapter in the powerful nature of discourse and co-teaching to create a twenty-first century classroom inclusive of all needs for all students.

The remaining sections of the chapter are grounded in the theory, practice, and examples of how to build inquiry-based instruction between co-teachers. The authors share examples of how two teachers can plan together. The authors expand each example by showing how the two teachers work together including how to build social norms, ensuring a community of learners, and facilitating "genuine" mathematical discourse. Throughout the examples, the authors share how scaffolding, universal design for learning, and some level of explicit instruction can be used to complement inquiry-based instruction. Overall, the focus on equity and engagement in inquiry-based learning is at the core of the chapter.

Today's classrooms are becoming more diverse. In response, many schools have implemented co-teaching as a means for promoting effective instruction in inclusive classrooms (Conderman & Johnston-Rodriguez, 2009; Gately & Gately, 2001; Malian & McRae, 2010; McKenzie, 2009). Although the delivery models of co-teaching can look different across settings, the benefits are significant. For example, when a co-teaching approach is used, more students with disabilities are included in general education classes, and this may reduce the stigma that students with disabilities experience. For educators, co-teaching offers opportunities for collaboration, while also providing professional support for one another.

The authors, Stephan and Dieker, present a close examination of the multiple steps in the co-teaching, co-planning, co-instructing, and co-assessing process, emphasizing that a cohesive partnership begins long before the teachers enter the classroom. Ideally, co-teachers share ownership in all aspects of the class and collaborate in all facets of the educational process. However, most teacher preparation programs are designed to prepare either general or special education teachers separately, rather than advancing a combined approach. Therefore, much of what teachers learn about co-teaching is acquired on the job. The authors offer clear examples of how teachers might use their specific training in mathematics or special education to support all students. For example, when describing the distinct teacher roles in the co-planning phase, the authors explain, "one teacher typically focuses on ensuring access to concepts,

peer dialogue, and conceptual understanding, while the other teacher ensures the trajectory of mastery of mathematical thinking is evolving at a pace needed while ensuring student autonomy in the process."

Research has illuminated a few areas to be cautious of when co-teaching. A study by Pearl and Miller (2007) noted that the individualized supports and accommodations for students in special education often are not used in the general education setting; creating a culture of failure without specialized instruction. The lack of preparation, administrative support (Dieker, 2001; Rea, 2005), and parity in the classroom (Dieker & Murawski, 2003; Spencer, 2005) also impact the quality of co-teaching.

The authors provide a critical description needed for the field of special education and general education to build a bridge focused on the concept of instructional autonomy. They provide a juxtaposition of the concept of implicit and explicit instruction, and the relationship to creating mathematical autonomy in students with disabilities through co-teaching. They argue that the two camps that "pit" implicit and explicit instruction against each other diminish the potential power of bringing together two teachers with different approaches (both found to be effective) to better meet the needs of all learners. They do emphasize the unfortunate land mine often found in special education, where the support is only given when a student is already behind. Direct instruction tends to be the answer, due to the implied efficiency, rather than scaffolding through concepts like UDL and ensuring the expertise of the second teacher is embraced to provide supports in the targeted areas of a students' disability.

Following their discussion of the philosophical differences across the fields, they provide a discussion on the current state of, need for, and positive outcomes found when the synergy of mathematics educator and special educator, who both are prepared to work in the field of mathematics, can make a direct impact on student learning and outcomes while potentially providing "real" time professional development for teachers across disciplines. They argue, from an equity perspective throughout the chapter, that students with disabilities deserve access to highly prepared mathematics teachers. Yet, at the same time, students deserve to have support from those who are prepared in their academic, social, or emotional needs. The power of the two teachers together provides rich outcomes for all.

The authors then provide a description of the anchors of co-teaching in the areas of co-planning, co-instructing, and co-assessing. The authors provide a convincing argument that the normative classroom is not favorable to all learners and that mathematics practices are often told directly to students instead of allowing for their own discovery of learning, which can eliminate the opportunity for students with disabilities to have a voice in discourse-based discussions. They provide an argument for lesson imaging, which allows both teachers to see the journey students will take over the course of the lesson. They discuss how teachers will provide supports, such as pictorial supports and use of mathematics of students, to give voice and discussion around mathematical concepts. The mathematics of students allows for building, by both teachers, on students' reasoning and individual and collective patterns of thinking to create roles for both teachers, focused on student-centered learning outcomes.

Much like the authors in the other chapters, a discussion is provided on the importance of social norms with buy-in, not only from the students, but also from both teachers. Stephan and Dieker focus on creating communities of learners, with the co-teachers serving more in the role of guiding student discussions, and only occasionally using more direct instruction, if deemed essential, by the specialists based on a clear understanding of the students' abilities and disabilities.

These authors provide examples of the process co-teachers can go through in discourse and the natural progression from small communities to whole class sessions, with the role of the co-teachers again being guides in the process. They emphasize a need for UDL approaches with a strong focus on the use of imagery as a way to have students build a deeper understanding of mathematical concepts. Throughout the discussion, Stephan and Dieker emphasize a key component of successful co-teaching is time for teachers to both construct and deconstruct their lesson imaging and outcomes. They conclude by reminding the field for students with disabilities, that procedural knowledge can often be practiced using software programs. However, the understanding of mathematical processes and procedures is harder to develop online and yet this type of understanding is essential to both build the foundation for the future and create students ready for the twenty-first century classroom. They argue that this type of understanding is easier, better, and more enjoyable to develop when two teachers are given time to work together.

The authentic, guided examples presented throughout each section of this chapter help put co-teaching into a usable framework. Because co-teaching is usually not taught explicitly, the engaging examples in the chapter, from the aliens eating our toast, to the analysis of how much pizza is needed for a party, helps teachers gain a better understanding of how this collaborative approach of teaching could work.

The research on co-teaching is still relatively limited and is primarily focused on teaching students with learning disabilities. There is minimal access to material involving students identified with other disabilities. There is still much to learn about preparing for co-teachers to work in inclusive settings. One challenge has been finding ways to deepen special educators' mathematical knowledge while also helping general education teachers increase their understanding of accessibility and accommodations. Stephan and Dieker give us hope for the future of co-teaching in mathematics by showing how a balanced partnership in teaching promotes equitable learning for all students

11.3   Collective Reflections Across Chapters

We need to engage all students in meaningful mathematics because over 2.4 million STEM-related jobs went unfilled in 2018 (SSEC, 2019). Currently, we see an aging STEM workforce (NASEM, 2018) while the supply and demand of those from culturally and neurodiverse backgrounds are limited (NSTC, 2018). As noted by these chapter authors, all students can learn mathematical skills, if the field embraces a range of options for both supports and access. Promoting diversity in the workforce is paramount (Sutton, 2017), yet the 2019 PISA highlights stagnant and declining performance in mathematics for many students at the margin. Addressing this gap is critical for the future of these students, as well as the workforce of the future, and the recommendations provided by these authors provide hope to align new outcomes to change this downward trajectory.

This section of the research-based book presents content focused on equity and access from researchers in special education, second language learners, and mathematics education. The content provided by the authors reflects both their own research and the collective nature of the status of the field. The high-level overview of these chapters provides case studies, research, practices, and discussions on how to ensure access for students from diverse cultural and neurodiverse backgrounds to advance their skills in mathematical practices, discourse, and thinking.

These authors provide a strong linkage between the current state of research and practice with an overall theme embodying the principles of UDL, which involves providing learners with multiple means of representation, multiple means of engagement, and multiple means of action/representation. The individual and collective works across the chapters remind both teachers and researchers about the critical nature of creating lessons using UDL, with access being a key concept in shaping mathematical thinking, much like the Americans with Disabilities Act (1990) shaped access to buildings. These authors do not necessarily frame their work in UDL principles, but collectively, their work is aligned with principles that ensure access at the onset of instruction to allow for maximum mastery, engagement, and learning for all students including those from economic, language, and neurodiverse backgrounds.

From our review of each individual chapter, we found collectively the chapters provide strong synergy for readers to consider how to impact diverse learners. The chapters collectively provide an extension of the future we hope to see for a wide range of learners in mathematics. What is so well said by these authors – that equity and access are essential for us to do better in the field of mathematics – frames our collective reflection of these chapters.

A current barrier to success for the field of special education is the lack of preparation on the part of special education teachers in understanding mathematical practices, and general education teachers in planning for and accommodating students with disabilities (Leko & Brownell, 2009). These same issues are prevalent for second language teachers and teachers who are often hired without certification in many schools of poverty. As noted by Stephan and Dieker in their chapter, a need for an understanding of concrete-representational-abstract (CRA) and approaches such as Realistic Mathematics Education (RME) can embrace a richer dialogue amongst an array of diverse learners (Freudenthal, 1973; Ball & Forzani, 2010). As emphasized by the work of Ball (1993), the focus of mathematics needs to allow students to accept responsibility for their own solutions and mathematical thinking. A student-centered approach, focused on students coming to their own mathematics understanding (Hoover et al., 2016), combined with specialized CRA is needed.

The research to practice gap remains prevalent in special education (Fuchs & Fuchs, 2001; Mastropieri et al., 2009; McLeskey & Billingsley, 2008), with few school and district leaders using evidence-based practices (EBP) to influence policy, perspective, and decision-making in areas such as mathematics. Narrowing this gap is critical to improving student outcomes (Bain et al., 2009; Sindelar & Brownell, 2016). Newer teachers need models of research-based validated practices with high standards of research and strong backgrounds in impacting learning outcomes as models (Brownell & Leko, 2018; Gilbert & Coomes, 2010), such as those provided in the cases studies within these chapters.

An example of EBP could be open-ended (OE) questions. Danielson and Axtell (2009) suggests that teachers prompt more authentic and thoughtful answers from students when they asked OE questions. OE questions prompt higher cognitive processing and critical thinking (Martinez, 1999) giving a more accurate understanding of what students have learned and their process behind the mathematical concept.

In a middle school mathematics classroom where teachers used more OE, students demonstrated greater retention of knowledge and understanding of a concept throughout the unit (Attali et al., 2016). A study of high school students with intellectual disabilities incorporated the use of OE questions in mathematics education to increase student vocabulary and comprehension. The use of OE questions prompted greater understanding and recall of the vocabulary (Roberts et al., 2019).

We applaud the focus of these authors on access. Other researchers have found that, with the appropriate tool, young students (PreK-second grade) with intellectual disabilities can develop basic coding skills using the Dash™ robot (Taylor et al., 2017). The authors found that participants could learn to code and build on their learning over a series of explicitly taught steps using UDL supports and guidance. Robotics and coding align with the essential components of highly effective mathematics outcomes of strengthening problem solving skills, collaboration, communication, and critical thinking; yet, if equity and access are not provided, the potential of robotics to influence higher achievements in mathematics will not occur. Simultaneously, if students with a range of learning differences are not given the opportunity to learn critical thinking through rich discourse and have access to manipulatives and classrooms where they can learn with their peers, then future careers with mathematical foundations will no longer be an option.

Collectively we applaud these authors by providing the field with clear examples of the state of access and equity in mathematics at this time. We challenge these authors and the field of mathematics, in general, to be early adopters of emerging technology tools in virtual reality and augmented reality (e.g., Pokémon Go) to start to consider how these tools might portray concepts. Yet, we must consider how we will ensure access for students with visual impairments or other sensory-related disorders. We also see a need for richer, more diverse, and dynamic opportunities for multisensory learning in mathematics. This type of Montessori approach often occurs at the elementary level, but often the ability to ask students to think about the smell, taste, or touch of more advanced mathematical concepts is needed to address those with limited pathways in the visual or auditory modality, which are often the default when teaching mathematics, even in rich discourse settings. We also encourage the field to consider co-teaching that moves beyond just general and special education teachers to be inclusive of physical therapists, occupational therapists, counselors (e.g., math anxiety), science coaches, reading coaches, physical education teachers, art teachers, music teachers, and anyone available in the building. Too often, others assume they do not have the background to contribute to mathematics, or mathematic teachers may assume these other professionals do not have a role in their content. However, those listed above add a clear dimension of UDL to the classroom and create pathways to learning based not only on auditory or visual input. The power of the future in mathematical outcomes is bringing synergy to both the content and access to the content in novel and UDL formats to ensure the development of richer and stronger neuropathways for deeper understanding in content, process, and outcomes of mathematical content.

11.4 Closing Summary

Magnitude and Severity of the Problem Osher et al. (2016) suggest a need for all students, but especially students with disabilities, to have targeted support to gain employability skills. Yet, teaching these discrete skills has proven challenging for practitioners working with students with disabilities (Rollins, 2016), especially if the student has language impairments, such as those with learning disabilities (LD), autism spectrum disorder, or intellectual disability (Charman et al., 2015).

Student needs can hinder employability and are aligned with increased risk of anxiety, depression, low self-esteem, and peer rejection (Lane et al., 2012), which has a direct impact on class participation, collaboration, and even access to general education settings. As Rose et al. (2013) note, the psychosocial implications of peer rejection can lead to bullying. Frequently, elementary teachers express concern in the areas of SEL and communication for students with disabilities, noting a direct negative impact for those who lack social, behavioral, and functional skills (Odluyurt & Batu, 2009).

UDL embraces learner variability through (1) multiple ways to engage students in learning, (2) multiple ways to present content, and (3) multiple ways for students to express what they are learning (Meyer et al., 2014; Rose et al., 2005; Rose & Meyer, 2002). This set of principles are critical in all settings but essential for those learners identified in these chapters; equally important to the neurodiversity of students with disabilities is use of the application of UDL in a variety of contexts and classrooms (e.g., Courey et al., 2013; Coyne et al., 2010; Kennedy et al., 2014; Marino et al., 2014). Most applications of UDL principles provide for considered variability in representation, engagement, and expression for students with a range of academic or motivational needs by removing barriers in the curriculum.

Universal Design for Learning is foundational to the P-20 educational infrastructure in all countries. The authors across these three chapters clearly addressed the principles of UDL (Rose & Meyers, 2008) of ensuring multiple means of representation, engagement, and action-expression in teaching to ensure the mathematical content was accessible to students from diverse backgrounds. Yet, what is important is these approaches support students with disabilities, but are helpful to all learners in mathematics classrooms. The evolution of education in mathematics, for students who are neurodiverse, could lie in the use of UDL. The introduction of UDL provides opportunities for students to engage in learning and demonstrate their knowledge, according to their strengths and interests. Through a UDL lens, including socializing, active learning, using concrete and abstract materials, communicating, and demonstrating their knowledge, is the pathway provided by these authors and a road we both advocate for personally and professionally. We hope the readers of these chapters will travel down these paths to ensure the success of each and every students in today's mathematics classrooms.

References

Attali, Y., Laitusis, C., & Stone, E. (2016). Differences in reaction to immediate feedback and opportunity to revise answers for multiple-choice and open-ended questions. *Educational and Psychological Measurement, 76*(5), 787–802.

Bain, A., Lancaster, J., Zundans, L., & Parkes, R. J. (2009). Embedding evidence-based practice in pre-service teacher preparation. *Teacher Education and Special Education, 32*(3), 215–225.

Ball, D. L. (1993). With an eye on the mathematical horizon: Dilemmas of teaching elementary school mathematics. *The Elementary School Journal, 93*(4), 373–397. https://doi.org/10.1086/461730

Ball, D. L., & Forzani, F. M. (2010). Teaching skillful teaching. *Educational Leadership, 68*(4), 40–45.

Brownell, M. T., & Leko, M. M. (2018). Advancing coherent theories of change in special education teacher education research: A response to the special issue on the science of teacher professional development. *Teacher Education and Special Education, 41*(2), 158–168.

Burgstahler, S. E. (Ed.). (2015). *Universal design in higher education: From principles to practice.* Cambridge, MA: Harvard Education Press

Charman, T., Ricketts, J., Dockrell, J. E., Lindsay, G., & Palikara, O. (2015). Emotional and behavioral problems in children with language impairments and children with autism spectrum disorders. *International Journal of Language & Communication Disorders, 50*(1), 84–93.

Conderman, G., & Johnston-Rodriguez, S. (2009). Beginning teachers' views of their collaborative roles. *Preventing School Failure, 53*(4), 235–244. https://doi.org/10.3200/PSFL.53.4.235-244

Courey, S. J., Tappe, P., Siker, J., & LePage, P. (2013). Improved lesson planning with universal design for learning (UDL). *Teacher Education and Special Education, 36*(1), 7–27. https://doi.org/10.1177/0888406412446178

Coyne, P., Pisha, B., Dalton, B., Zeph, L. A., & Smith, N. C. (2010). Literacy by design: A universal design for learning approach for students with significant intellectual disabilities. *Remedial and Special Education, 33*(3), 162–172. https://doi.org/10.1177/0741932510381651

Danielson, C., & Axtell, D. (2009). *Implementing the framework for teaching in enhancing professional practice.* ASCD.

Dieker, L. A. (2001). What are the characteristics of "effective" middle and high school co-taught teams? *Preventing School Failure, 46*(1), 14–25.

Dieker, L. A., & Murawski, W. W. (2003). Co-teaching at the secondary level: Unique issues, current trends, and suggestions for success. *High School Journal, 86*(4), 1–13. https://doi.org/10.1353/hsj.2003.0007

Freudenthal, H. (1973). What groups mean in mathematics and what they should mean in mathematical education. *Developments in Mathematical Education*, 101–114.

Fuchs, L. S., & Fuchs, D. (2001). Principles for sustaining research-based practice in the schools: A case study. *Focus on Exceptional Children, 33*(6), 1.

Gately, S. E., & Gately, F. J. (2001). Understanding coteaching components. *Teaching Exceptional Children, 33*(4), 40–47.

Gilbert, M. J., & Coomes, J. (2010). What mathematics do high school teachers need to know? *The Mathematics Teacher, 103*(6), 418–423.

Hoover, M., Mosvold, R., Ball, D. L., & Lai, Y. (2016). Making progress on mathematical knowledge for teaching. *The Mathematics Enthusiast, 13*(1), 3–34.

Kennedy, M. J., Thomas, C. N., Meyer, J. P., Alves, K. D., & Lloyd, J. W. (2014). Using evidence-based multimedia to improve vocabulary performance of adolescents with LD: A UDL approach. *Learning Disability Quarterly, 37*(2), 71–86. https://doi.org/10.1177/0731948713507262

Lane, K. L., Menzies, H. M., Oakes, W. P., Lambert, W., Cox, M., & Hankins, K. (2012). A validation of the student risk screening scale for internalizing and externalizing behaviors: Patterns in rural and urban elementary schools. *Behavioral Disorders, 37*(4), 244–270.

Leko, M. M., & Brownell, M. T. (2009). Crafting quality professional development for special educators: What school leaders should know. *Teaching Exceptional Children, 42*(1), 64–70.

Malian, I., & McRae, E. (2010). Co-teaching beliefs to support inclusive education: Survey of relationships between general and special educators in inclusive classes. *Electronic Journal for Inclusive Education, 2*(6), 1–18.

Marino, M. T., Gotch, C. M., Israel, M., Vasquez, E., III, Basham, J. D., & Becht, K. (2014). UDL in the middle school science classroom: Can video games and alternative text heighten engagement and learning for students with learning disabilities? *Learning Disability Quarterly, 37*(2), 87–99. https://doi.org/10.1177/0731948713503963

Martinez, M. E. (1999). Cognition and the question of test item format. *Educational Psychologist, 34*(4), 207–218. https://doi.org/10.1207/s15326985ep3404_2

Mastropieri, M. A., Berkeley, S., McDuffie, K. A., Graff, H., Marshak, L., Conners, N. A., Diamond, C. M., Simpkins, P., Bowdey, F. R., Fulcher, A., Scruggs, T. E., & Cuenca-Sanchez, Y. (2009). What is published in the field of special education? An analysis of 11 prominent journals. *Exceptional Children, 76*(1), 95–109.

McKenzie, R. G. (2009). Elevating instruction for secondary-school students with learning disabilities by demystifying the highly qualified subject matter requirement. *Learning Disabilities Research and Practice, 24*, 143–150.

McLeskey, J., & Billingsley, B. S. (2008). How does the quality and stability of the teaching force influence the research-to-practice gap? A perspective on the teacher shortage in special education. *Remedial and Special Education, 29*(5), 293–305.

Meyer, A., Rose, D. H., & Gordon, D. (2014). *Universal design for learning: Theory and practice.* CAST Professional Publishing.

National Academies of Sciences, Engineering, and Medicine. (2018). *Aging and disability: Beyond stereotypes to inclusion: Proceedings of a workshop.* The National Academies Press.

National Research Council. (2015). *Guide to implementing the next generation science standards.* National Academies Press.

National Science and Technology Council. (2018). *Charting a course for success: America's strategy for STEM education.* https://www.whitehouse.gov/wpcontent/uploads/2018/12/STEM-Education-Strategic-Plan-2018.pdf

Odluyurt, S., & Batu, E. S. (2009). Determining the preparatory skills of preschools based on the opinions of teachers and literature review. *Educational Sciences: Theory and Practice, 9*(4), 1841–1851.

Osher, D., Kidron, Y., Brackett, M., Dymnicki, A., Jones, S., & Weissberg, R. (2016). Advancing the science and practice of social and emotional learning. *Review of Research in Education, 40*(1), 644–681. https://doi.org/10.3102/0091732X16673595

Pearl, C. E., & Miller, K. J. (2007). Co-taught middle school mathematics classrooms: accommodations and enhancements for students with specific learning disabilities. *Focus on Learning Problems in Mathematics, 2*, 1–20.

Rea, P. (2005). 20 ways to engage your administrator in your collaboration initiative. *Intervention in School and Clinic, 40*, 312–316.

Roberts, C. A., Kim, S., Tandy, J., & Meyer, N. (2019). Using content area literacy strategies during shared reading to increase comprehension of high school students with moderate intellectual disability on adapted science text. *Education and Training in Autism and Developmental Disabilities, 54*(2), 147–160.

Rollins, P. R. (2016). Words are not enough: Providing the context for social communication and interaction. *Topics in Language Disorders, 36*(3), 198–216.

Rose, C. A., Forber-Pratt, A. J., Espelage, D. L., & Aragon, S. R. (2013). The influence of psychosocial factors on bullying involvement of students with disabilities. *Theory Into Practice, 52*(4), 272–279.

Rose, D. H., & Meyer, A. (2002). *Teaching every student in the digital age: Universal design for learning.* Association for Supervision and Curriculum Development.

Rose, D. H., & Meyer, A. (2008). *A practical reader in Universal Design for Learning.* Cambridge, MA: Harvard Education Press.

Rose, D. H., Meyer, A., & Hitchcock, C. (2005). *The universally designed classroom: Accessible curriculum and digital technologies*. Harvard Education Press.

Sindelar P. T., & Brownell M. T. (2016). Preparing and retaining effective special education teachers: Systemic solutions for addressing teacher shortages. Ed Prep Matters. https://edprepmatters.net/2016/03/preparing-and-retaining-effective-special-education-teachers-systemic-solutionsfor-addressing-teacher-shortages/

Smithsonian Science Education Center. (2019). *The STEM imperative*. Retrieved from https://ssec.si.edu/stem-imperative

Spencer, S. A. (2005). An interview with Dr. Lynne Cook and Dr. June Downing: The practicalities of collaboration in special education service delivery. *Intervention in School and Clinic, 40*, 296–300.

Sutton, H. (2017). Students with disabilities as likely to enter STEM fields as those without disabilities. *Disability Compliance for Higher Education, 22*(9), 9–19. https://doi.org/10.1002/dhe.30292

Taylor, M. S., Vasquez, E., & Donehower, C. (2017). Computer programming with early elementary students with Down syndrome. *Journal of Special Education Technology, 32*(3), 149–159. https://doi.org/10.1177/0162643417704439

Thomas, C. N., Van Garderen, D., Scheuermann, A., & Lee, E. J. (2015). Applying a universal design for learning framework to mediate the language demands of mathematics. *Reading & Writing Quarterly, 31*(3), 207–234.

Part IV

Chapter 12
Counting

Helen Thouless, Caroline Hilton, and Tim Webb

Abstract In this chapter, we start by defining what it means to be a counter and then explain what children need to be able to do to be successful at counting. Additionally, we define subitizing, explain its importance, and how it differs from counting. We complete the first section by discussing the importance of counting as a prerequisite to other aspects of arithmetic. In the following section, we discuss several challenges that counting poses for children experiencing mathematical difficulties, illustrated by examples from our research. One of these challenges includes the language of counting. In English, aspects of the language and sequencing of counting can be difficult for children, especially the teen number words, decade boundaries, counting above 100, and sequencing of numbers. Many of these characteristics are common to other languages too. Other aspects of counting that prove difficult for many children are maintaining one-to-one correspondence while counting unfamiliar sequences, understanding cardinality, finger gnosis, and using the formal symbolic notation for counting. In the final section of this chapter, we discuss and outline the activities that we have found to be useful to enhance the counting skills of children with mathematical difficulties.

Keywords Counting · Arithmetic · Language of counting · Subitizing · Dyscalculia · Assessment of number sense · Finger gnosis · Working memory

H. Thouless (✉)
St Mary's University, London, UK
e-mail: helen.thouless@stmarys.ac.uk

C. Hilton
Institute of Education—UCL, London, UK

T. Webb
New River College Outreach Service, Islington, London, UK

© The Author(s), under exclusive license to Springer Nature Switzerland AG 2022
Y. P. Xin et al. (eds.), *Enabling Mathematics Learning of Struggling Students*,
Research in Mathematics Education,
https://doi.org/10.1007/978-3-030-95216-7_12

12.1 The Importance of Counting and Subitizing

12.1.1 Counting

It's as easy as one, two, three

Despite its reputation as something that is easy, counting is a complicated process that takes children years to master. In order to become a proficient counter, a child has to master and coordinate a number of different concepts. Each of these concepts is more complicated than they seem, as each is comprised of several skills. In this section, we discuss the foundational research on counting. There are newer studies that build on this research, but we consider the ones cited here as too important to omit.

The most commonly used model for describing what successful counting looks like is the model proposed by Gelman and Gallistel (1978). Gelman and Gallistel proposed five principles that include three "how to count" principles and two "what to count" principles.

The "how to count" principles are:

- The one-to-one principle (each object is counted once and once only)
- The stable-order principle (the number words are said in a stable order, even if a child cannot remember all the words – e.g., one, two, three, five, six)
- The cardinal principle (the last word in the count indicates the numerosity of the set)

The "what to count" principles are:

- The abstraction principle (any group of objects can be counted)
- The order-irrelevance principle (objects can be counted in any order and the answer will always be the same)

Each of these principles is more complicated than it appears on the surface. For example, even when children understand the importance of the one-to-one principle, they have to develop reliable strategies to ensure that they do count each object exactly once. One useful strategy that many children use is tagging, which is touching the objects as they count. Alibali and DiRusso (1999) found that preschool children counted more accurately when tagging than when they gestured toward the objects, but even gesturing resulted in a more accurate count than when they used no active means for coordinating the correspondence between the number words and the objects to be counted. When counting involves less familiar number sequences, children are likely to make mismatches between their gestures and speech, and thus lose track of the counting process (Graham, 1999).

As children develop the stable-order principle, they go through a series of developmental steps to develop the canonical order for numbers. At first, children appear to learn the number word sequence as one chunk, "onetwothreefourfive". Then they have to recognize that the number sequence is composed of distinct words, each of which has a unique meaning and comes in a specific order in the list (Baroody &

Wilkins, 1999). In English, children have to learn by rote the words up to 20 (Callahan & Clements, 1984), because the English language has no discernible pattern to the counting words up to 20. Once a child can count beyond 20, they can begin to perceive patterns in the counting words: the single-digit sequence 1–9 is repeated in each decade; decade transitions are signaled by nine; the decade names are related to single digit names and so they can use their prior knowledge of the single-digit patterns to generate the rest of the sequence to one hundred (Baroody & Wilkins, 1999; Callahan & Clements, 1984). Once children can count to one hundred, they have to learn how knowledge of the patterns in the numbers and place value can be utilized to count above one hundred. Another related number sequencing skill is being able to name the number that comes next in the count without starting at one or zero, which is a skill that most typically developing children master at around 5 years of age (Baroody & Wilkins, 1999).

There is no specific sequence for when children develop each of the counting principles. They may show different aspects of their understanding of counting as their knowledge of the counting principles develops and as they learn to coordinate these different principles (Johnson et al., 2019). Even children as young as 3 can have implicit knowledge of the counting principles with set sizes greater than the range that they can count successfully (Gelman & Meck, 1983). Johnson et al. (2019) found that more than half of the 3–5-year-old children in their sample understood the cardinality principle (i.e., the last number they said in their count), and used this principle even if other aspects of the count were wrong and they therefore said the wrong answer.

Although Gelman and Gallistel's (1978) counting principles is a commonly cited model, another helpful model to consider was proposed by Steffe et al. (1983). This model is based on three key abilities:

- The ability to say the number word sequence
- The ability to identify (isolate) the countables (the objects that are to be counted)
- The ability to match (one-to-one) the counting words to the countables

These two models differ in a number of ways. An important difference is that Gelman and Gallistel (1978) are interested in "principles", while Steffe et al. (1983) focus on "abilities" in terms of the child's experiences.

Accordingly, Steffe et al. (1983) defined five levels of counting:

1. Perceptual counters – in order to count and calculate children need to have physical objects, whether they are the original objects or representations, such as fingers.
2. Figural counters – children can count without the need for the actual physical objects. For example, if four counters are covered by a cloth, a child would be able to say how many counters there were altogether if one more was added.
3. Counters of motor unit items – physical actions, such as moving fingers, become the objects that are counted, rather than the perceptual or figural items themselves.
4. Counters of verbal unit items – children no longer need any form of physical action, but can use the number words as countable objects in their own right.

5. Counters of abstract unit items – children can perform calculations without relating their actions to any actual objects or physical actions.

According to Steffe et al. (1983), children often move between these different types of counting and may even fall into more than one category. One of the issues that Steffe et al. acknowledged with this categorization, is the fact that it is very difficult to observe the different stages with confidence. They later argued that "at best an observer can make educated guesses, taking into account – as does any experienced diagnostician – several indications collected over an extended period of observation" (Steffe & Cobb, 1988, p.19).

Children need time to learn that counting is used to enumerate a set of objects and that unequal counts provide information about relative numerosities (Fuson & Hall, 1983). Children also need time to become confident that counting is more reliable than estimating for providing exact numerosities of groups of objects (Cowan, 1987). This may partially explain Wynn's (1990) observation that some children were able to count sets of objects that were placed in front of them, but when the same children were engaged in a *give a number* task, for example "give me three counters", they would often fail to count and would instead grab a handful of objects. *Give a number* tasks are more complicated than counting sets because the child needs to remember the requested number, label each object taken with the correct number word, monitor their counting, and stop when they have the right number of objects (Johnson et al., 2019).

As has been shown, learning to count requires complex coordination of multiple skills. As a result, assessing children's counting skills is no easy task, especially as these skills change over time. Fuson (1988) attempted to resolve this problem by identifying a developmental path for children from 3½ to 6 years of age. She observed that:

> Both action parts of counting immovable objects – pointing and saying number words – undergo progressive internalization with age. Pointing may move from touching to pointing from a distance to using eye fixation. Saying number words moves from saying audible words to making readable lip movements to making abbreviated and unreadable lip movements to silent mental production of words. (Fuson, 1988, p. 85)

This model of skill development is useful, but should be used with caution, given that different children may develop along different trajectories and that counting is used in different situations for different purposes.

12.1.2 Subitizing

A related but separate skill to counting is subitizing. Subitizing is when we instantly recognize the number of objects in a small group without counting (Sayers, 2015). Subitizing is a capacity that we are born with, as even very young babies can differentiate between sets of different sizes for small numbers of objects (Starkey & Cooper, 1980).

There are two types of subitizing: perceptual and conceptual subitizing. *Perceptual subitizing* is when a person automatically and instantly recognizes small numerosities without having to count (Sayers et al., 2016). At 2 ½ years old, children can subitize two or three objects (Schaeffer et al., 1974), while adults can subitize (perceptually) up to four or five objects arranged randomly.

Sometimes common structures of numbers such as dice and dominoes enable quick recognition of patterns of dots. This is known as *conceptual subitizing*, which extends the ability to recognize numerosities beyond our perceptual subitizing range. This may be done by partitioning these large groups into smaller groups that can be perceptually subitized (Sayers et al., 2016), but as time goes on, patterns such as those on dice and dominoes, become easily recognizable because of their arrangements. Conceptual subitizing can be a mid-point between enactive and symbolic understanding, because children can use the joining of images of numbers in specific patterns (moving beyond counting by ones) to build confidence and competence with number bonds before they move on to purely symbolic knowledge (Sayers, 2015).

12.1.3 Counting as a Prerequisite to Other Aspects of Arithmetic

There is much discussion about "Number sense" in relation to numerical understanding. In everyday life, "number sense" is often used to refer to a set of skills that are believed to be important for people to manage tasks, such as budgeting, estimating, and using timetables. The term "number sense", as it relates to education in early number and arithmetic, has received a lot of scrutiny and has been defined in many different ways. Berch (2005) identified 30 listed components of number sense taken from a wide range of definitions. For brevity, we will use the seven key themes identified by Back et al. (2013, p.1836):

- Awareness of the relationship between number and quantity
- Understanding of number symbols, vocabulary, and meaning
- Systematic counting, including notions of ordinality and cardinality
- Awareness of magnitude and comparisons between different magnitudes
- An understanding of different representations of number
- Competence with simple arithmetical operations
- An awareness of number patterns, including recognizing missing numbers

With these in mind, it can be seen that counting experiences are important for developing number sense. For example, rote counting helps children learn that numbers come in a specific sequence, and therefore that different numbers are smaller or larger than others (Franke, 2003). As children count, they see patterns in the numbers, for example the ones that repeat every decade and a similar pattern in the tens place (Callahan & Clements, 1984). For an example of a child who explicitly uses her knowledge of single digit patterns to support her crossing decade boundaries you may watch the video of Anna counting (https://prek-math-te.stanford.edu/counting/anna-counts (Ginsburg, 2020)). As children count beyond one hundred, they get a stronger sense of the relationship between ones, tens, and hundreds, in that ten ones make ten, ten tens make one hundred, ten hundreds make one thousand, etc. (Franke, 2003), which aids with the formation of place value concepts (see Chap. 14 for a more detailed discussion of this). Object counting and *Give me a number* tasks also help children develop a concrete understanding of numerical relationships, allowing them to understand what the abstract number names mean in reality and that these meanings can be used to count many different types of things and to understand cardinality (Gelman & Meck, 1983).

These counting skills also impact children's ability to calculate and solve problems. Thouless (2014) found that two-thirds of her 9–10-year-old students with dyslexia made counting errors when solving word problems. They had understood the word problems and were utilizing strategies that should produce correct answers, but due to counting errors gave the wrong answer. Some of these errors were due to miscounts at the decade boundaries (e.g., from 29 to 30) and some were due to one-to-one correspondence errors. Importantly, in the simple counting tasks the children had not shown these one-to-one correspondence errors, but with the added difficulty of making sense of the word problem, these procedural errors appeared. In general, if counting is difficult the amount of attention that is required to count will preclude sufficient attention being allocated to following the procedure correctly or to solving the problem (Dowker, 2005).

Finally, we need to consider the role of counting in addition and subtraction. Carpenter and Moser (1984), in a longitudinal study of 88 children aged 6–9 years, observed the changes in strategy use when children engaged in activities involving addition and subtraction. For addition, they identified five distinct strategies:

- Count-all (e.g., for 4 + 5, counting out four objects, then counting out five objects, and finally counting all the objects: 1–2–3–4; 5–6–7–8–9)
- Count on from the first number (e.g., for 4 + 5, counting on from four while keeping track of the five items being added: 4; 5–6–7–8–9)
- Count on from the larger number (e.g., for 4 + 5, counting on from five while keeping track of the four items being added: 5; 6–7–8–9)
- Use known number facts retrieved from memory (e.g., I know that 4 + 5 = 9)
- Use derived facts (e.g., using the fact that 4 + 4 = 8 to work out that 4 + 5 must be equivalent to 8 + 1)

For children to move from the count-all strategy, they need to be able to count on from any number in the number sequence. For children to be able to count on from the larger number, they need to be able to quickly identify the greater of two numbers. These skills are all underpinned by knowledge and understanding of the counting system. The same can be seen with subtraction.

For subtraction, Carpenter and Moser (1984) again observed a series of five distinct strategies:

- Separating from (e.g., for 6–2, count out six objects and take two away to leave four objects)
- Adding on (e.g., for 6–2, count out two objects and then add on objects until six objects have been produced. Count out the four objects that have been added)
- Matching (e.g., for 6–2, line up the six objects and the two objects so that they are matched one-to-one. Count the remaining four objects in the unmatched group)
- Counting down from (e.g., for 6–2, count back two from six by keeping track both of the counting words five and four and the number of words counted, to get the answer four)
- Counting up from given (e.g., for 6–2, count up from two to six, keeping track of the number of counting words used, to get the answer four)

Unlike addition, the strategies for subtraction do not appear to have a developmental hierarchy. For a detailed view on learning trajectories, it may be worth exploring the "Learning and Teaching with Learning Trajectories" website, created by Clements and Sarama (2017/2019) (https://learningtrajectories.org/index.php/learning_trajectories). These learning trajectories are not proposed as developmental, but rather as a scaffold that supports learning.

12.2   The Challenges of Counting

12.2.1   The Language of Number

The skills of counting that we use are often taken for granted. In fact, there are remote communities all over the world where there is very little need for counting and so the language of counting is minimal, or may not exist at all (Gelman & Butterworth, 2005; Frank et al., 2008). Thus, counting is very much a cultural artifact, as are the very words we use to count. The first challenge is to ensure that the counting opportunities we provide for children are purposeful and meaningful. The next is to support children's understanding of foundational patterns in the number words themselves.

12.2.1.1   Does It Matter What Number Words We Use?

Children learn to use number words for counting real objects long before they are able to use number words as abstract entities to solve purely mathematical problems (Hughes, 1986). For example, Hughes (1986) found that 3–4-year-old children were able to solve numerical problems with cubes, but could not perform similar calculations when they were asked using the abstract language of mathematics (e.g., "If I have three cubes in my box and I take one out, how many will I have left?" as opposed to "What's three take away one?"). Children need a lot of practice with counting real objects, before they can move onto the abstract world of mathematics.

When solving problems with real objects, children often model the problems using their fingers as concrete referents. Hughes (1986, p.51) argues that "Fingers can thus play a crucial role in linking the abstract and the concrete, because they can be both representations of objects and objects in their own right." This is supported by recent evidence from neuroscience (Berteletti & Booth, 2015).

In order to count objects, children need to become familiar with our base-10 number system. This system is the result of thousands of years of human development. It is a very sophisticated system, so it is not surprising that it takes time for children to learn how to work with it. A particular challenge when learning to count, is to understand the links between our spoken and written number systems (transcoding) (Nunes & Bryant, 1996). In the English language, this is particularly evident with the teen numbers, which do not seem to follow the same rules as for numbers between 21 and 99 (and beyond). For example, look at the numbers "19" and "91". When we say the numbers, they both start with "nine" and so it is easy to get them confused. In addition, it is easy to confuse numbers such as "forty" with "fourteen", because the last syllables are very similar (this is a particular challenge for a child with a hearing impairment). To properly count in the base ten system, children have to learn about these inconsistencies in our language of numbers; they have to learn that the teen numbers do not follow the rules of the tens numbers. This is important because children learn to use numbers verbally before they start to read and write numbers. When children are in the early stages of learning to read and write numbers, problems are often observed due to transcoding errors caused by the confusion between the spoken and the written number words (Nunes & Bryant, 1996).

In addition to difficulties with transcoding, children who are schooled in languages, such as English, which have irregular counting systems, may struggle with remembering the number words and internalizing the meaning of numbers. A regular base-10 counting system requires knowledge of the words zero to nine, with the former being a challenging word for "nothing to count." All numbers in this system are made from combinations of these. For example, in a regular counting system, such as Welsh, the number "47" is "pedwar deg saith" (which translates literally as "four-ten-seven"). This means that children learning this number system can make all the numbers up to 99 using only ten number words. In addition, the relationship between the number words and the actual numbers they represent expresses the multiplicative relationship that is implicit in our base-10 place value system (see Chap. 14 for further discussion of this). For example, in Welsh, the number 80 is identified by the words "wythdeg" ("eight-ten"). This is very different from the irregular number words in the English language where the number "47" is "forty-seven". This places additional demands on the number of words (in a fixed sequence) that need to be learned and also the meaning of those words within the base-10 number system. For example, a child learning number words in English has to learn all the words from zero to ten as well as the teen words and words for the "tens" numbers such as "twenty, thirty, forty" and so on. It has been shown that children who learn in a language (e.g., Chinese) that uses a regular number system do better in arithmetic tasks (Fuson & Kwon, 1992), due to their increased compatibility with the place value system, and find it easier to learn the correspondence between oral and written representations of numbers in that system (Dowker & Roberts, 2015). In addition, Dowker and Roberts (2015) have suggested that children who learn to use more regular counting systems, develop a better understanding of numerical relationships, especially for numbers beyond 20.

12.2.2 The Role of Working Memory

Children with poor working memory (see Chap. 2) are likely to have difficulty with arithmetic, particularly when solving problems that require counting on or counting back mentally (Gathercole & Alloway, 2008). This is because working memory overload often results in errors being made. The demands on children to learn number facts and perform mental calculations, without concrete or visual aids to support them, can place a significant demand on children with poor working memory. This can lead to a lack of confidence and lack of progress.

In general, children with poor working memory score poorly in all areas of mathematics (Gathercole & Alloway, 2008). In assessments of working memory, however, children with special educational needs and disabilities often have an uneven profile. For example, children with motor coordination difficulties tend to be more successful with tests of verbal working memory than visuo-spatial working memory, whereas children with language impairments tend to be more successful with tasks of visuo-spatial working memory than tasks involving verbal working memory (Gathercole & Alloway, 2008). Consequently, it is important to know the specific working memory issues that children have, in order to know how best to support them.

12.2.3 Mathematical Learning Disabilities and Dyscalculia

The terms Dyscalculia and Mathematical Learning Disabilities (MLD) are highly contentious and have different meanings in different countries (see Chap. 1 for further discussion of this). The unifying consensus on the topic is that the children identified as having MLD or being dyscalculic have difficulties learning number facts both procedurally and conceptually (Scherer et al., 2016). Procedural difficulties with number fact knowledge revolve around limited fact retrieval when problem solving, whereas conceptual difficulties revolve around place value issues such as grouping and segmenting, the relationship of numbers to the place value system, verbal counting, counting by groups, counting principles, and transferring word problems into mathematical expressions. Children identified as being dyscalculic also have difficulties with subitizing and will be slow to work out which of two numbers is larger (especially presented symbolically) (Reeve et al., 2012).

Of particular relevance to this chapter are the studies that suggest that young children with MLD have persistent difficulties in understanding counting concepts

(Geary et al., 2007). As a consequence, children may be less likely to move beyond laborious counting-based strategies when engaging in arithmetic problem solving. When compared with their typically developing peers, or with children who are performing at a lower level, children with MLD have been described as those that:

- Rely on developmentally immature strategies
- Frequently commit counting errors
- Use immature counting procedures [they often use counting-all rather than counting-on]
- Have difficulties retrieving basic facts from long-term memory

(Geary & Hoard, 2005, p.258)

It has been suggested that children with MLD have a specific difficulty with number representation or processing, rather than with more general mathematical problem-solving skills.

While the notion of MLD may be problematic, the recommended strategies to support children who are identified as having MLD are similar to those that are recommended for other low-achievers and will be discussed later in this chapter.

12.2.4 The Role of Fingers

> Whenever a counting technique, worthy of the name, exists at all, *finger counting* has been found either to precede it or accompany it. (Dantzig, 2007, p.9)

It has been recognized across the disciplines of education, psychology, and neuroscience that fingers play an important role in the development of children's early number concepts. It can be no coincidence that we have ten fingers (including thumbs) and that we use a base-10 number system. It also seems likely that our tally system is linked to our hands, with the sticks being fingers and the tied group being a hand or fist. In fact, the word for "five" in many languages has its root in the word for "hand" or "fist". Nevertheless, the use of fingers is a learned, not a spontaneous activity (Crollen et al., 2012). In most cases, children begin to use their fingers in response to those around them modeling their use (Hughes, 1986).

While English and many other cultures base their counting on base-10, there are other cultures that use other bases. For example, Pular, a West African language, essentially uses base-5 so that three is tati, five is jowi, and eight is jeetati (five-three). Although this counting system is in a different base than English, it is still related to how many fingers we have and children are still taught to count on their hands.

When learning to use their fingers, a child often uses them in three ways: to represent numerosities (e.g., by holding up three fingers to show that she has three strawberries); to keep track during a count (e.g., by counting on three to find out how many strawberries she has in total); and to point at objects in one-to-one correspondence activities.

To use one's fingers as a tool to help with counting, requires the ability to identify each individual finger. For example, when we are asked to find the fourth character in our password, or to help with skip counting or group counting for multiplication, it is common practice to use fingers. What is amazing is that we know when we have got to the fourth finger – we do not need to count to check. The ability to do this so reliably is due to our finger gnosis. Finger gnosis (or finger awareness) is concerned with knowing one's fingers and, for example, being able to identify one's own fingers in response to touch. In typically developing children, finger gnosis develops quickly up to the age of 6 years and then continues to develop at a slower rate up to the age of 12 years, at which point it should be fully developed (Strauss et al., 2006). When finger gnosis is disrupted in some way, young children often have difficulty with tasks involving numbers, because they try to work out calculations mentally, rather than using their fingers as a counting tool to help (Hilton, 2019).

Research suggests that if children have poor finger gnosis (Noël, 2005), or poor finger-counting skills (Jordan et al., 1992), they are likely to be delayed in their acquisition of early arithmetic knowledge and understanding. In fact, the relationship between finger gnosis and achievement in mathematics is stronger than the relationship between tests of general cognitive ability and achievement in mathematics (Noël, 2005).

Use of fingers for calculation is an embodied activity, so it can quite quickly become a part of our practice. Fuson and Secada (1986) conducted an intervention study comparing the use of diagrams and fingers for solving arithmetic problems. They found that the children who were taught to use their fingers continued to use them after the intervention, whereas the other children did not independently choose to continue using the diagrams.

Given the significance of fingers for our number development, we need to consider how to support children who do not easily use them. It has been found that interventions that support children in developing their finger gnosis are very effective (Gracia-Bafalluy & Noël, 2008; Hilton, 2019). A series of activities based on the research done by Gracia-Bafalluy and Noël (2008), entitled "Activities for Finger Training", can be downloaded from the website "Youcubed" (https://www.youcubed.org/wp-content/uploads/2017/03/Finger-Activities-vF.pdf).

12.3   Assessing Counting Skills

In this section, we will be exploring three different research-based assessment approaches: the use of a more formal assessment tool (Jordan et al., 2008), the use of effective frameworks based on foundational number sense (FoNS) (Andrews & Sayers, 2015), and assessment through play based on the work of Wager and Parks (2016). As a teacher, you have to decide which approach will give you the best information about your children's knowledge, understanding, and skills. The examples we have provided are just examples, and if these approaches are not suitable for you feel free to adapt them or find other approaches.

It has been shown that children from backgrounds of lower socioeconomic status (SES) often start school with fewer skills in early number knowledge (Ramani & Siegler, 2008). Poor number sense puts children at risk for mathematical difficulties (Jordan et al., 2008). Therefore, we need to have useful assessment tools to identify children's areas of strengths and weaknesses in order to provide effective interventions.

12.3.1   Number Sense Assessment Tool

Jordan et al. (2008) developed a number sense assessment tool that is specifically aimed at identifying the early number skills that underpin later arithmetic and numerical problem-solving. The components in the test include counting and number recognition, number knowledge, as well as aspects of numerical operations. We describe the questions asked and provide some supplementary questions to support a wider range of skills that may help to provide useful additional information.

For counting, children are asked to count as far as they can, with the intention of stopping them if they get to ten. Children can restart only once, but they can self-correct as often as they need. In the next activity, a puppet counts an arrangement of alternating blue and yellow counters (or dots). The puppet may count the yellow or the blue counters first. The puppet will sometimes make mistakes and will sometimes count correctly. The child needs to say whether the puppet has counted correctly. We think it would be useful to add to this an example of "give me......counters", with a small number of counters (say three or four). This is because while a child is counting, they may find it hard to hold in mind the number of items that were requested. This has been shown by Wynn (1990) to be a more challenging task and is distinguished from a demonstration of cardinality using Gelman and Gallistel's (1978) counting principles.

The more abstract number knowledge aspect of the assessment focuses on knowledge of symbolically represented numbers, as can be seen below. The tasks have been described so that they can be adapted by teachers and support staff.

In the first part of the assessment, children are shown a numeral (e.g., 4) and asked what number comes next and then what number comes two numbers after. We suggest that it would also be useful to ask what numbers come before, in order to determine how well children know the number sequence backwards. This assessment will tell you how well children can count on and back from any given number, or whether when calculating the answers, they need to start from one.

Next, children are shown two numbers (e.g., 5 and 8) and are asked which number is bigger, or which number is smaller. This assessment is to see if children recognize the symbols, can attach meaning to them, and can identify their position in relation to each other on a mental number line (Siegler & Booth, 2004). For children who are at a lower developmental stage, this could be adapted and piles of different colored counters could be placed in front of a child for them to say which pile is bigger or smaller. While this task relies more on visual perception, it will let you know whether the child understands the language of relative size.

Finally, children are presented with three numbers, each placed at one corner of an equilateral triangle (e.g., 7, 11, and 13). The triangle is shown with the base horizontal and with the middle number at the apex of the triangle. The child is asked which number is closer to the one at the apex (in this case, "Which is closer to 11, 7 or 13?").

To assess number operations, you can start with questions involving counters and a box (or other appropriate resource) to hide the counters. Both the assessor and the child have a set of counters. A small number of counters are shown to the child. The counters are then hidden in the box and the child is asked how many counters have been hidden. The child can either say the number or put out the same number of counters from their collection. The task is then repeated, but this time a small number of counters are added or removed, one at a time. The child then has to say or demonstrate how many counters there are in the box. We suggest leaving the box open, so that the child can check. The final questions are focused on some "real life" problems, such as "If you have three crayons and I have four crayons, how many crayons do we have altogether?"

If this test is used, we recommend that a note is made of the strategies the child uses when solving the problems. If possible, it may be helpful to video record the child, to enable a more in-depth analysis of the strategies used and to help plan for next steps.

12.3.2   Fundamental Number Sense (FoNS)

Fundamental number sense is based on the idea that children need number sense in order to be able to access the mathematics curriculum (Gersten & Chard, 1999). In order to identify the characteristics of a child with FoNS, Andrews and Sayers (2015) have identified an eight-point observation schedule. While their schedule has been designed for use in classroom observations, we believe that it could equally well be used as an assessment (and planning) tool. The schedule is exemplified in Table 12.1 (from Andrews & Sayers, 2015, p.262).

The FoNS framework has been trialed in Hungary, Sweden, and England and initial findings suggest that the framework can be used across cultural boundaries (Andrews & Sayers, 2015). By identifying children's counting characteristics, this framework can provide a useful tool for observation, assessment, and planning, making it very versatile and supportive to classroom practitioners.

Table 12.1 Exemplifying the FoNS framework

FoNS characteristic	What children might be doing
Number recognition	Identify a particular number symbol from a collection of number symbols and name a number when shown its symbol
Systematic counting	Count systematically, both forwards and backwards, from arbitrary starting points
Relating number to quantity	Understand the one-to-one correspondence between a number's name and the quantity it represents
Quantity discrimination	Compare magnitudes and deploy language like "bigger than" or "smaller than"
Different representations	Recognize, work with, and make connections between different representations of number
Estimation	Estimate, whether it be the size of a set or the size of an object
Simple arithmetic	Perform simple addition and subtraction operations
Number patterns	Recognize and extend number patterns and, in particular, identify a missing number

12.3.3 Using Learning Stories

Learning stories involve four actions of assessment: describing, documenting, discussing, and deciding (Carr, 2001). Describing is the process of identifying learning opportunities for a child, which appeal to their interests and ways of learning. When describing, the observer creates a narrative of what the child is trying to do and how they are doing it. Documenting is the process of gaining evidence through, for example, photographs, work samples, or video recordings. Discussing involves the process of reflection, interpretation, and discussion with others, such as parents, care givers, or other staff, in order to better understand what has been observed. Deciding is the process of planning next steps, based on the findings.

By using learning stories to assess mathematical knowledge and understanding, Wager and Parks (2016) proposed a method that relies on in-the-moment decision making. The focus of this approach to assessing children's mathematical understanding relies on school staff being able to notice the moments during play, when there is an opportunity to observe children and support them in exploring and communicating their mathematical experiences. During their study, Wager and Parks (2016) identified occasions where school staff found opportunities to explore children's mathematical knowledge and understanding through observations, questioning, and scaffolding. They also noted occasions where the staff member had tried to understand the children's thinking by engaging with the activities and sustained shared thinking.

One potential disadvantage of this method is that it relies on a thorough understanding of the pedagogical aspects discussed in this chapter. However, it enables school staff to be much more responsive to the children and to capitalize on those opportunities that happen spontaneously and are motivating and interesting for the children. The vignette below, taken from Wager and Parks' (2016) study illustrates how this assessment tool can work:

Birdie first noticed when, unprompted, Tommy counted the four children at the art table and then counted out one marker for each child. After observing Tommy's spontaneous count-ing, Birdie prompted him to count several more times that morning, each time with a higher number. When another friend joined the art table (where the markers and colored pencils were now superheroes instead of drawing utensils), Birdie asked him "how many friends were at the art table?" and "can you count them?" She then followed him to the block area where he was playing with six cars and asked "how many?" and asked again at snack time when nine snacks were handed out. Birdie described what she learned about Tommy from these interactions, "When counting people he remembered to count himself. When he lost his place or got confused he started all over. He pointed to each thing when he counted, but he sometimes touched them. Sometimes he starts counting from the right side, but he usu-ally counted from the left side. He also lined things up in a row to count them. (p. 996)

As can be seen from this episode, the teacher not only learned about the child's knowledge and understanding of counting but was also able to articulate how the child understood their mathematics within a meaningful context (for further discus-sion of this, see Chap. 9).

12.4 Activities to Enhance Counting Skills

12.4.1 Ten Nice Things (Skinner, 1997)

Ten Nice Things is an appropriate way for assessing children between 2 and 4 years of age. Each child selects ten nice things from a larger collection. The first child rolls a die with dots on it, counts the dots, and gives that number of objects to the child next to them. That child then rolls the die, counts the dots, and gives that num-ber of objects to the next child, and so on.

As an example, we present Emmanuel, a 4-year-old, who was attending a nurs-ery school class for children with special educational needs. It had been difficult to assess Emmanuel because of his speech, language, and communication needs. However, playing the Ten Nice Things game with him allowed me to assess his counting skills in a fun, nonthreatening environment.

At first, it was just Emmanuel and I playing the game, because I knew that he had dif-ficulties with turn-taking. We each started with ten objects and a dotted die. Emmanuel started off subitizing up to three dots, but with practice he could subitize all six of the dice patterns. He could also count out a set of ten objects. When Emmanuel became confident with this version of the game, I increased the number of objects to fifteen and used two dotted dice. Counting out 15 objects and adding the sums of the two dice were both more difficult tasks for him, but with my support he was able to engage in these more difficult versions of the game. Simultaneously, I also started to get him to play the ten objects ver-sion of the game with one of his classmates. Importantly, playing this higher-level version of the game provided me with ample opportunities to assess his counting skills beyond ten.

This rather simple but engrossing game supports children to develop their one-to-one correspondence, stable order principle, cardinality, counting sets, making sets, subitizing, sharing, and turn-taking skills. Simultaneously, it allows a teacher to assess those skills in a playful context.

12.4.2 Making Sense of Counting

The use of a regular counting system can be very helpful for children who have difficulty remembering all the number words and for those who get confused with the meaning of number words. By adopting a regular number system (see below), that uses fewer words, children are able to access the mathematics without the language getting in the way, because the language demands are reduced and the base-10 meanings of the words above ten, are explicit in the words used. Children who use a home language where place value is transparent (for example, Japanese, Mandarin, Welsh, and Swahili), could be encouraged to use their home language to count and maybe share this knowledge with others (Table 12.2).

A system such as this can be adapted to meet the needs of a class or individual child and as long as parents/care givers are notified of this alternative number system, children should be able to access numbers and numerical activities more easily both at school and at home.

I used this system with a group of children who understood the principles of addition and subtraction, but frequently made errors because they had difficulty remembering the words for the "teen" numbers. Providing the children with this alternative counting system enabled them to access the mathematics and demonstrate their understanding with confidence.

12.4.3 Counting Collections

Counting Collections is an instructional activity that focuses on object counting while encouraging children's engagement in the activity (Franke et al., 2018). In Counting Collections, the teacher gives each child a bag containing a number of objects appropriate for the number range within which the child needs to practice counting. The child counts the objects in the bag. If the number of objects are small

Table 12.2 In English, a possible set of number words could be

1 – one	11 – ten-one	21 – two-ten-one
2 – two	12 – ten-two	22 – two-ten-two
3 – three	13 – ten-three	23 – two-ten-three
4 – four	14 – ten-four	24 – two-ten-four
5 – five	15 – ten-five	25 – two-ten-five
6 – six	16 – ten-six	26 – two-ten-six
7 – seven	17 – ten-seven	27 – two-ten-seven
8 – eight	18 – ten-eight	28 – two-ten-eight
9 – nine	19 – ten-nine	29 – two-ten-nine
10 – ten	20 – two-ten	30 – three-ten

the child can count by ones, but with larger numbers the child may group the objects to make the counting more efficient. Once the child finishes counting, they write the number of objects they counted and represent how they counted them. This activity supports beginning counters to develop all five of the counting principles (Gelman & Gallistel, 1978), but it also supports more advanced counters to develop unitizing, skip-counting, and experiences with place value.

Sekio was an 8-year-old boy who attended a self-contained special education class. At the beginning of the school year, he could count up to 12 objects accurately, had inconsistent cardinality, and wrote "7" when asked to represent this count. By the end of the school year, he could reliably count up to 37 objects accurately, consistently demonstrated cardinality, and knew that he should represent every block counted.

Sekio had already spent 4 years in the education system and yet was unable to count beyond 12 accurately. During the year, he was given regular experiences of counting beyond the troublesome teens. From the beginning, he was always given bags with at least 20–30 objects in them, and these numbers increased as his confidence increased. The bags contained interesting objects, such as colorful erasers, cars, frogs, beads, pennies, etc. (see Chap. 9 for more discussion of this).

Sekio was given organizers to help his accurate counting, such as a number line, so that he could match one object with each number, giving him the visual cue of the number to help him count accurately. He was also given organizers to help him represent his count, such as a blank number grid to write his count in.

Sekio showed better counting skills and growth in his coordination of the counting principles when he was working with larger sets of objects and with sets of objects that were motivating and interesting. These findings match those of Johnson et al. (2019), who found that children can show understanding of more complicated tasks that are not shown on simpler tasks. Therefore, we should not restrict children to counting small sets of objects because of our preconceived understanding of their capabilities.

12.4.4 Choral Counting

Choral Counting is an instructional activity that both incorporates appropriate mathematical content for young children and provides an opportunity to engage in mathematical discussions (Franke et al., 2018). There are two sections to Choral Counting: (1) the number sequence and (2) pattern identification and expansion. It is during the second section of the activity that children engage in mathematical discourse by expressing their own mathematical ideas.

In Choral Counting, the teacher has to first choose an appropriate counting sequence for the children. For younger children, appropriate counting sequences are by ones, twos, fives, tens, or backwards by ones. Once the counting sequence has been introduced to the children, the class counts together while the teacher strategically records the count on the board so that certain patterns emerge. After writing three or more rows or columns, the teacher stops the count and asks the children to identify patterns in the numbers. Once a child has stated a pattern they can be asked to extend, compare, or justify their pattern, and other children can be asked to build on what the first child has said. This activity supports children to develop a stable number sequence, both by ones and when skip-counting, and to promote an understanding of place value.

Ali was a 6-year-old boy who had speech, language, and communication needs. At the beginning of the school year, Ali had difficulties counting by ones above 12, but could reliably count ten objects, and recognized the symbols: 1, 2, 3, 8, and 10. When Ali started participating in Choral Counting, he could recognize that a count by 2 s, arranged in a 5x6 grid, would always have a four in the ones place in the second column. Later in the year, he recognized that the last two digits in the first row of a count by ten would always be ten. He not only recognized the regularity in the number system in columns but could also see regularity across columns and could predict that the number to the right of 310 would be 410. Although these numbers were well above his counting range, he could use the regularity in the number system to predict what would come next. By the end of the year, Ali could rote count to 100 with the visual support of a hundreds-chart and recognize many of the numbers up to 100. Through Choral Counting, Ali had made significant progress and, importantly, he then understood that there is a system and pattern to numbers.

12.4.5 Quick Images—Subitizing

Quick Images is an instructional activity designed to promote conceptual subitizing and to help children develop flexibility with numbers (Parrish, 2010). The teacher shows an image that has dots or objects arranged with some sort of order, e.g., eight dots within a Tens Frame. The image is only seen by the children for 2–3 seconds. Then the teacher either asks the children how many dots they thought they saw or asks them to draw what they saw. Then the teacher tells the children that they will get a chance to revise their answer before showing the image for another 2–3 seconds. Once the children are sure about how many dots they saw, the teacher shows the image again and asks the children to explain how they saw the image while the teacher records their thinking.

The teacher in a primary self-contained class laminated a number of fives and tens-frames with a variety of numbers of dots on them and carried them with her to the bus line. In this way, she could use every moment of the school day for learning. There were a number of questions she asked with these cards, including: "How many dots are there?", "How do you see the dots?", and "How many more dots do I need to make ten?" With these questions she was addressing conceptual subitizing and number bonds up to ten. With her most advanced children she could also show two cards at the same time to help them think about adding strategies above ten (Hollister, 2009).

12.4.6   Children's Literature

Books that require and develop counting skills are very helpful. For example, the book "One is a Snail, Ten is a Crab" (Pulley Sayre & Sayr, 2004), supports children's understanding of part-part-whole relationships. This is done by exploring the different numbers of legs that can be obtained by combining different animals. For example, you can make five legs by putting together a snail (one leg) and a dog (four legs). It then moves on to help children understand about counting in tens. Other books, such as, "Five Little Monkeys Jumping on the Bed" (Christelow, 1989) provide children with opportunities to count backwards as well as forwards.

Children's literature can provide very powerful ways to engage children in counting activities and in problem solving. When choosing number books to read, it is helpful to know that some researchers have found that children learn more about counting when numbers are embedded in a story (Carrazza & Levine, 2019).

When I have used children's literature, it has helped to engage children by providing a meaningful context. The great thing about children's literature is that it engages children's imagination, so the activities can be directly related to the book and the book becomes a stimulus for further mathematical activities.

12.4.7   Music

Music can provide very powerful ways for children to "feel" a rhythm and so practice counting as they move to the music. For example, children can be engaged in moving round the room to a march such as the music from "The Great Escape" (Ear8002, 2012) for practicing counting in twos and exploring multiplicative relationships. This can be developed further by exploring music with different meters. Activities like these enable children to explore and count in a dynamic, multisensory, and embodied context.

12.4.8   The Seven Principles of Working Memory Intervention

As many children who have difficulty with counting also have poor working memory, we thought it would be useful to include some tips on how to support children with poor working memory to access the curriculum. These "principles" are based on the work of Gathercole and Alloway (2008).

1. Recognize working memory failures
2. Monitor the child
3. Evaluate working memory loads
4. Reduce working memory loads when necessary

5. Repeat important information
6. Encourage use of memory aids
7. Develop the child's own strategies to support working memory

12.5   Conclusion

While counting appears to be a simple mathematical activity, it is both more involved than it seems and fundamental to number sense and arithmetic. The complexity of counting means that it causes many pitfalls to the young learner of mathematics: depending on the child's home language, the language of counting may make the sequence and structure of the number system more or less obscure; a more obscure counting system (such as the one in English) may put more of a burden on children's working memory and make it harder for them to allocate sufficient resources to other aspects of arithmetic, such as accurate tagging (one-to-one correspondence) or stopping a count down at the appropriate spot. Also, children without finger gnosis or a good sense of the relative size of numbers, will have difficulties counting accurately or knowing that their count is reasonable.

Given the issues caused by poor counting that we discussed in this chapter, it is vital to know how skillful young children are at counting and which aspects of counting they find difficult. Fortunately, there are a wide variety of assessments that can give this type of information as well as many engaging, educational activities that will give children the practice they need to become more expert counters.

References

Alibali, M. W., & DiRusso, A. A. (1999). The function of gesture in learning to count: More than keeping track. *Cognitive Development, 14*, 37–56.

Andrews, P., & Sayers, J. (2015). Identifying opportunities for grade one children to acquire foundational number sense: Developing a framework for cross cultural classroom analyses. *Early Childhood Education Journal, 43*(4), 257–267.

Back, J., Sayers, J., & Andrews, P. (2013). The development of foundational number sense in England and Hungary: A case study comparison. In *Eighth Congress of European Research in Mathematics Education (CERME 8), Antalya, Turkey, 6th to 10th February, 2013* (pp. 1835–1844).

Baroody, A. J., & Wilkins, J. L. (1999). The development of informal counting, number, and arithmetic skills and concepts. In J. V. Copley (Ed.), *Mathematics in the early years* (pp. 48–65). National Council of Teachers of Mathematics; National Association for the Education of Young Children.

Berch, D. B. (2005). Making sense of number sense: Implications for children with mathematical disabilities. *Journal of learning disabilities, 38*(4), 333–339.

Berteletti, I., & Booth, J. R. (2015). Perceiving fingers in single-digit arithmetic problems. *Frontiers in Psychology, 6*, 226.

Callahan, L. G., & Clements, D. H. (1984). Sex differences in rote-counting ability on entry to first-grade: Some observations. *Journal for Research in Mathematics Education, 15*(5), 378–382.

Carpenter, T. P., & Moser, J. M. (1984). The acquisition of addition and subtraction concepts in grades one through three. *Journal for Research in Mathematics Education, 15*(3), 179–202.

Carr, M. (2001). *Assessment in early childhood settings: Learning stories.* Sage.

Carrazza, C., & Levine, S. C. (2019). *How numbers are presented in counting books matters for children's learning: A parent-delivered intervention.* Conference talk: Society for Research in Child Development, Baltimore, MD.

Christelow, E. (1989). *Five little monkeys jumping on the bed.* Clarion Books.

Clements, D. H., & Sarama, J. (2017/2019). *Learning and teaching with learning trajectories [LT]².* Retrieved from Marsico Institute, Morgridge College of Education, University of Denver.

Cowan, R. (1987). When do children trust counting as a basis for relative number judgments? *Journal of Experimental Child Psychology, 43*(3), 328–345.

Crollen, V., Seron, X., & Noël, M. P. (2012). Is finger-counting necessary for the development of arithmetic abilities? In F. Domahs, L. Kaufmann, & M. H. Fischer (Eds.), *Handy numbers: Finger counting and numerical cognition.* Frontiers E-books.

Dantzig, T. (2007). *Number: The language of science* (4th ed.). Penguin.

Dowker, A. (2005). *Individual Differences in Arithmetic: Implications for Psychology, neuroscience and education.* Psychology Press.

Dowker, A., & Roberts, M. (2015). Does the transparency of the counting system affect children's numerical abilities? *Frontiers in Psychology, 6,* 945.

Ear8002. (2012, August 20). *"Elmer Bernstein": "The Great Escape" from Athens (X 2)* [Video]. Retrieved from https://www.youtube.com/watch?v=eeOoh2i7VCw

Frank, M. C., Everett, D. L., Fedorenko, E., & Gibson, E. (2008). Number as a cognitive technology: Evidence from Pirahã language and cognition. *Cognition, 108*(3), 819–824.

Franke, M. L. (2003). Fostering young children's mathematical understanding. In C. Howes (Ed.), *Teaching 4- to 8-year olds.* Paul H. Brookes Publishing.

Franke, M. L., Kazemi, E., & Turrou, A. C. (2018). *Choral counting & counting collections: Transforming the PreK-5 math classroom.* Stenhouse Publishers.

Fuson, K. C. (1988). *Children's Counting and Concepts of Number.* Springer.

Fuson, K. C., & Hall, J. (1983). The acquisition of early number word meanings: A conceptual analysis and review. In H. P. Ginsburg (Ed.), *The development of mathematical thinking.* Academic Press.

Fuson, K. C., & Kwon, Y. (1992). Korean children's understanding of multidigit addition and subtraction. *Child development, 63*(2), 491–506.

Fuson, K. C., & Secada, W. G. (1986). Teaching children to add by counting-on with one-handed finger patterns. *Cognition and Instruction, 3*(3), 229–260.

Gathercole, S., & Alloway, T. P. (2008). *Working memory and learning: A practical guide for teachers.* Sage.

Geary, D. C., & Hoard, M. K. (2005). Learning disabilities in arithmetic and mathematics: Theoretical and empirical perspectives. In J. I. D. Campbell (Ed.), *Handbook of mathematical cognition* (pp. 253–267). Psychology Press.

Geary, D. C., Hoard, M. K., Byrd-Craven, J., Nugent, L., & Numtee, C. (2007). Cognitive mechanisms underlying achievement deficits in children with mathematical learning disability. *Child development, 78*(4), 1343–1359.

Gelman, R., & Butterworth, B. (2005). Number and language: How are they related? *Trends in Cognitive Sciences, 9*(1), 6–10.

Gelman, R., & Gallistel, C. R. (1978). *The child's understanding of number.* Harvard University Press.

Gelman, R., & Meck, E. (1983). Preschoolers' counting: Principles before skill. *Cognition, 13,* 343–359.

Gersten, R., & Chard, D. (1999). Number sense: Rethinking arithmetic instruction for students with mathematical disabilities. *The Journal of Special Education, 33*(1), 18–28.

Ginsburg, H. (2020). *Anna counts,* DREME TE. https://prek-math-te.stanford.edu/counting/anna-counts

Gracia-Bafalluy, M., & Noël, M. P. (2008). Does finger training increase young children's numerical performance? *Cortex, 44*(4), 368–375.

Graham, T. A. (1999). The role of gesture in children's learning to count. *Journal of Experimental Child Psychology, 74,* 333–355.

Hilton, C. (2019). Fingers matter: the development of strategies for solving arithmetic problems in children with Apert syndrome. *Frontiers in Education, 4,* 131.

Hollister, A. (2009). Focused instruction on Quick Images: A guided math group video featuring Michael Flynn. In J. Storeygard (Ed.), *My kids can: Making math accessible to all learners, k-5* (pp. 38–42). Heinemann.

Hughes, M. (1986). *Children and number: Difficulties in learning mathematics.* Wiley-Blackwell.

Johnson, N. C., Turrou, A., McMillan, B., Raygoza, M., & Franke, M. (2019). "Can you help me count these pennies?": Surfacing preschoolers' understandings of counting. *Mathematical Thinking and Learning, 21*(4), 237–264. https://doi.org/10.1080/10986065.2019.1588206

Jordan, N. C., Glutting, J., & Ramineni, C. (2008). A number sense assessment tool for identifying children at risk for mathematical difficulties. In A. Dowker (Ed.), *Mathematical difficulties* (pp. 45–58). Academic Press.

Jordan, N. C., Huttenlocher, J., & Levine, S. C. (1992). Differential calculation abilities in young children from middle-and low-income families. *Developmental Psychology, 28*(4), 644.

Noël, M. P. (2005). Finger gnosia: A predictor of numerical abilities in children? *Child Neuropsychology, 11*(5), 413–430.

Nunes, T., & Bryant, P. (1996). *Children doing mathematics.* Blackwell Publishers Ltd..

Parrish, S. (2010). *Number Talks: Helping children build mental math and computation strategies.* Math Solutions.

Pulley Sayre, A., & Sayr, J. (2004). *One is a snail, ten is a crab: A counting by feet book.* Walker Books Ltd..

Ramani, G. B., & Siegler, R. S. (2008). Promoting broad and stable improvements in low-income children's numerical knowledge through playing number board games. *Child Development, 79*(2), 375–394.

Reeve, R., Reynolds, F., Humberstone, J., & Butterworth, B. (2012). Stability and change in markers of core numerical competencies. *Journal of Experimental Psychology: General, 141*(4), 649–666.

Sayers, J. (2015). Building Bridges—Making connections between counting and arithmetic: Subitising. *Primary Mathematics, 13,* 22–25.

Sayers, J., Andrews, P., & Boistrup, L. B. (2016). The role of conceptual subitising in the development of foundational number sense. In T. Meaney, O. Helenius, M. L. Johansson, T. Lange, & A. Wernberg (Eds.), *Mathematics education in the early years* (pp. 371–394). Springer.

Schaeffer, B., Eggleston, V. H., & Scott, J. L. (1974). Number development in young children. *Cognitive Psychology, 6,* 357–379.

Scherer, P., Beswick, K., DeBlois, L., Healy, L., & Opitz, E. (2016). Assistance of students with mathematical learning difficulties: How can research support practice? *ZDM, 48,* 633–649.

Siegler, R. S., & Booth, J. L. (2004). Development of numerical estimation in young children. *Child Development, 75*(2), 428–444.

Skinner, C. (1997). *Board games for the nursery (Set A).* BEAM.

Starkey, P., & Cooper, R. G. (1980). Perception of numbers by human infants. *Science, 210,* 1033–1035.

Steffe, L., & Cobb, P. (1988). *Construction of arithmetical meanings and strategies.* Springer.

Steffe, L., von Glasersfeld, E., Richards, J., & Cobb, P. (1983). *Children's counting types: Philosophy, theory, and application.* Praeger.

Strauss, E., Sherman, E. M., & Spreen, O. (2006). *A compendium of neuropsychological tests: Administration, norms, and commentary*. Oxford University Press.

Thouless, H. (2014). Counting difficulties for students with dyslexia. In S. Pope (Ed.) *Proceedings of the 8th British Congress of Mathematics Education* (pp. 319–326). University of Nottingham.

Wager, A. A., & Parks, A. N. (2016). Assessing early number learning in play. *ZDM, 48*(7), 991–1002.

Wynn, K. (1990). Children's understanding of counting. *Cognition, 36*(2), 155–193.

Chapter 13
Additive Reasoning and Problem-Solving

Yan Ping Xin and Signe Kastberg

Abstract In this chapter we illustrate how a computer-assisted intervention program, COMPS-A that integrates a constructivist view of learning and explicit teaching of mathematical model-based problem-solving, can help students with learning disabilities or difficulties in mathematics. Building on students' development of fundamental ideas such as "number as the composite unit" (which naturally leads to the part-part-whole additive relationships), COMPS-A emphasizes students' understanding and representation of mathematics relations in algebraic equations and, thus, supports growth in generalized problem-solving skills. Findings from empirical studies indicate that elementary students with learning difficulties in mathematics can be expected to move beyond concrete operations and toward thinking symbolically or algebraically. Algebraic conceptualizations of mathematical relations and model-based problem-solving can be taught through systematic strategy instruction. Introducing symbolic representation and algebraic thinking in earlier grades may facilitate a smoother transition from elementary to higher level mathematics and improve middle- and high school mathematics performance.

Keywords Learning disabilities · Learning difficulties · Elementary mathematics · Problem-solving · Computer-assisted instruction · CAI · Instructional strategies · Intervention · Word problem-solving · Model-based learning · Mathematical models · Mathematical relations · Additive reasoning · abstract level of representation · algebra readiness.

Y. P. Xin (✉)
Department of Educational Studies, Purdue University, West Lafayette, IN, USA
e-mail: yxin@purdue.edu

S. Kastberg
Department of Curriculum and Instruction, Purdue University, West Lafayette, IN, USA

During the past decade (2007–2017) the mathematical performance of students with disabilities or learning difficulties has not shown significant improvement. According to the most recent National Assessment of Educational Progress (NAEP, 2019) in mathematics, while 14% of fourth-graders without disabilities in the U.S. scored below the basic level, 50% of fourth-graders with disabilities scored below the basic level. While 25% of eighth-graders without disabilities scored below the basic level, 68% of eighth-graders with disabilities scored below the basic level. The gap between students with disabilities and their same-age peers grew. Note that students who scored at the basic level exhibited only partial mastery of prerequisite skills (NAEP, 2019). That is, over 50% fourth-graders and 68% eighth-graders with disabilities have limited prerequisite skills for fourth- or eighth-grade mathematics. The teaching of mathematics to students with diverse needs challenges teachers. Additive word problem-solving and reasoning is a foundation for learning many mathematics concepts and a core part of elementary mathematics. Many primary school students with learning disabilities/difficulties in mathematics (LDM) have limited proficiency in additive problem-solving that results in lifelong challenges with mathematical problem-solving. Quotes from our interviews (Xin et al., 2018) with elementary mathematics teachers who worked with students with diverse needs suggest that these teachers face significant challenges in their work with children with LDM.

"The biggest challenge school teachers encountered when working with students with learning difficulties is concept building." (RH, fifth-grade teacher)

"The challenge is to help students build the concept of one-to-one correspondence - This builds up the background knowledge for place value." (HS, third-grade teacher)

"They [students] do not come to us retaining what they learn from second grade [the year before]. The amount retained from [second grade] can set them back and be a difficulty for us." (AE, third-grade teacher)

"Using regular math programs, teachers do not know how to adapt them to students with special needs." (JB, fourth-grade teacher/worked with students with LDM)

Teachers who worked with a range of students at different levels of academic and problem-solving performance asserted that *"Children are not ready to learn something that I teach."* There is a need of intervention programs that focus on building conceptual understanding of fundamental mathematical ideas that will enable students with LDM to make sense of grade-level mathematics and advance at the rate of their same-age peers.

This chapter first briefly describes common practices in the United States in teaching mathematics word problem-solving. We then introduce two paradigms in solving word problems: Operational versus Relational. Next, we review recent mathematics interventions in additive reasoning and problem-solving involving students who are struggling in mathematics, followed by a description of a research-based intervention program, Conceptual Model-based Problem Solving (COMPS). Finally, we showcase a web-based mathematics problem-solving computer tutor developed on the basis of research from mathematics education and special education.

13.1   Common Practice in Teaching Elementary Word-Solving

Experienced elementary teachers who teach additive or multiplicative word problem-solving to students in the inclusive classrooms or resources classrooms often focus on *choice of operation* in teaching word problem-solving. For example, Mrs. Smith described how she taught arithmetic word problem-solving to her fourth-grade students (Xin, 2016).

Excerpt 1 (May 29, 2012):

> *"Each problem that I went through with the children I began by having a student read aloud the problem. From that point on, we always had a conversation about whether or not we should be using multiplication or division."*

Choice of operation is a distinctive feature of traditional arithmetic word problem-solving in the U.S. elementary classrooms. To choose operations to use in problem-solving, it is common to see students rely on the "keyword" strategy. Table 13.1 illustrates some of the keywords recommended by an online resource published by a commercial mathematics textbook series publisher.

As shown in Table 13.1, the resource recommends that words such as "altogether," indicate an operation, such as addition, be used; words such as "take away" indicate an operation of subtraction; and words such as "time(s)" signal an operation of multiplication; etc. The keyword strategy, which has been the practice in the United States for generations (Sowder, 1988), directs students' attention toward isolated "cue" words in the problem. The keyword strategy might be a "quick trick" that sometimes results in selecting correct operations; however, it has nothing to do with mathematics. In particular, the keyword strategy does not orient students' attention to the underlying mathematical relations necessary for constructing mathematical models that support problem-solving. Mathematical relations are patterns or structures between quantities referred to in a problem situation.

Table 13.1 Key words and choice of operations (adapted from an online resource for enVisionMath)

Addition	Subtraction	Multiplication	Division
plus	dropped /take away	product of	per
more than	lost/ fell	times	out of
combined	change	twice	sharing
altogether/in all	difference	multiplied by	each/every
total	less/less than /fewer than		

Applying the keyword strategy might contribute to students being prone to "reversal operation" errors when encountering the so-called inconsistent language problems (Lewis & Mayer, 1987; Mayer, 1999). For instance, in the problem, "Pat has 7 books. She has 4 more books than Amy. How many books does Amy have?", students might mistakenly add due to the presence of the key word "more," when they need to subtract to find the solution (Xin, 2007; Xin et al., 2011). Yet the target of this problem is a relationship between the quantity of books Pat has and the quantity of books Amy has.

Even when teachers have the opportunity to consider teaching approaches other than the keyword strategy, they may continue to rely on teaching key words to improve performance with additive word problems. A teacher enrolled in an online special education methods course as part of a master's program exhibited this perspective in a weekly reflection post after reading about research-based word problem-solving intervention strategies (e.g., cognitive heuristic strategies, model-based problem-solving [see later in the chapter for model-based problem-solving]): "*after all, at the end of the day, it is still the keyword strategy!*"

Unfortunately, reliance on key words as the most efficient instructional strategy for word problem-solving is common. Many pre-service and in-service teachers firmly believe that the "keyword strategy" is THE strategy to teach mathematics word problem-solving. It may take generations to undo or shatter this "robust" yet ill-rooted strategy (Cathcart et al., 2006). This chapter seeks to address this reliance by providing an alternative instructional strategy shown to support students' additive word problem-solving performance and conceptions.

Another strategy commonly used in teaching word problem-solving includes "draw a picture." Students are encouraged to draw objects or use cubes /blocks to represent the actual objects described in the word problem. Students then use counting by ones of the drawn objects as the primary strategy for solving the problem. While, there are a range of "pictures" that students could draw, not all pictures help students correctly solve problems. For instance, Walker and Poteet (1988–1989) compared a "picture drawing" intervention condition with the traditional word problem-solving instructional condition (Polya's four-step problem-solving process: Read/understand the problem, develop a plan, solve, and look back and check, Polya, 1957). The picture drawing condition asked students to draw a diagrammatic representation of the story, write a number sentence from the representation, and solve the number sentence. After seventeen 30-minute instructional sessions students' mathematics word problem-solving performance decreased from pretest to posttest. Results further indicate that if the student-drawn pictures do not illustrate mathematical relations described in the word problem, the pictures may not help students generate solution plans or mathematics equations for accurate problem-solving (Gersten et al., 2009).

Yet another strategy commonly used by teachers and students is called "guess and check." Greer (1992) described this strategy as: "look at the numbers; they will tell you which operations to use. Try all the operations and choose the most reasonable answer" (p. 28). Excerpt 2, from a survey of elementary teachers' approaches to teaching of problem-solving strategies, illustrates teachers' use of guess and check.

Excerpt 2 (May 29, 2012):

- Edwin received a total of $374 to buy basketballs for the basketball team. Each basketball costs $34. How many basketballs can he buy?

"Since this problem needed to use division that involved two-digits, it posed quite a challenge for my fourth-graders. We did a lot of *guessing and checking* as we worked through a division problem of this sort. ......" (Xin, 2016)

"Another thing I try to stress to them is whether or not our numbers for our answer are going to be increasing (multiplication) or decreasing (division)." (Xin, 2016)

According to existing literature in mathematics education, children's use of "guess and check" strategies to solve problems can be a foundation for more efficient strategies. Yet teachers like those cited in Xin (2016) seem to encourage students to use inefficient approaches and even "rules that expire" (Karp et al., 2014). Students using guess and check, beyond as a means to build intuition about operations, will work more slowly than peers using more efficient methods.

Often, when the numbers in the problem are small, it might be manageable for students to correctly solve the problem using the "draw a picture" or "guess and check" strategy. However, when the numbers become large, such problem-solving process may become cumbersome or inefficient. More importantly, in isolation and without further classroom dialogue, none of these strategies focuses on an analysis of underlying mathematical relations in the word problem and representing such relations in mathematical models at a symbolic level for solution.

To summarize, the illustrations of the above-described strategies share a common feature: selection of operation was the focus of the problem-solving reasoning and process. According to Polotskaia and Savard (2018), there are two paradigms in mathematics problem-solving, the "Operational Paradigm" and the "Relational Paradigm" (p. 70). The strategies teachers used in excerpts 1 and 2 were aligned with the operational paradigm as they were choosing an operation to use with the numbers in the problem to produce a solution. Below we define and discuss the two distinctive paradigms based on the work by Polotskaia and Savard (2018).

13.2 Two Paradigms in Solving Word Problems: Operational Versus Relational

Operational Paradigm The *operational paradigm* focuses on semantic analysis of the word problem involving transforming the words of the problem into an arithmetical operation (Polotskaia & Savard, 2018). "This approach sees a word problem as numerical data connected by a semantic link in a story to be transformed into numbers and connected by an operation." (p. 72).

Historically, it has been commonly believed that selecting and applying an appropriate arithmetic operation is crucial to success in problem-solving. Commonly used teaching practices including the "*keyword*" strategy focus on translating "cue" words into operation signs (see Table 13.1). Most intervention strategies developed for helping students with LDM to solve word problems focus on translating word problem "storyline" into numbers connected with an operation sign. For instance, stories such as "making purchases" or "spending money" would be associated with or translated into a "minus" sign in a mathematics sentence or equation. The operational paradigm relies on learners' development of understanding that is *driven by arithmetical operations* rather than mathematical relationships (Polotskaia & Savard, 2018). According to scholars in mathematics education (Thompson, 1993), focusing on the operation (whether to "add" or "subtract," for instance) relevant to calculation is likely to distract learners from mathematical relations depicted in the word problem. In contrast, it is the mathematical relation, on which the mathematical model is conceptualized, that contributes to generalized problem-solving (Jonassen, 2003).

Relational Paradigm The *relational paradigm* is built on the law of composition in that any number can be written as a combination of any two smaller numbers, and "the relation between two elements determines a unique third element as a function" (Davydov, 1982, p. 229). This law of composition is the basis for addition and subtraction (Polotskaia & Savard, 2018). Davydov (1982) argues that the law of composition is the root of relationships between quantities.

With the relational paradigm, learners need to read the *entire* story to understand the underlying mathematical relation in the word problem. Given an additive word problem situation, for instance, "Rachel had some flowers in a big vase. Then, 9 of the flowers wilted, so she took those flowers out. There were 27 flowers left in the vase. How many flowers were in the vase in the beginning?" learners need to understand that the story is about the part-part-whole relation between the sets or units described in the word problem context (i.e., # of flowers wilted plus # flowers left in vase = total # of flowers in the vase). Under the relational paradigm, the mathematical reasoning process can be described as moving from understanding relationships between physical objects (without focusing solely on numbers) toward a *holistic* understanding of the additive structures present in word problems (Polotskaia & Savard, 2018).

To summarize, it is certainly important for students to understand the word problem story or task in the real-world context. It is also important for the learners to represent and describe the underlying mathematical relations (e.g., additive or multiplicative relations) before applying an operation to solve the problem.

13.3 Recent Intervention in Additive Reasoning and Problem-Solving

Literature syntheses and meta-analyses have explored instructional interventions for teaching students with LDM mathematics problem-solving strategies. For instance, Zheng et al. (2012) reviewed reports of seven group and eight single-subject design studies (1986–2009) in teaching word problem-solving to students with mathematics disabilities. They identified key instructional elements common across effective intervention programs, including (a) clear instructional objectives, (b) advanced organizers, (c) providing needed instructional scaffolding and fading the prompts gradually, (c) explaining underlying concepts with elaborations, (d) modeling the skills to be taught, (e) engaging students in discourse-oriented mathematics instruction (encourage questioning, engaging students in dialogues), (f) sequencing instructional/learning activities, and (g) breaking the task into smaller components or steps. These instructional elements provide support (such as introducing and fading prompts) when used appropriately in concert with emphasis on student activity and reasoning.

Zhang and Xin (2012) conducted a meta-analysis of reports of word problem-solving intervention studies for children with LDM published from 1996 to 2009. This study analyzed 29 group comparison and 10 single-case design studies that investigated word problem-solving interventions for students with LDM. This study addressed the relative effects of three categories of word problem-solving instructional strategies: problem structure representation techniques, cognitive strategy training, and technology. Findings indicated that problem structure representation resulted in the highest *effect size*[1] (ES = 2.64, calculated from group design studies), compared to cognitive strategy training (ES = 1.86) and assistive technology (ES = 1.22). As indicated by the authors of this meta-analysis study, the key instructional techniques for representing problem structure were explicit instruction and conceptual modeling.

Hughes et al. (2014) conducted a meta-analysis of algebra interventions involving students with LDM. This study evaluated 12 mathematics intervention studies (1983–2013) involving elementary students and secondary students with LDM. It examined instructional-related theoretical frameworks in the context of learning theories, which include cognitive learning theory, model-based interventions, as well as other instructional elements such as co-teaching, concrete-representational-abstract (CRA) instructional sequences, graphic organizers, and technology-based interventions. The findings from this meta-analysis study indicate that the interventions were effective overall at elementary and secondary levels. In particular, results indicate that applied cognitive or model-based interventions had the strongest effectiveness (ES = 0.68) in improving students' performance on solving algebra mathematics problems. The meta-analysis study also indicated that CRA, which included a gradual instructional sequence moving from concrete to semi-concrete and abstract level of representations, resulted in a moderate effectiveness (ES = 0.52). This finding is significant since Common Core State Standards for Mathematics (CCSSM, 2010) has emphasized the importance of algebraic thinking in the elementary grades.

---

[1] Effect size (ES) is used to measure the effectiveness of a treatment. EF is defined as the standardized mean difference between the experimental and the control group (Glass et al., 1981; Hedges & Olkin, 1985). An ES of 2.00 means 97.5% of students in the experimental group (e.g., received instruction in mathematical model-based representation and solving) performed better than the control group.

More recently, Kim and Xin (2022) conducted a comprehensive review of reports of empirical research studies published between 1981 and February 2019 in which computer-assisted mathematics word problem-solving interventions were used to facilitate the learning of mathematical word problem-solving of elementary and secondary school students with LDM. This study examined 13 studies including ten group comparison design studies and three single subject design studies, guided by three theoretical frameworks: behavioral-oriented instructional strategies, cognitive learning theory-based strategies, and model-based learning approaches. Findings from this review indicate that computer-assisted intervention programs designed on the basis of model-based learning approaches demonstrated the largest effect size (ES = 1.25), when compared to cognitive strategy instructional programs (ES = 1.02) or behavioral-oriented interventions (ES = 0.32).

In summary, findings from these research syntheses and meta-analyses support the use of model-based intervention programs. The problem structure representation strategy, which was found most effective in Zhang and Xin's (2012) meta-analysis, is similar to the model-based interventions classified in the Hughes et al. (2014) study. Both emphasize representing mathematical relations in model equations. To conclude, the findings from meta-analyses conducted by Kim et al. (2022), Hughes et al. (2014), and Zhang and Xin (2012) favor model-based interventions for word problem-solving instruction for students with LDM.

13.4 Model-Based Learning

According to *Principles and Standards for School Mathematics* (National Council of Teachers of Mathematics, 2000):

"Instructional programs from prekindergarten through grade 12 should enable each and every student to—

- Understand patterns, relations, and functions
- Represent and analyze mathematical situations and structures using algebraic symbols
- Use mathematical models to represent and understand quantitative relationships
- Analyze change in various contexts" (https://www.nctm.org/Standards-and-Positions/Principles-and-Standards/algebra/)

Central to this recommendation is the use of mathematical models to represent and understand quantitative relationships. In a more recent related effort to situate K-12 STEM education more centrally, model-based learning has been emphasized (Seel, 2017). According to Seel (2017), "helping students to develop powerful models should be among the most important goals of mathematics instruction" (p. 932). Indeed, mathematical modeling is emphasized as an essential practice in the U.S. Common Core Mathematical Practice Standards (Common Core State Standards Initiative, 2012).

13.4.1   Mathematical Modeling

Blum and Leiss (2005) provide a framework for modeling (see Fig. 13.1). In this modeling cycle, one must (a) read and understand the task, (b) structure the task and develop a real situational model, (c) connect it to and/or represent it with a relevant mathematical model, (d) solve and obtain the mathematical results, (e) interpret the mathematical results in a real problem context, and (f) validate the results (either end the task or re-modify the mathematical model if it does not fit the situation). In light of research in mathematics education, many students have difficulty making the transition from a real *situational model* to a *mathematical model*. This is often a weak area in students' mathematical understanding (Blomhøj, 2004).

Modeling involves translation or representation of a real problem situation into a mathematical expression or model. Mathematical models are an essential part of all areas of mathematics, including arithmetic, and should be introduced to all age groups including elementary students (Dundar et al. 2012). It should be noted that engaging students in the modeling process does not necessarily mean they become engaged in the discovery or invention of mathematical models or complex notational systems. According to Lesh et al. (2003), however, it does mean that when such models or systems are provided by the teacher, "the central activities that students need to engage in is the unpacking of the meaning of the system" (p. 216), representation of the real problem situation in a mathematical expression or model, and the flexible use of the model to solve real-world problems.

In this chapter, we briefly introduce the Conceptual Model-based Problem-Solving approach (Xin, 2012) which supports the relational paradigm. Then we showcase a computer-assisted intervention tutor to nurture additive reasoning and problem-solving through constructivist-oriented learning activities and conceptual model-based problem-solving.

Fig. 13.1 Modeling cycle (Blum & Leiss, 2005)

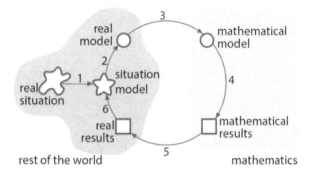

13.4.2 Conceptual Model-Based Problem-Solving (COMPS)

COMPS (Xin, 2012) has emerged as an effective intervention approach that promotes mathematical model-based problem-solving. With the COMPS approach, students develop a structural understanding of mathematical relationships in word problems and construct a cohesive mathematical model by creating and working with meaningful representation (Brenner et al., 1997).

Representations that model an *underlying mathematical structure* serve to facilitate solution planning and accurate problem-solving. For instance, Part + Part = Whole (PPW) is a cohesive mathematical model that generalizes various additive word problem situations. Table 13.2 illustrates sample representations of a range of additive word problems in the PPW mathematical model equation.

The COMPS program reflects a shift away from traditional problem-solving instruction that focuses on the *choice of operation* to compute solutions, toward a mathematical model-based problem-solving approach that emphasizes an understanding and representation of mathematical relations in algebraic equations. As seen in Table 13.2, the COMPS approach connects elementary arithmetic to algebra learning by promoting mathematical model-driven reasoning and problem-solving. That is, rather than "gambling" on what operation to use for the solution, the COMPS approach asks that students represent the word problem in the PPW mathematical model equation (e.g., $P + P = W$, the unknown quantity will be represented by a letter). Then the problem will be solved through finding the unknown quantity in the equation. Now, the challenge is, how to help students construct the PPW mathematical model. As we have shared in the beginning of this chapter, many students do not come to class ready to construct this abstract PPW model. Before we get into the "how" part, we first dissect the additive word-problem structure.

13.5 Additive Word-Problem Structure and Its Variations

The additive problem structure includes a range of *Part-Part-Whole* and *Additive Compare* problem structures. A *Part-Part-Whole* (PPW) problem describes an additive relation between multiple parts and the whole, that is, "Part and Part make up the Whole." It includes problems such as combine (e.g., *Christine has 5 apples. John has 4 apples. How many apples do they have together?*), change-join (e.g., *Christine had 5 apples. John gave her 4 more apples. How many apples does Christine have now?*), and change-separate (e.g., *Christine had 9 apples. Then she gave away 4 apples. How many apples does she have now?*) (Van de Walle, 2004). The unknown quantity can be either the *part* or the *whole*. An *Additive Compare (AC)* problem compares two quantities and involves a compare sentence that describes one quantity as "more" (AC-more) or "less" (AC-less) than the other quantity (e.g., *"Christine has 9 apples. She has 5 more apples than John. How many apples does John have?"* or *"Christine has 9 apples. John has 4 less apples than Christine. How many apples does John have?"*). The unknown can be the *big*, *small*, or *difference* quantity (see variations of additive word problems in Table 13.2).

Table 13.2 *Additive word problem situations and corresponding model representation* (adapted from Xin, 2012)

Part-Part-Whole (PPW) Problem Situations	Sample Model Representation
1. Amy has 3 books. Pat has 4 books. How many books do they have all together? *(Combine)* 2. Amy had 3 books. Pat gave Amy 4 more books on her birthday. How many books does Amy have now? *(Change-join)* 3. Amy had some books. She gave Pat 4 books, and found that she only has 3 books left. How many books did Amy have in the beginning? *(Change-separate)*	Part Part Whole $3 + 4 = a$
4. Amy and Pat have found that together they have a total of 7 books. Pat says that she has 4 books. How many books does Amy have? *(Combine)* 5. Amy had 7 books in her collection. She gave away some books. Now she has only 4 books left. How many book did she gave away? *(Change-separate)* 6. Amy had some books. Pat gave Amy 4 more books. Now Amy has 7 books. How many books did Amy have in the beginning? *(Change-join)* 7. Amy had 7 books. Then she lost 4 books. How many books does Amy have now? *(Change-Separate)*	Part Part Whole $a + 4 = 7$
8. Amy and Pat have found that together they have a total of 7 books. Amy says that she has 3 books. How many books does Pat have? *(Combine)* 9. Amy had 3 books. Then Pat gave her some books. Now Amy has 7 books. How many books did Pat give Amy? *(Change-join)*	Part Part Whole $3 + a = 7$
Additive Compare Problem Types	*Additive Compare*
10. Amy has 3 books. Pat has 4 more books than Amy. How many books does Pat have? *(Compare-more)* 11. Amy has 3 books. She has 4 fewer books than Pat. How many books does Pat have? *(Compare-less)*	Part Part Whole $3 + 4 = a$ Smaller Difference Bigger
12. Pat has 7 books. She has 4 more books than Amy. How many books does Amy have? *(Compare-more)* 13. Pat has 7 books. Amy has 4 fewer books than Pat. How many books does Amy have? *(Compare-less)*	Part Part Whole $a + 4 = 7$ Smaller Difference Bigger
14. Pat has 7 books. Amy has 3 books. How many more books does Pat have than Amy? *(Compare-more)* 15. Pat has 7 books. Amy has 3 books. How many fewer books does Amy have than Pat? *(Compare-less)*	Part Part Whole $3 + a = 7$ Smaller Difference Bigger

As seen in Table 13.2, *part*, *part*, and *whole* are the three basic quantities in the PPW model equation (e.g., P + P = W). It should be noted that the three basic elements in the PPW model equation will have unique denotations when a specific problem subtype applies. For example, in a *combine* problem type, "*Amy has 3 books. Pat has 4 books. How many books do they have all together?*" (see Table 13.2, problem 1), the number of books Amy has and number of books Pat has are the two *parts*; these two parts make up the combined amount (i.e., "all together") or the *whole*. In contrast, in a *change* problem type, "*Amy had 7 books. Then she lost 4 books. How many books does Amy have now?*" (see Table 13.2, problem 7), the number of books Amy had in the beginning is the *whole* (7), whereas the number of books Amy lost (4) and the number of books she has now (*a*) are the two *parts* that make up the whole or the beginning amount (7).

Similarly, in an *Additive Compare* problem (AC-More), *"Amy has 3 books. Pat has 4 more books than Amy. How many books does Pat have?"* (see Table 13.2, problem 10), the number of books Amy has (the smaller quantity "3") and the difference between Pat and Amy (difference quantity "4") make up the total, that is, the number of books Pat has (the bigger quantity which is the unknown "*a*," $3 + 4 = a$). Or in other words, the number of books Pat has is the same as the number of books Amy has plus the difference quantity ($a = 3 + 4$). In contrast, in an AC-Less problem, for instance, *"Amy has 3 books. She has 4 fewer books than Pat. How many books does Pat have?"* (virtually the same problem but worded differently, see Table 13.2, problem 11), the number of books Pat has (the bigger quantity which is the unknown "*a*") equals the number of books Amy has (the smaller quantity "3") plus the difference quantity (i.e., 4).

13.6 Instructional Phases

After we understand the additive word problem structure, we are ready to talk about the instructional phases in promoting students' construction of the mathematical model, that is, *Part and Part makes up the Whole* ($P + P = W$). Instruction in the COMPS program focuses on: *mathematical model representation* and word *problem-solving*. During the instruction of *mathematical model representation*, word stories with no unknowns are used to help students understand the mathematical structure and relations among the quantities. Specifically, students learn to decontextualize the word problem story and identify the three key elements i.e., *part, part, and whole*. Students then map the three elements from the word problem story to its corresponding COMPS diagram equation (see sample representations in Table 13.2, last column). During that stage, as all the quantities are given in the story (no unknowns) students can check the truth value of the equation to see if they have mapped the three elements correctly onto the mathematical model equation. This process also shapes and/or reinforces the concept of "equality" and the meaning of the equal sign.

13.6.1 Mathematical Model Representation

To facilitate students' representation of a word problem story onto the PPW mathematical model equation, Xin (2012) has created a set of word problem story grammar questions as linguistic scaffolding to support student's construction of the PPW mathematical model.

While the core of the COMPS program is its explicit teaching of mathematical model-based word problem-solving, it also provides linguistic scaffolding that draws students' attention to "word problem (WP) story grammar" (Xin et al., 2008) that is directly linked to the mathematical model representation. According to literature in reading comprehension (Rand, 1984), story grammar addresses the elements of a story. It involves a set of expectations or knowledge about the internal structure of stories that makes both comprehension and recall more efficient. Consistent use of the same questions about stories (e.g., by asking a series of story grammar questions regarding who, what, where, when, and why) equips students with a framework they can apply on their own (Gurney et al., 1990). Research demonstrates that explicit instruction in both story grammar and story mapping (organizing and representing the internal structures of stories) has positive effects on the reading comprehension of students with/without learning disabilities (e.g., Boulineau et al., 2004). Borrowing the concept of story grammar from reading comprehension literature, *WP Story Grammar* (Xin et al., 2008), denotes mathematical structure and relations that are common across a range of word problem situations. As shown in Fig. 13.2b, WP story grammar prompting questions, *"Which sentence describes one quantity as "more" or "less" than the other? Who has more, or which quantity is the bigger one? Who has less, or which quantity is the smaller one? Which sentence tells about the bigger quantity? Which sentence tells about the smaller quantity?"*, are used as a linguistic scaffold to facilitate students' comprehension of the word problem and the mathematical relations involved. These questions promote a correct representation of information from the word problem onto the mathematical model equation that is then used to solve the problem.

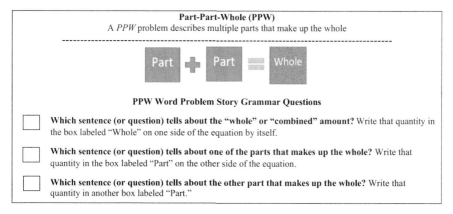

Fig. 13.2 (**a**) Conceptual model of PPW word problems. (Adapted from Xin, 2012). (**b**) Conceptual model of *AC* word problems. (Adapted from Xin, 2012)

Fig. 13.2   continued

13.6.2   *Mathematical Problem-Solving Using COMPS*

The problem representation instructional phase will be followed by the mathematics word problem-solving phase. During the problem-solving phase, students will be presented with real-world problems with an unknown quantity. When representing a problem with an unknown quantity in the mathematical model equation, students can choose to use a letter (any letter they prefer) to represent the unknown quantity. Xin (2012) created a DOTS checklist (see Fig. 13.3) to guide the problem-solving process for students with LDM. The DOTS checklist serves as a meta-cognitive strategy to help students regulate or monitor the problem-solving process to curtail students jumping to circling keywords and/or using guess and check strategies with numbers in the problem and gambling on the operation to use. This DOTS checklist can be modified depending on the type of word problems students are learning to solve. As this chapter only deals with additive word problems, the DOTS checklist presented here focuses on additive mathematical structure and problem-solving.

For students with LDM, explicit strategy modeling and explanation may be important. Below we provide a sample teacher–student interaction when the teacher is guiding the student through the process of solving an additive *compare* problem with *referent* unit as the unknown (involving "inconsistent language"— that is, the word "fewer" in the problem does not signify the operation of subtraction).

DOTS Checklist
• **D**etect the additive problem structure and mathematical relations.
• **O**rganize the information in PPW mathematical model equation
• **T**ransform the equation, as needed, to isolate the unknown quantity.
• **S**olve for the unknown quantity in the equation and check your answer.

Fig. 13.3 DOTS checklist. (Adapted from Xin, 2012)

Kaylin has 18 candies. She has 6 fewer candies than Melody.
How many candies does Melody have?
(Students read the problem together.)
Teacher: What is this problem about?
Students: This problem compares the number of candies Kaylin has to the number of candies that Melody has.
Teacher: Correct! It is a *comparison* problem that describes one quantity as *more* or *less* than the other quantity.
(Teacher presents the Additive Compare *Word Problem Story Grammar* Poster, see Fig. 13.2b) Can you tell which sentence tells one quantity is more or less than the other?
Students: Kaylin has 6 fewer candies than Melody.
Teacher: That is correct. This comparison sentence tells us that the number of candies Kaylin has is *less* than the number of candies Melody has. Let's underline this comparison sentence!
Students: Underlines the sentence in the problem.
Teacher: What is the difference between the number of candies Kaylin has and the number of candies Melody has?
Students: Kaylin has 6 candies less… The difference between the two is "6."
Teacher: Let's write the *difference* amount 6 in the PPW diagram, the box labeled as "*difference.*"
Students write 6 in the box labeled as *difference.*
Teacher: From the comparison sentence: "Kaylin has 6 fewer candies than Melody," can you tell me who has more, and who has less?
Students: "Kaylin has 6 fewer ……," so Kaylin has less, and Melody has more.
Teacher: Superb! Please help me name the box for "smaller" and name the box for "bigger" on the PPW diagram. (Note: The "name tags" are used as visual scaffolds/anchors to help student organize the information)
Students: Names the box for "smaller" as Kaylin and names the box for "bigger" as Melody (See below sample diagram equations. Based on student's prior knowledge, teacher may consider different ways of representing the information in the PPW diagram equation to promote mental flexibility).

Teacher: What number we will write in the box for "Kaylin" in the PPW diagram equation?

Students: "Kaylin has 18 candies," so we will write "18" in the box for Kaylin.

Teacher: Great! Now what number we will write in the box for "Melody"?

Students: ……

Teacher: Do we know the number of candies Melody has?

Students: No.

Teacher: Correct, that is the unknown quantity we are asked to find out.

Let's write letter "*a*" to represent the unknown quantity in the box for Melody in the diagram equation.

Students: Write letter *a* in the box for Melody.

Teacher: Now we have completed the mapping of information onto the PPW diagram equation. (Teacher points to diagram equation. See below) "It tells that, the number of candies Kaylin has PLUS the difference amount (6) EQUALS the bigger quantity—that is, the number of candies Melody has."

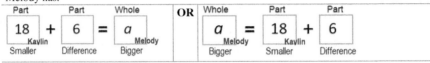

Teacher: To solve for the unknown quantity *a*, we will simply rewrite the math equation as:

$a = 18 + 6$

Can someone help solve for the unknown quantity *a*? 18 plus 6 equals?

Students: 24.

Teacher: 24 what?

Student: 24 candies.

Teacher: Correct! What is your complete answer to the problem? (Teacher may remind students to check back the word problem and provide prompt as needed, such as "what are we asked to solve for?")

Students: Melody has 24 candies.

Teacher: Great! please write down your answer in your worksheet. I will write the answer "24" above letter "*a*" in the equation. Does our answer "24 candies" make sense?

Students: The underlined sentence (in the problem) says "Kaylin has 6 fewer candies than Melody," so Melody has more candies than Kaylin. We know Kaylin has 18 candies, and our answer 24 is more than 18, so our answer "Melody has 24 candies" seems to make sense.

Teacher: Good thinking! That is correct, "Melody has 24 candies; 24 is *6 more* than 18, the number of candies that Kaylin has. (Teacher should point to corresponding numbers in the diagram equation while confirming students' thinking), or 24 equals 18 plus 6!

Depending on the needs of each individual student with LDM, the teacher can decide on proper instructional scaffoldings supported by explicit strategy explanation, guided practice, performance monitoring with corrective feedback, and independent practice. During the beginning of the instruction, concrete bar models (made of unifix cubes or unit blocks) and then semi-concrete bar models (on paper) can be used to demonstrate the part-part-whole relation and help students make the transition to the abstract PPW model equation. Figure 13.4 illustrates the bridging between the bar model and the part-part-whole model equation. Existing research has identified the importance of movement from physical models to mathematical ones in supporting algebraic thinking and learning (Witzel, 2005).

Finally, after students' acquisition of solving word problems with a specific semantic structure, such as PPW problem structures, it is important to present students with a mixture of problems providing them with opportunity to apply proper

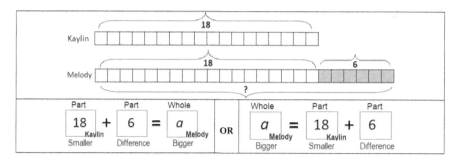

Fig. 13.4 Use bar model (upper panel) as a scaffold to facilitate students' understanding of the Part-Part-Whole model equation (lower panel)

strategies/diagram equations to solve a range of additive word problems (e.g., PPW as well as additive compare word problems). Please refer to the Appendix for sample problems with mixed problem structures.

13.7 Computer-Assisted Intelligent Tutor: COMPS-A

In this section, we will showcase a web-based mathematical problem-solving tutor, COMPS-A (Xin et al., 2015) developed through the support from U.S. National Science Foundation. The goal of the COMPS-A program was to help elementary students construct mathematical model-based reasoning and problem-solving. The program includes two modules. Module A nurtures children to develop more sophisticated ways of counting, as a spring board to the construction of part-part-whole relations. Module B systematically promotes student's construction of the PPW model equation for generalized problem-solving.

13.7.1 Module A: Nurturing the Fundamental Mathematical Idea of Number as a Composite Unit

The goal of Module A is to nurture students' construction of the composite unit and additive double counting, which serve as the foundation for the development of additive reasoning and problem-solving.

The primary cognitive structure that underlies a child's conceptualization of additive part-part-whole problem situations is what Steffe et al. (1983) termed the *composite unit*. Additive reasoning requires students to operate with numbers as composite units (Tzur et al., 2013). Composite units (unit composed of unit items) are the result of mental operations developed through challenges to counting acts. A child develops a progressively more sophisticated system of counting behaviors (Olive, 2001) to address challenges to their counting acts. Through counting by ones children iterate the unit of one to create larger units (e.g., $1 + 1 + 1 = 3$). Gradually, the nested nature of the resulting, composed quantity becomes explicit (e.g., $(1 + 1 + 1) + 1 = 4$; $4 + 1 = 5$, etc.). When number is conceived of as a composite unit, a child can join a unit of 3 and a unit of 4, for instance, to compose a larger unit of 7; they can also break or decompose 7 into 3 and 4 (or 1 and 6, 2 and 5, 4 and 3, 5 and 2, and 6 and 1). These actions create relationships between quantities and numbers. Eventually, a child can think of $7–3 = ?$ as $3 + ? = 7$ for instance. That is, a composite unit of 7 (whole) is comprised of a unit of (3) and another unit that the child can find. With composing and decomposing of quantities, a key aspect of additive reasoning, the child can understand the part-part-whole (PPW) conceptual model (Fuson & Wills, 1988).

In particular, for Module A, the curricular sequence was developed based on the work of Steffe et al. (1983), Wright et al. (2007), and Cobb and colleagues (Cobb & Wheatley, 1988; Cobb & Merkel, 1989). The focus of the curriculum in the module is the development of the composite unit for whole numbers to 100. The goal of the curriculum in module A is to challenge children's counting acts and provoke changes to the children's mental operations resulting in the development of the composite unit. Most children in grades 2 and 3 will have achieved some or all of the stages described by Olive (2001) through "internalization" (p. 5) of the results of their counting acts. To provoke development, the curriculum in Module A uses a series of uncovered and covered collections with and without structure (e. g. dots in ten frames versus random dots).

The child can work with collections where one quantity is covered and can generate a total when given more units demonstrating her development of numerical composites. Numerical composites allow children to take the numeral 7 as a representative of a collection of seven objects. Finding a total using a numerical composite is not yet evidence of a composite unit in which three quantities are related. For example, students with numerical composites can find a total when three additional units are added to a covered collection of seven units, but may not yet have the flexibility to know that the numerical composite of ten contains the composites of three and seven.

Module A is guided by a constructivist view of mathematics learning and includes a series of activities involving the use of virtual manipulatives, such as unifix cubes and coverings. Module A consists of seven lessons, the first lesson is used to determine where a student should start based on their needs. Lessons A2–A5 involve the development of composite units from 2 to 100 using ones, fives, and tens. Lesson A2 focuses on movement from perceptual to figurative counting to support the student to create mental images of quantities less than 10. Lesson A3 focuses on building a visual structure for quantities up to 10 using five and ten frames. For example, students are shown 6 cookies and the quantity is covered. Students are then asked how many cookies are needed to make 10 cookies. Lesson A4 focuses on building counting using covered collections that promote counting on. Lesson A5 focuses on building tens and ones structure for numbers up to 20 using ones, fives, and tens. Ten is presented as two collections of five to support counting by fives toward building mental images of ten.

Lesson A6 and A7 support students to use the tens and ones structure of a number as a quantity. Lesson A6 presents quantities as multiples of ten and asks students to determine the result of adding or subtracting a multiple of ten. For example, 50 is presented as five rods of ten. Students are then asked how many units are added to generate a quantity of 85. Numbers in problem situations are represented as either numerals or quantities to encourage learners to build connections between mental representations of quantities, number words, and numerals. Finally, in Lesson A7 quantities presented are not multiples of ten, although still less than 100, and are presented as either rods and singles, numerals, or using number words. This lesson provides opportunities to work from a number not at a decade (such as 74) and determine the result when a given quantity is added or subtracted or an unknown quantity is added or subtracted and the resulting quantity is provided.

Lessons A1–A5 are situated in a bakery context (see sample screenshots in Fig. 13.4). Lessons A6 and A7 involve building connections between quantities and numerals as representations of tens and ones. To support student transition from the bakery context-based addition and subtraction and the PPW models used in module B, lessons A6 and A7 are presented in quantities of tiles and numerals rather than the bakery context.

To illustrate, in Fig. 13.5, the assessment items in lesson A1 demonstrate the computer environment that asks the student to create different quantities of biscuits using the quantities in the stock room. Once the stock room is stocked with boxes of biscuits, challenges are presented in which a number of biscuits is given and then covered. Students are asked to bring additional biscuits and asked how many biscuits will be in the box.

For example, in lesson A2 in Fig. 13.5, three biscuits are given and the student is asked to bring two more. The process of bringing quantities to the display case to create new quantities provides opportunities for children to use counting strategically (see Fig. 13.5). Covers for the quantities can be removed by the students, but will only be shown for a few seconds to encourage the development of mental representations of the quantities. Structured and unstructured collections are also provided to support students to build visual models of small collections with the same numerosity and assign numerals to them as in the example from lesson A3 in Fig. 13.5.

A series of uncovered and covered collections are used to develop numerical composites. Problems with a covered collection are given to students for them to figure out a total when given more units. The development of composite units includes further restricting countable items by covering both collections of units. The lessons A6 and A7 involve the development of composites units using the tens and ones structure. Lesson A6 and A7 involve building connections between quantities and numerals as representations of tens and ones. The tools in Lessons A6 and A7 are building blocks that are used to explore changes from a given quantity or number to another given quantity or number.

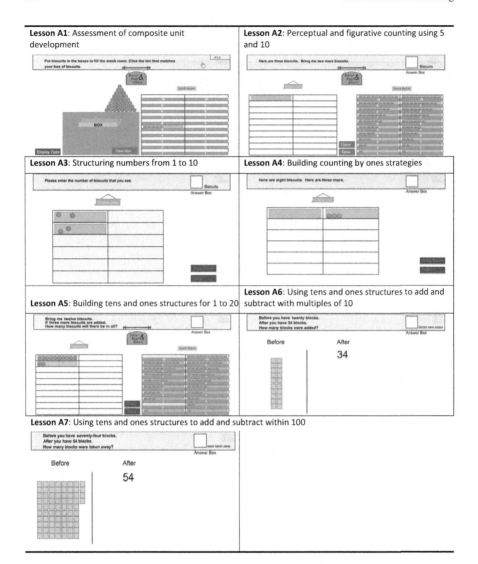

Fig. 13.5 Sample screenshots Lessons A1 to A7 of Module A in the COMPS-A computer tutor (©COMPS-RtI, Xin et al., 2015)

13.7.2   *Module B: Additive Word Problem Representation and Solving*

Module B, developed based on Xin (2012), links the two ideas: the concept of composite unit, from Module A, to the additive mathematical model, "*Part* and *Part* makes up the *Whole.*" As students encounter real-world mathematical word problems, the tutor program also provides them with bar models to facilitate their

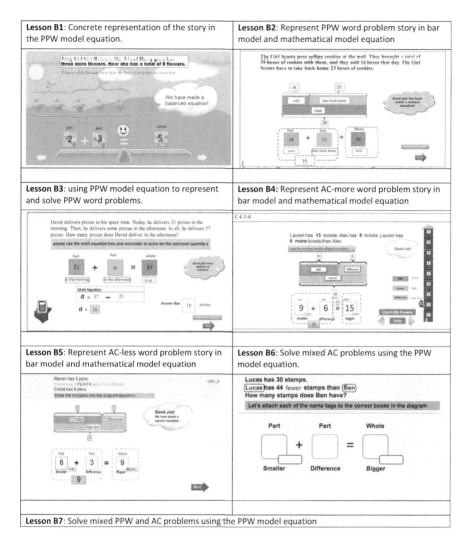

Fig. 13.6 Sample screenshots Lessons B1 to B7 of Modul B in the COMPS-A computer tutor (©COMPS-RtI, Xin et al., 2015)

conceptual transition from the isolated unit of ones to the composite unit (CU) that involves multiple individual ones; the bar model also facilitates students' conceptual transition from a semi-concrete to a symbolic level of operation when using mathematical model equations to represent and solve simple part-part-whole (PPW) word problems. Figure 13.5 lesson B2 illustrates the connection between the bar-model representation and the PPW mathematical model equation representation.

In Lesson B1, the computer tutor provides the learner with concrete picture representations (see Fig. 13.6 for Lesson B1 screenshot) to help students understand the mathematical relations described in the word problem story. This relation is then expressed by the Part + Part = Whole (PPW) model equation. Emphasis is also given to the meaning of the equal sign in the mathematics equation; a balancing scale is used to help students understand that the combined numerical value from the left side of the equation should be the same as that on the right side of the equation.

In Lesson B2, bar models are used to facilitate students' transition from semi-concrete bar model representation to abstract PPW model equation representation. Students represent the word problem stories (with no unknown quantities) in the Part-Part-Whole diagram equation (i.e., P + P = W). After that, students also check the "balance" of the equation to verify the accuracy of their representation (see Fig. 13.6 for Lesson B2 sample screenshot).

In Lesson B3, students use the PPW diagram equation to represent and solve PPW word problems where one of the three quantities is missing (see Fig. 13.6 for Lesson B3 sample screenshot). The tasks involved in this lesson include "*change-join*," "*change- separate*," and "*combine*" problem structure (Van de Walle, 2004). Students represent the word problem in the Part-Part-Whole model equation, and then solve for the unknown quantity.

Before solving the problem, students learn to "label" each of the three boxes (i.e., "Part," "Part," and "Whole/Total") based on the specific context of the word problem story (see Fig. 13.6 for Lesson B3 sample screenshot). The "labeling" of the three boxes serves as a visual scaffold to help students organize the information and correctly represent the quantities in the PPW model equation based on their understanding of the relations in the word problem.

Lessons B4, B5, and B6 extend students' knowledge base pertinent to PPW mathematical structure as they learn to represent (Lesson B4 and Lesson B5) and solve a range of *additive compare* problems (Lesson B6) including the "more than" and the "less than" situations. In Lesson B4 students will work with *additive compare* problems that involve "*more than*" relations (see Lesson B4 for sample screenshot); whereas in Lesson B5 students will work with problems that involve "*less than*" relations (see Lesson B5 for sample screenshot). As shown in Fig. 13.6, Lessons B4 and B5 screenshots, students will engage in "labeling" the three boxes that are associated with the three quantities (i.e., "smaller," or "bigger," or "difference") in the PPW model equation. The labeling activity helps *anchor* correct mapping of numbers from the word problem to the PPW model equation on the basis of comprehending the underlying mathematical relations.

After students engage in representing variously constructed additive compare problem situations onto the PPW model equation, in Lesson B6 (see Fig. 13.6 Lesson B6 sample screenshot), they will be presented with real-world additive problems in which they need to solve for one unknown quantity (either the "smaller," "bigger," or the "difference" amount).

In Lesson B7, students will be provided with opportunities to represent and solve a range of additive word problems (see Table 13.1), with the unknown quantity (represented by letter a) being either one of the "parts," or the "whole." Guided by the DOTS checklist (see Fig. 13.3), students will engage in decoding various problem contexts and decontextualizing mathematical relations from a real-world story problem, and then representing it in the mathematical model equation, P + P = W, for solution.

13.8 Bridging Arithmetic and Algebra

As we all know, many students, in the United Stated at least, experience tremendous difficulty in learning algebra. Many students have developed algebra "phobia." Based on existing literature, at least two reasons are associated with students' phobia of algebra: (a) a lack of exposure to algebraic ideas and thinking, and (b) students' tendency to learn algebra as mere symbol manipulation (Chappell & Strutchens, 2001). The COMPS program serves to extend students' arithmetic experiences, connecting arithmetic learning with algebra learning.

As evidenced by empirical research studies (e.g., Xin et al., 2020, 2008, 2011), COMPS is particularly helpful for students with LDM, who are likely to experience disadvantages in working memory and information organization. Using the COMPS program, students do not need to gamble on the operations (e.g., whether to add or subtract). Because one mathematical model equation (e.g., Part + Part = Whole) represents and bridges a range of additive word problem situations; and the model equation *drives* the solution plan. With the advancing of technology, the web-based computer-assisted COMPS-A program provides students with animations to illustrate how three units of one, for instance, can be merged together to become one unit of three (a bar that is measured as three), which helps students construct and reinforce the mathematical idea, "part and part makes up the whole." This process is expected to help students make the mental leap toward mathematical model-based problem-solving at the symbolic level. The COMPS-A tutor program will prepare students for the CCSSM and promotes generalized problem-solving skills through bridging the arithmetic problem-solving and algebra learning.

Acknowledgement ©All screenshots from the COMPS-A computer tutor presented in this chapter are copyrighted by the COMPS-RtI project ([i]Xin et al., 2015-2018). All rights reserved. Therefore, reproduction, modification, storage, in any form or by any means is strictly prohibited without prior written permission from the project director (yxin@purdue.edu) and the authors.

[i]This research was funded by the National Science Foundation, under grant1503451. The opinions expressed do not necessarily reflect the views of the Foundation.

Appendix

Mixed Additive Word Problem-Solving Worksheet (Adapted from Xin, 2012)

1. Kelsie said she had 82 apples. If Lee had 32 fewer apples than Kelsie, how many apples did Lee have?
2. Selina had some video games. Then, her brother Andy gave her 24 more video games. Now Selina has 67 video games. How many video games did Selina have in the beginning?
3. Taylor and her friend Wendy collect marbles. As of today, Taylor has 93 marbles. Taylor has 53 more marbles than Wendy. How many marbles does Wendy have?

4. Dana has 28 goldfish in her aquarium. She has 32 fewer goldfish than her friend Gesell. How many goldfish does Gisela have in her aquarium?
5. Gilbert had 56 paperback books. Then his brother, Sean, gave him some more paperback books. Now Gilbert has 113 paperback books. How many paperback books did Sean give Gilbert?
6. Adriana has 70 cows. Michelle has 35 fewer cows than Adriana. How many cows does Michelle have?

References

Blum, W., & Leiss, D. (2005). Modellieren mit der "Tanken"-Aufgaben. *Mathematic Lehren, 128*, 18–21.

Blomhøj, M. (2004). Mathematical modelling – a theory for practice. In B. Clarke et al. (Eds.), *International perspectives on learning and teaching mathematics* (pp. 145–159). National Center for Mathematics Education.

Boulineau, T., Fore, C., Hagan-Burke, S., & Burke, M. D. (2004). Use of story-mapping to increase the story-grammar text comprehension of elementary students with learning disabilities. *Learning Disability Quarterly, 27*(2), 105–121.

Brenner, M. E., Mayer, R. E., Moseley, B., Brar, T., Duran, R., Reed, B. S., et al. (1997). Learning by understanding: The role of multiple representations in learning algebra. *American Educational Research Journal, 34*, 663–689.

Cathcart, W. G., Pothier, Y. M., Vance, J. H., & Bezuk, N. S. (2006). *Learning mathematics in elementary and middle schools: A learner-centered approach* (4th ed.). Pearson Merrill Prentice Hall.

Chappell, M. F., & Strutchens, M. E. (2001). Creating connections: Promoting algebraic thinking with concrete models. *Mathematics Teaching in the Middle School, 7*(1), 20–25.

National Governors Association Center, & for Best Practices & Council of Chief State School Officers. (2010). *Common core state standards for mathematics.* Authors.

Common Core State Standards Initiative. (2012). *Introduction: Standards for mathematical practice.* Retrieved from http://www.corestandards.org/the-standards/mathematics.

Cobb, P., & Merkel, G. (1989). Thinking strategies as an example of teaching arithmetic through problem solving. In P. Trafton (Ed.), *New directions for elementary school mathematics: 1989 yearbook of the National Council of Teachers of Mathematics* (pp. 70–81). NCTM.

Cobb, P., & Wheatley, G. (1988). Children's initial understandings of ten. *Focus on Learning Problems in Mathematics, 10*(3), 1–28.

Davydov, V. V. (1982). Psychological characteristics of the formation of mathematical operations in children. In T. P. Carpenter, J. M. Moser, & T. A. Romberg (Eds.), *Addition and subtraction: Cognitive perspective* (pp. 225–238). Lawrence Erlbaum Associates.

Dundar, S., Gokkurt, B., & Soylu, Y. (2012). Mathematical Modelling at a Glance: A Theoretical Study, *Procedia – Social and Behavioral Sciences, 46*, 3465–3470.

Fuson, K. C., & Willis, G. B. (1988). Subtracting by counting up: More evidence. *Journal for Research in Mathematics Education, 19*, 402–420.

Gersten, R., Chard, D. J., Jayanthi, M., Baker, S. K., Morphy, P., & Flojo, J. (2009). Mathematics instruction for students with learning disabilities: A meta-analysis of instructional components. *Review of Educational Research, 79*, 1202–1242. https://doi.org/10.3102/0034654309334431

Glass, G. V., McGaw, B., & Smith, M. L. (1981). *Meta-analysis in social research.* Sage Publication.

Greer, B. (1992). Multiplication and division as models of situations. In D. Grouws (Ed.), *Handbook of research on mathematics teaching and learning* (pp. 276–295). Macmillan.

Gurney, D., Gersten, R., Dimino, J., & Carnine, D. (1990). Story grammar: Effective literature instruction for high school students with learning disabilities. *Journal of Learning Disabilities, 23*(6), 335–342, 348. https://doi.org/10.1177/002221949002300603

Hedges, L. V., & Olkin, I. (1985). *Statistical methods for meta-analysis*. Academic Press.

Hughes, E. M., Witzel, B. S., Riccomini, P. J., Fries, K. M., & Kanyongo, G. Y. (2014). A meta-analysis of algebra interventions for learners with disabilities and struggling learners. *The Journal of the International Association of Special Education, 15*(1), 36–47.

Jonassen, D. H. (2003). Designing research-based instruction for story problems. *Educational Psychology Review, 15*, 267–296.

Karp, K. S., Bush, S. B., & Dougherty, B. J. (2014). 13 rules that expire. *Teaching Children Mathematics, 21*(1), 18–25.

Kim, S. J., & Xin, Y. P. (2022). A Synthesis of Computer-Assisted Mathematical Word Problem-Solving Instruction for Students with Learning Disabilities or difficulties. *Learning Disabilities: A Contemporary Journal, 20(1).* https://www.ldw-ldcj.org/images/Kim__Xin_2022.pdf

Lesh, R., Doerr, H. M., Carmona, G., & Hjalmarson, M. (2003). Beyond constructivism. *Mathematical Thinking and Learning, 5*(2&3), 211–233.

Lewis, A. B., & Mayer, R. E. (1987). Students' miscomprehension of relational statements in arithmetic word problems. *Journal of Education & Psychology, 79*, 361–371.

Mayer, R. E. (1999). *The promise of educational psychology Vol. I: Learning in the content areas.* Merrill Prentice Hall.

National Council of Teachers of Mathematics. (2000). Principles and standards for school mathematics. Author.

National Assessment of Educational Progress result. (NAEP, 2019). Extracted from https://www.nationsreportcard.gov/mathematics/nation/scores/?grade=4

Olive, J. (2001). Children's number sequences: An exploration of Steffe's constructs and an extrapolation to rational numbers of arithmetic. *Mathematical Education, 11*(1), 1–9.

Polotskaiaa, E., & Savardb, A. (2018). Using the Relational Paradigm: Effects on pupils' reasoning in solving additive word problems. *Research in Mathematics Education, 20*(1), 70–90. https://doi.org/10.1080/14794802.2018.1442740

Polya, G. (1957). *How to solve it* (2nd ed.). Doubleday.

Rand, R.H. (1984). Computer algebra in applied mathematics: An introduction to MACSYMA. Boston: Pitman.

Seel, N. M. (2017). Model-based learning: a synthesis of theory and research. *Educational Technology Research and Development, 65*(4), 931–966. https://doi.org/10.1007/s11423-016-9507-9

Sowder, L. (1988). Children's solutions of story problems. *The Journal of Mathematical Behavior, 7*, 227–238.

Steffe, L. P., von Glasersfeld, E., Richards, J., & Cobb, P. (1983). *Children's counting types.* Praeger.

Thompson, P. W. (1993). Quantitative reasoning, complexity, and additive structures. *Educational Studies in Mathematics, 25*(3), 165–208. https://doi.org/10.1007/BF01273861

Tzur, R., Johnson, H. L., McClintock, E., Kenney, R. H., Xin, Y. P., Si, L., Woodward, J., Hord, C., & Jin, X. (2013). Distinguishing schemes and tasks in children's development of multiplicative reasoning. *PNA, 7(3), 85–101.*

Van de Walle, J. A. (2004). *Elementary and middle school mathematics: Teaching developmentally* (5th ed.). Allyn & Bacon.

Walker, D. W., & Poteet, J. A. (1989–90). A comparison of two methods of teaching mathematics story problem-solving with learning disabled students. *National Forum of Special Education Journal, 1*, 44–51.

Witzel, B. S. (2005). Using CRA to teach algebra to students with math difficulties in inclusive settings. *Learning Disabilities: A Contemporary Journal, 3*(2), 49–60.

Wright, R., Stanger, G., Stafford, A., & Martland, J. (2007). *Teaching number in the classroom with 4–8 year olds.* Sage.

Xin, Y. P. (2012). Conceptual model-based problem solving: Teach students with learning difficulties to solve math problems. *Sense*. https://doi.org/10.1007/978-94-6209-104-7_1

Xin, Y. P. (2016). Conceptual model-based problem solving. In P. Fermer, J. Kilpatrick, & E. Pehkonen (Eds.), *Posing and solving mathematical problems: Advances and new perspectives* (pp. 231–254). Springers.

Xin, Y. P. (2007). Word-problem-solving tasks presented in textbooks and their relation to student performance: A cross-curriculum comparison case study. *The Journal of Educational Research, 100*, 347–359.

Xin, Y. P., Kim, S. J., Lei, Q., Wei, S., Liu, B., Wang, W., Kastberg, S., Chen, Y., Yang, X., Ma, X., & Richardson, S. E. (2020). The Impact of a Conceptual Model-based Intervention Program on math problem-solving performance of at-risk English learners. *Reading and Writing Quarterly: Overcoming Learning Difficulties, 36*(2), 104-123. https://www.tandfonline.com/doi/full/10.1080/10573569.2019.1702909

Xin, Y. P., Kastberg, S., & Chen, V. (2015). *Conceptual Model-based Problem Solving (COMPS): A response to intervention program for students with LDM*. National Science Foundation (NSF) funded project.

Xin, Y. P., Liu, J., & Zheng, X. (2011). A cross-cultural lesson comparison on teaching the connection between multiplication and division. *School Science and Mathematics, 111*(7), 354–367.

Xin, Y. P. (PI)., Wang, W., Van Nahmen, M. A., & Sanders, D. (2018). *Conceptual Model-based Mathematics Intelligent Tutors (COMMIT)*, NSF I-Corps funded project.

Xin, Y. P., Wiles, B., & Lin, Y. (2008). Teaching conceptual model-based word-problem story grammar to enhance mathematics problem solving. *Journal of Special Education, 42*(3), 163–178. https://doi.org/10.1177/0022466907312895

Zhang, D., & Xin, Y. P. (2012). A follow-up meta-analysis of word problem solving interventions for students with learning problems. *The Journal of Educational Research, 105*(5), 303–318.

Zheng, X., Flynn, L. J., & Swanson, L. (2012). Experimental intervention studies on word problem solving and math disabilities: A selective analysis of the literature. *Learning Disability Quarterly, 36*(2), 97–111. https://doi.org/10.1177/07319487124442

Chapter 14
Nurturing Multiplicative Reasoning with Whole Numbers

Ron Tzur and Yan Ping Xin

Abstract While central to a person's life and study of mathematics, learning to reason and operate multiplicatively with whole numbers presents challenges and difficulties for students and teachers. To support teachers' design and implementation of instruction that nurtures multiplicative reasoning in students with learning disabilities or difficulties in mathematics (LDM), in this chapter we present a learning progression (trajectory) comprised of six schemes. The first four open the way to reasoning in a place value number system and to establishing the distributive property of multiplication over addition. The latter two open the way for reasoning about division as the inverse of multiplication. We describe how the understanding of these schemes paves the road to mathematical model-based problem representation and generalized problem-solving skills. We illustrate our intervention programs in the context of a computerized, web-based tutor program (abbreviated PGBM-COMPS) that we have used, and teachers can use, to nurture multiplicative reasoning and problem-solving by students with LDM. We also bridge children's learning of elementary arithmetic with geometry through the big ideas of multiplicative reasoning. Finally, we culminate with a few studies that show the impact of our intervention program on students' learning and problem-solving performance (including transfer to novel situations).

Keywords Multiplicative reasoning · Problem-solving · Learning difficulties · Instructional strategies · Intervention · Mathematical models

R. Tzur
University of Colorado Denver, Denver, CO, USA
e-mail: RON.TZUR@UCDENVER.EDU

Y. P. Xin (✉)
Purdue University, West Lafayette, IN, USA
e-mail: yxin@purdue.edu

14.1 Introduction

To present our purpose in this chapter, we begin with a strong, imperative statement for anyone working to promote children's multiplicative reasoning with whole numbers: *repeated addition is not multiplication* (Simon et al., 2018; Tzur et al., 2021). It is addition, being repeated—and is rooted in additive reasoning that involves unit preserving operations (Schwartz, 1988). Yes, children can and often are taught to use repeated addition to find answers to multiplicative situations (Baroody, 2006; Kling & Bay-Williams, 2015; Whitacre & Nickerson, 2016). However, such a focus seems to be at the heart of the challenges they then face in situations that require to reason multiplicatively. We started with this statement because we believe the teaching of multiplication as repeated addition is a major source of many students' struggles when instruction advances, roughly around third grade, from additive to multiplicative operations on whole numbers.

To illustrate the difference between additive and multiplicative reasoning, we introduce a key construct about numerical thinking, namely, number as a composite unit (Steffe, 1992)—a single unit created from smaller units. For example, the number six can be thought of as made of six units of 1 (hereafter referred to as ones, or 1s), of 4 + 2, of 2 + 2 + 2, or of three copies of the number 2. The latter two, when considered in a word problem situation, can help understand the difference between additive and multiplicative reasoning, respectively.

Consider a person asked how many cookies are in three bags, each containing two cookies. When adding 2 cookies + 2 cookies + 2 cookies that person's cognitive operations are on the same unit throughout and the result is still the same unit (e.g., cookies). In contrast, reasoning multiplicatively in such a situation would involve considering two types of units that produce a third type:

- Three *bags* (first composite unit made of three single bags);
- Two *cookies per bag* (a second composite unit relating 1s with a composite unit);
- *Total of 1s* (a composite unit, made of six 1s, which *differs from the two other units*).

Cognitively, multiplicative operations are unit transforming (Davydov, 1992; Schwartz, 1988; Simon et al., 2018), because they involve distribution (coordination) of copies of an item of one composite unit (e.g., 2-per-bag) into items of another composite unit (3 bags), in service of a goal of figuring out items of a third unit (e.g., a composite unit of six cookies). A related way to distinguish multiplicative reasoning is in focusing on the many-for-one coordination, as opposed to one-for-one in additive reasoning (Clark & Kamii, 1996). With this meaning of multiplicative reasoning, and a sharp distinction from repeated addition, our purpose in this chapter is to portray a way of nurturing multiplicative reasoning with whole numbers in all children.

This purpose resides in a stance on multiplicative reasoning (MR) as foundational for more advanced mathematical thinking (Harel & Confrey, 1994; Simon et al., 2018) and for being able to solve multiplicative problem situations meaningfully and efficiently. Specifically, this foundation would afford, or constrain, children's progress to meaningful reasoning in fractional situations (Hackenberg, 2007; Hackenberg & Tillema, 2009; Thompson & Saldanha, 2003) and later in algebraic situations (Hackenberg & Lee, 2015). In this regard, Vergnaud (1983) sorted multiplicative problem structures into three types: *isomorphism of measures*, *product of measures*, and *multiple proportions*. The first involves operating with equal groups (EG), such as the three bags with two cookies each, and asking about a total of cookies. It could also involve givens of six cookies, placed two per bag, and asking about the number of bags. The second structure, multiplicative compare (MC) of products of measure, pertains to the coordination of a given quantity and an operation on it, such as, Pat has two cookies and Sam has three times as many cookies, and asking about the amount of cookies Sam has. In addition to the EG and MC problems that are commonly seen in elementary mathematics curriculum, the second structure (product of measures) pertains to *Cartesian products* and *rectangular area* problems. The third structure (multiple proportions) pertains to problems such as, a bag of two apples costs 80 cents, and asking about the cost of three bags.

The problem situations distinguished in the foregoing are by no means exhaustive. As Greer (1992) asserted, "Extension of the concepts of multiplication and division can be continued indefinitely to encompass directed numbers, complex numbers, matrices, and so on" (p. 279). In this chapter, we focus on multiplicative reasoning underlying solutions to EG and MC problem types, because they are needed for promoting the conceptual foundations of multiplicative reasoning. To familiarize teachers with ways of nurturing such reasoning, we organize the chapter around the modules of a web-based tutoring program, *PGBM-COMPS* (see next section). That is, we share a tool developed and successfully used to nurture multiplicative reasoning in students designated as having learning disabilities and difficulties in mathematics (Lei et al., 2020; Xin et al., 2016, 2017a, b, 2020). This tutor was designed to support conceptual, model-based learning that builds on what students know by promoting the construction of fundamental mathematical ideas necessary for the development of multiplicative reasoning and problem-solving.

14.2 Nurturing Multiplicative Reasoning through the PGBM-COMPS Computer Tutor

Supported by the National Science Foundation (Xin et al., 2008a), the PGBM-COMPS computer tutor draws primarily on two research-based frameworks: a constructivist view of learning from mathematics education (Simon et al., 2004; Steffe & Cobb, 1988; Tzur, 2019; von Glasersfeld, 1995), and Conceptual Model-based Problem Solving (COMPS) (Xin, 2012) that generalizes the underlying structures of word problems. In particular, the PGBM-COMPS tutor (Xin et al., 2017b) includes five modules (A, B, C, D, and E) that integrate two components: (a) the

"Please Go Bring Me …" (PGBM; Tzur et al., 2020) anchor tasks that promote students' development of six schemes (Tzur et al., 2013) that are fundamental to multiplicative reasoning, and (b) the COMPS (Xin et al., 2008b, 2011), which emphasizes understanding and symbolic representation of word problems in mathematical model equations.

Next, we describe the six concepts (schemes) of multiplicative reasoning along with their relevant modules of the tutor. A succinct summary of all six schemes, including sample of assessment tasks for each, are found in Table 14.1. With the integration of PGBM and COMPS, the tutoring program always moves students from operations on concrete objects (e.g., cubes and towers), through operations on figural items (e.g., one's fingers standing for a set of composite units and/or 1s is covered), to abstract items (numbers and symbols).

Table 14.1 Six MR schemes and sample tasks to assess student's development of them

Problem type	Sample problems situations
Multiplicative Double Counting	Pretend that you have made many towers, each made of 7 cubes. How many cubes are in every tower? Your Answer: _____ How many cubes are in the first 4 towers? Your Answer: _____ So we can count those by seven, "7, 14, 21, 28 …" Do you think you will say the number 70 if you continue counting cubes in the towers? Your Answer: _____ Why? _____ Do you think you will say the number 84 if you continue counting cubes in the towers? Your Answer: _____ Why? _____
Same Unit Coordination	Rachael has built 13 towers with 2 cubes in each. Mary has built 7 towers with 4 cubes in each. Who has more towers, Rachael or Mary? Your Answer: _____ How many more towers does she have?
Unit Differentiation and Selection	Tom's father bought 6 pizzas. Each pizza had 4 slices. Tom's mother bought a few more pizzas. Then, there were 9 pizzas. How many more slices did Toms' mother bring?
Mixed Unit Coordination	Maria made birthday bags. She wants each bag to have 6 candies. After making 3 bags, she still had 12 candies left. How many bags will she have altogether after putting these 12 candies in bags?
Quotitive Division	There are 28 students in Ms. Franklin's class. During reading, she puts all students in groups of 4. She asked a student (Steve): "How many groups will I make?" Steve said: "32. Because 28 + 4 is 32." Do you think that Steve is correct? Your Answer: _____ Why? _____
Partitive Division	After an art class, there were 78 crayons out on the tables. There are 6 boxes for the crayons. Ms. Brown puts the same number of crayons in each box. How many crayons would she put in each box?

Module A₁: Multiplicative Double Counting (mDC) In this PGBM-COMPS tutor program, the first scheme students will learn is *multiplicative double counting* (mDC) (Steffe, 1992, Steffe & Cobb, 1998, Tzur et al., 2013). This scheme is constructed through the child's reflection on activities of building and enumerating composite units and 1s that constitute these composite units. For example, a child may be asked (in class or by the tutor program) to go and bring seven towers, each containing two cubes (see Fig. 14.1, left panel, *PGBM 7T₂*; note: At this point the tutor shows neither the fingers component nor the symbolic one as shown in the Figure on the right). In the tutor program, the child does this by clicking twice on the pile of cubes (see Fig. 14.1, left panel, yellow "pyramid" at the bottom-center of the screen). This will create a tower of two cubes placed within the pink line to the left of the ruler. The child then copies that tower seven times into the pink (lined) "mat" (e.g., here we see a step in the process consisting of five such towers, made of different color cubes). The left most frame in the figure shows the four key questions the tutor program (or a partner in the classroom) then asks of the child, which appear one-by-one:

1. How many cubes are in each tower?
2. How many towers did you bring?
3. How many cubes are there in all?
4. How did you figure out that answer?

These four key questions were designed to promote reflection (see Tzur & Hunt, this volume) on the activity sequence of *first* building each tower (i.e., producing a composite unit from one), *then* iterating that composite unit a given number of times (e.g., creating seven copies of the tower of 2), *then* accounting for the total of 1s in all towers the child produced. That is, a child's activity includes separating the composite unit (unit rate, 2-per-tower) from the composite unit of units (seven towers of two cubes each), leading to the distribution (coordination) of those two unit types to produce a third composite unit (e.g., fourteen 1s). The tutor program follows the child's answers with responses about their correctness, hence also strengthening

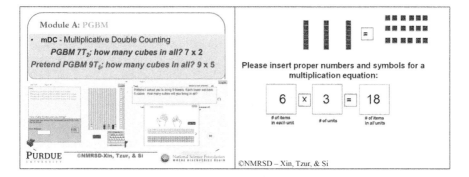

Fig. 14.1 Sample screenshot of Module A of PGBM-COMPS—multiplicative double counting

short-term memory (e.g., between the activity of producing seven towers and the answer to that question).

Indeed, the most crucial question is the fourth, as it promotes the child's reasoning of how they processed the given numbers to arrive at the answer—whether correct or incorrect. For example, one child may explain, "I counted all cubes in the picture: 1–2; 3–4; … 13–14." Another child may explain, "I skip-counted by 2s: 2–4–6–8–10–12–14 and stopped when there were no more towers to count." Note that those two children are using premultiplicative reasoning. Another child, moving into multiplicative double reasoning (here, mDC), may explain: "The first tower is 2, two-is-4, … 7-is-14," indicating simultaneous operation on both the pairs of cubes and the towers.

When the tutor program (or a teacher) determines that the child is properly operating on and solving problem situations with the concrete cubes, it moves to the next phase—solving problems with given numbers that are either (a) produced as towers and then being *covered* or (b) not being produced at all ("for pretend" task types). In Fig. 14.1, an example of such a problem is given in the red font (e.g., *Pretend PGBM 9T$_5$*). At this point, the tutor presents the finger component, in which a child can select the number of cubes per each tower and then click on fingers that stand for towers (with accrual numbers of cubes appearing, too). Outside the tutor, a teacher would explicitly encourage children to use their fingers to count, *simultaneously*, the accrual of 1s (5–10– … 40–45) and the accrual of composite units (e.g., raising 1 finger at a time until reaching 9 fingers, 1–2– … 8–9). When established, the mDC scheme includes a child's anticipation that a total number of items (say, 12 cubes, or apples) is a composite unit made of another composite unit (4 baskets, or towers), each of which is a composite unit itself (3 apples, or cubes).

When the tutor program (or a teacher) determines that the child is properly using the simultaneous count of figural items, it moves to the next phase—solving problems while using equation models of the situation. That is, the last part of every module involves the COMPS component to move from operating on concrete or figural items to operating on abstract, symbolic items. This is shown in Fig. 14.1 (right panel), using a problem situation of 3 towers, each made of 6 cubes. The child sees the picture of towers, and/or given the numbers in a word problem and learns (gradually in the tutor) to fill in the different types of units involved. Critically, the word "*times*" and the symbol for multiplication are introduced in direct, relevant linkage with the child's activity (e.g., going 6 *times* to bring, each time, a tower of 3 cubes). Once a child seems facile with using the COMPS equation for solving towers-and-cubes tasks, the tutor program proceeds to tasks in which the model equation is used to solve realistic word problems for the unknown product. At that point, the COMPS component presents the equation with general terms: # of items in each unit × # of units = total # of items in all units (OR, simply: unit rate × # of units = product; see Xin, 2012).

Module A$_2$: Same Unit Coordination (SUC) In the PGBM-COMPS tutor program, the second scheme is termed *Same Unit Coordination* (SUC). We did not include a separate figure of the tutor program for this scheme, but a situation for it can be seen in Fig. 14.2 (left panel). It should be noted that "same unit" refers here only to composite units—*not* to 1s. For example, a child may be given a situation involving two compilations of composite units of eleven 1s, one comprised of 4 towers with 11 cubes each and the other with 3 towers each made of 11 cubes. A same unit coordination task may ask either how many towers are there altogether or how many more *towers* are on the left side of the screen (or fewer towers on the right side). That is, the child's goal is to figure out the *sum or difference of composite units* (not of 1s). Like in the mDC scheme, in the tutor program the child proceeds from working on SUC tasks with concrete compilations of towers and cubes, through figural items, to abstract and symbolic items.

Teachers often, and rightly, ask why is the SUC scheme considered within the progression of multiplicative reasoning, as it essentially involves additive operations on composite units. A response to this question can engage the teachers (and here, you—the reader) in thinking of typical errors many children were found to make while working on such tasks (see Ding et al., 2019). For example, if asked about the sum of towers in the above situation, many children respond with 77 towers. They mistakenly multiply to find the total of 1s in each compilation ($4 \times 11 = 44$ and $3 \times 11 = 33$) then add the totals ($44 + 33 = 77$). Teachers soon realize that SUC is crucial as it involves focusing on composite units while *not losing sight of the composite nature* of those units. That is, a child needs to conceptualize the composite units as units in their own right—made of a given number of 1s.

Cognitively, the SUC scheme entails applying the nested nature of units of 1 (e.g., four 1s and seven 1s are nested in eleven 1s) also to composite units (e.g., 4 towers of 11 and 3 towers of 11 are nested in a larger compilation of 7 towers of 11 cubes each). This latter way of reasoning is crucial, for instance, when reasoning in a place value, base ten number system. For example, consider the task: "Pat had four coins of 10 cents each and received seven more coins of 10 cents each for her birthday; how many coins of 10 cents each did Pat have then?" A typical error would be 110 (coins). An established SUC scheme will lead to correctly reasoning there must

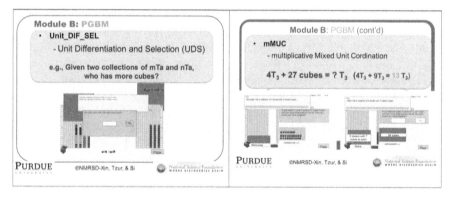

Fig. 14.2 Core tasks and sample screenshots of Module B of the PGBM-COMPS tutor: UDS and MUC

be 11 coins of 10 cents each that, later on (see mixed-unit coordination below), would serve as a basis for also reasoning about the equivalent amount of 1 dollar and 10 cents. When established, the SUC scheme provides a basis for operating on specific composite units such as 10s, 100s, and 1000s, with contexts including distance, weight, and money (Fuson et al., 1997a, b).

Module B_1: *Unit Differentiation and Selection (UDS)* In the PGBM-COMPS tutor program, the third scheme children learn is termed *Unit Differentiation and Selection* (UDS; see Tzur et al., 2013). It involves explicitly distinguishing operations on composite units from operations on 1s, and operating multiplicatively on the difference of 1s between two compilations of composite units. Typical tasks include, "You have 4 towers of 11 cubes each ($4T_{11}$) and I have 3 towers of 11 cubes each ($3T_{11}$). How are our collections similar? Different? How many more cubes do you have?" This task is presented in Fig. 14.2, left panel. Before those three questions are asked of the children, they build the situation by producing and placing the respective compilations of towers on the pink mats of each side of the screen. (Note: the gray rectangles are covers that could be placed over some or all of the towers/cubes in each side to encourage operation on figural or abstract items.) Another situation could differ not in the number of towers but rather in the number of 1s in each composite unit (e.g., $4T_{11}$ and $4T_5$).

The child's goal in the task is to specify the similarities and differences, and to figure out the difference in 1s between the two compilations. The child's activity can include (a) operating multiplicatively on each compilation separately to find its total of 1s (e.g., $4 \times 11 = 44$; $3 \times 11 = 33$), and then find the difference ($44 - 33 = 11$ cubes). This is termed a *Total-First* strategy. Its advantage is the routine procedure involved and available to children who have constructed mDC. However, this strategy becomes cumbersome as soon as numbers are large. For example, consider the task: "Store A has 302 six-packs of juice and Store B has 298 six-packs of juice; How many more cans of juice does Store A have than Store B?" To solve this by multiplying 302×6 and 298×6, then subtracting the products seems futile.

Instead, the goal we would set for the children's learning is to construct a scheme underlying a *Difference-First Strategy*. In that scheme, the child first set the goal of finding the difference in the number of composite units between the two compilations (e.g., $302 - 298 = 4$ six-packs). The follow-up subgoal would be to find the total of 1s only in that portion of the two quantities, that is, the difference in composite units between the two compilations (e.g., 4 six-packs $\times$ 6 cans/pack = 24 cans). Such a strategy (and scheme) underlies a meaningful understanding of the distributive property of multiplication over addition (e.g., $302 \times 6 - 298 \times 6 = (302 - 298) \times 6$).

We promote children's use and coordination of both strategies as a means to nurture their construction of the UDS scheme. This scheme's name—unit differentiation and selection—is complex. But the terms it consists of were set to remind the teacher of the mental operations involved. First, a child needs to differentiate (distinguish) between the unit rate (e.g., six cans per pack) and the number of composite units in each of the given compilations Then, the child needs to select the units on which they operate in the first and second step of each strategy. In the Total First Strategy, the child selects to first operate on the totals of 1s in each compilation (using mDC), then on the difference in 1s (using additive reasoning). In the Difference First Strategy, the child selects to first operate additively on the compilations of composite units (using SUC), then on the 1s in that difference (using mDC). This last explanation helps clarify why, conceptual, mDC and SUC are necessary to establish the UDS scheme with a Difference-First Strategy, and thus in nurturing a cognitive foundation for the distributive property. Most importantly, all three schemes serve a crucial role in constructing the fourth scheme, which constitutes a cognitive foundation for operating in place value number systems.

Module B_2: Mixed-Unit Coordination (MUC) In the PGBM-COMPS tutor program, the fourth scheme children learn is termed *Mixed Unit Coordination* (MUC; see Tzur et al., 2009, 2013). The term (mixed unit) connotes a child now construct operations involving both composite units and 1s (e.g., 10s and 1s). Figure 14.2 right panel presents core tasks/activities and screenshots of a typical task requiring this scheme: "Under the gray cover the computer placed $4T_3$. And on the pink mat (on the right) it placed 27 more cubes. If we put all 27 cubes in towers of 3 cubes each and move under the gray cover—how many **towers** will we have in all?" For teachers, this situation is presented symbolically in the light-green frame as "$4T_3 + 27$ cubes $= ? \ T_3$."

After the UDS scheme has enabled distinguishing composite units from 1s, the MUC scheme involves operating on 1s to answer questions about composite units in two or more compilations. The child's goal when solving such tasks is to figure out the number of composite units in a global compilation that would combine both given quantities—composite units and 1s (e.g., four towers of 3 cubes and 27 single cubes). To this end, the child's activity includes selection and coordination of the unit rate (e.g., 3) that characterizes the given compilation of towers with a segmenting operation (Steffe, 1992) on the given number of 1s (i.e., a reversal of mDC, as in 3-is-1-tower, 6-is-2, … 27-is-9). This coordinated operation first yields the additional number of composite units (9 towers) in the collection of 1s and opens the way to adding it to the initially given number of composite units (e.g., $4T_3 + 9T_3 = 13T_3$). An MUC task may also ask about the total of 1s (instead of composite units), which a child would answer by applying mDC only to the compilation of composite units, then adding it with the given number of 1s (e.g., $4 \times 3 = 12$ cubes; 27 cubes $+12$ cubes $=39$ cubes). The latter could be shown in a place value context: "A school prepares vans for a field trip; each van carries 10 people. They already have 4 full vans but 29 more people that need to go on the trip. How many vans in all does the school need for all people to be able to go?" Notice that such a task requires a more sophisticated way of reasoning, as the last van needed will not be full.

The example of a school's field trip above helps accentuate why MUC is a cognitive basis for meaningfully operating in a place value base ten system (or other bases). Later, the MUC scheme would also serve as a basis for meaningfully operating on algebraic equations, such as in cases of "collecting like-terms" (e.g.,

$3x^2 + 5x - 40 + 3x - x^2 + 10 = 2x^2 + 8x - 30$). Teachers need to be mindful that advancing from 10s and 1s to larger units (e.g., 100s, 10s, and 1s) is anything but trivial. The reason is that a child would then have to conceptualize the larger unit as a 3-level unit (ten units of ten units of 1s). Our research work with students (grades 3, 4, and 5) in the US and China (forthcoming) has shown how challenging such situations can be. For example, ~75% of US and ~50% of Chinese students incorrectly solved the task: "A bag contains 10 apples and a box contains 10 bags. A school has 1 box, 16 bags, and 11single apples that are not packed. How many single apples does it have in all?" Typical *errors* we found included adding $1 + 16 + 11 = 28$ or $100 + 16 + 11 = 127$.

Besides the key role MUC serves in place value operations, we stress another, critical aspect of the progression of the three schemes past mDC (i.e., SUC, UDS, and MUC) in supporting children's construction of divisional schemes. Many teachers with whom we worked decided, initially, to skip those three schemes. Instead, they opted for proceeding from mDC directly into the first (of two) divisional schemes, namely, quotitive (measurement) division. After all, they said, division is but a reversal of mDC. At issue is that such a reversal may be known to the teacher, whereas for children it is yet to be established. Those teachers attested to students' "hitting a wall" of conceptual understanding in division, which led the teachers to go back and nurture the entire progression from mDC to MUC.

Cognitively, the reason this progression supports division has to do with the segmenting operation involved in MUC tasks (e.g., segmenting 27 cubes into nine segments of three 1s each). Key here is that, having established UDS before MUC, a child would have easier, independent access (without being told or shown) to the operation of selecting the characteristic unit rate (e.g., 3 cubes per tower) from one of the givens and imposing it on the segmenting operation. That is, MUC supports "measuring" the given total of 1s by a segment of size of the unit rate (e.g., measuring 27 cubes by segments of three 1s each to find 9 such segments constitute 27). It is in this sense that MUC—with mDC, SUC, and UDS established prior to it—can serve as a cognitive precursor for constructing the two divisional schemes—to which we turn next.

Module C: Quotitive Division (QD)   In the PGBM-COMPS tutor program, the fifth scheme children will learn is termed *Quotitive Division* (QD; see Tzur et al., 2013). Figure 14.3 left panel presents a typical task and the screenshots of the tutor. The givens here include a total of 1s ("containing" composite unit, e.g., 40 cubes) and the number of 1s in each (unit rate, e.g., 8 cubes per tower). The question, and the child's goal, is to figure out how many units of the given size would constitute the larger composite unit. Initially, a child brings forth the segmenting operating as explained above. Through reflection on such activities, the child constructs an anticipation of the effect of iterating those segments, which leads to advancing from a segmenting to a partitioning operation.

The difference between the two operations is subtle, yet essential. In segmenting, the child cannot yet anticipate the possibility and effect of iterating a composite unit; they literally find such effect in action (Hackenberg & Tillema, 2009; Steffe,

1992; Steffe & Cobb, 1998; Steffe & Olive, 2010). In our example, they have to iterate the unit of 8 while keeping track of how many times it is being iterated (e.g., 8 is 1, 16 is 2, ... 40 is 5). In partitioning, they have interiorized this operation by constructing composite units (not just 1s anymore) as iterable—a unit that can be iterated without actually carrying out that operation. It is the advance from segmenting to partitioning, with interiorized, iterable composite units, which we use as a criterion for attributing to a child the construction of division (the quotitive, or measurement type).

It is only when the child establishes the QD scheme at the anticipatory stage (see Tzur & Hunt, this volume) that we recommend introducing it as a reversal of mDC. The reason is straightforward: anticipating the reversal linkage entails coordinating mDC with something the child first has to have in place. Once established, the QD scheme provides a basis for conceiving of division as an inverse operation of multiplication, and thus for using multiplication fact families to solve division problems in which the total (or product) and the size of each group (i.e., unit rate) is given. Accordingly, once all six schemes were learned using the PGBM game, the program also connects the PGBM game with mathematical model expressions (e.g., $8 \times (1 + 1 + 1 + 1 + 1) = 40$), which leads to the COMPS component of the program (seen in Fig. 14.3 right panel).

The purpose of the COMPS part is to provide children with opportunities to represent and solve a range of real-world QD tasks in a mathematical model equation. That is, we use the COMPS-QD part of Module C to promote children's modeling of solving QD tasks at the symbolic level: Unit Rate (UR) × # of Units [unknown quantity] = Given Total or Product (Xin, 2012). In our example, a child is given the total of 40 cubes, and asked to find the number of towers can be made if each tower is made of 8 cubes. We note that this part of the program consists of two phases. First, the child becomes familiar with *problem representation* in an equation. Only then, the child proceeds to using the equation for *problem-solving*. During the phase of *problem representation*, word stories with no unknowns are used to help children understand the problem structure and the mathematical

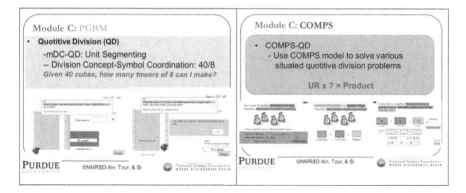

Fig. 14.3 Core tasks and sample screenshots of Module C of the PGBM-COMPS tutor

relations involved (e.g., equal groups problem situation are represented in the equation: Unit rate × # or units = Total # of items across all units, or the *product*). Specifically, the goal is for children to decontextualize the mathematical relations involved in the problem and represent it in the corresponding model equation (Xin, 2015). Accordingly, during this phase of instruction, as all quantities are given in the story (no unknowns), children can successfully learn to "map" all the given information onto the model equation. They then use the equation to check it is "balanced," that is, to reinforce an equation represents an equivalence (see Xin, 2015) between the left and right sides.

Once the child becomes facile with representing all three quantities involved does the tutor moves on to the problem-solving phase. In this phase, children are presented with word problems that involve an unknown quantity. When representing a problem with an unknown quantity in the mathematical model equation, the tutor enables children to choose and insert any letter they prefer for representing the unknown quantity. This activity can, later on, contribute to their construction of a concept of variable as being independent of the specific letter used to represent it.

Module D: Partitive Division (PD) In the PGBM-COMPS tutor program, the sixth scheme children will learn is termed *Partitive Division* (PD; see Tzur et al., 2013). Figure 14.4 left panel presents a typical task and the screenshots of the tutor. The givens here include a total of 1s (the global composite unit, e.g., 30 cubes) and the number of towers (# of units, e.g., 3 towers). The question, and the child's goal, is to figure out how many 1s would constitute each of the given composite units (e.g., 10 cubes). Initially, children bring forth an activity of distributing the given 1s into each group, one-by-one. This activity is predivisional as they are not focusing and operating on composite units, yet. Through reflection on this activity, which we consider as a precursor for division, they begin noticing and constructing an anticipation that each "round" of distribution can also be thought of as a composite unit (e.g., giving 1 to each of three towers creates a "layer" of the first three cubes, giving 1 to the next layer is another group of three 1s, and the total of cubes distributed is six, etc.). Eventually, they interiorize the operation of distributing such composite units onto the given number of groups at once, and begin thinking about the effect of such distribution in terms of "how many such 'layers' could be produced before I run out of cubes." They may show this by experimenting with smaller numbers (e.g., "Let me try 4 layers; that's only 12 cubes; so let me try 7 layers; that's only 21 cubes; so what about 10 layers—that's 30 cubes and I'll be done"). A key transition would be for the child to then relate the number of layers of 1s distributed to each composite unit (e.g., three cubes distributed among three towers produce one layer of "height" equal to 1 cube), so the total number of layers is equal to the number of 1s that constitute each group (e.g., 10 layers across three towers means that each tower is made of 10 cubes).

Introducing prompts and constraints to the child's activity (e.g., "Do you think there would be more than one cube in each tower? Will 3 cubes work? Why?"), children will begin anticipate that each round of distribution of 1s would yield a composite unit (e.g., a 'layer'). They then could reorganize their coordinated-counting act to figure out the end result (# of cubes in each tower or the *unit rate*)

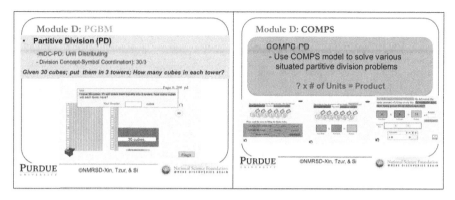

Fig. 14.4 Tasks/activities and sample screenshots from Module D of the PGBM-COMPS tutor

without carrying out the distribution, that is, to establish the PD scheme. In this process of trial and error, children would likely also notice they could use the multiplication/division facts learned earlier for mDC and QD. This could allow a teacher to later nurture the children's linkage between the two types of division into a single (twofold) construct we call division. It would also support their transition to using the same model equation—only this time the given and the unknown switch roles (e.g., instead of being given a total of 1s, say, 30, and a unit rate, say 3 cubes-per-tower—a task will continue giving the total but the other given will be the number of groups, say, 3 towers).

Similar to Module C, Module D proceeds from the PGBM context for nurturing PD to the COMPS context (Fig. 14.4 right panel). The purpose of the latter is to provide children with opportunities to represent and solve a range of real-world problems in a mathematical model equation (Xin, 2012). That is, to promote their modeling and solving of partitive division problems at the abstract, symbolic level (i.e., UR [unknown quantity] × # of Units = Given total"). Again, like in Module C, they will first become facile with representing the given information in the equation (no unknowns, yet) and only then the COMPS program engages them in representing and solving problems while inserting their letter of choice, for the unknown quantity, into the equation.

Module E: Using COMPS to solve a range of multiplicative word problems When children complete the work through the first four modules of this tutor program, they arrive at the final module where they learn to use COMPS to solve a range of multiplicative word problems building on their previously constructed schemes. Table 14.2 presents the range of word problem the child would solve. Problems that students encounter include measurement division (or quotitive division), fair share (partitive division), *Rate × Quantity* (product unknown), as well as various multiplicative compare problem situations. The left and right columns are presented for teachers to see sample problem situations and how it would be represented in a model equation. We believe this table could help teachers create their own problems, using contexts they recognize as familiar and favorable to the children.

Table 14.2 Multiplicative word problem situations and model equation representations (adapted from Xin, 2012)

Problem variations	Sample problem situations	Sample model equations
Equal groups		
Partitive division/ *Fair share*	A school arranged a visit to the museum in Lafayette. It spent a total of $667 buying 23 tickets. How much does each ticket cost?	Unit Rate a × # of Units 23 = Product 667
Quotitive division/ *Measurement division*	There are a total of 575 students in Centennial Elementary School. If one classroom can hold 25 students, how many classrooms does the school need?	Unit Rate 25 × # of Units a = Product 575
Product unknown/ *Rate × quantity*	Emily has a stamp collection book with a total of 27 pages, and each page can hold 13 stamps. If Emily filled up this collection book, how many stamps would she have?	Unit Rate 13 × # of Units 27 = Product a
Multiplicative compare		
Compared set unknown	Isaac has 11 marbles. Cameron has 22 times as many marbles as Isaac. How many marbles does Cameron have?	Unit 11 × Multiplier 22 = Product a
Referent unit unknown	Gina has sent out 462 packages in the last week for the post office. Gina has sent out 21 times as many packages as her friend Dane. How many packages has Dane sent out?	Unit a × Multiplier 21 = Product 462
Multiplier unknown	It rained 147 inches in New York one year. In Washington D.C., it only rained 21 inches during the same year. The amount of rain in New York is how many times the amount of rain in Washington D.C. that year?	Unit 21 × Multiplier a = Product 147

Figure 14.5 presents the scaffolds provided by the tutor (described in the next paragraph) as students solve multiplicative compare problems (left panel) and an organizing mnemonic DOTS (similar to the DOTS checklist described in Chap. 13) to facilitate problem representation and solving: (a) Detect the problem structure, (b) Organize the information using the diagram, (c) Transform the diagram into a meaningful math equation, and (d) Solve for the unknown quantity in the question and check your answer.

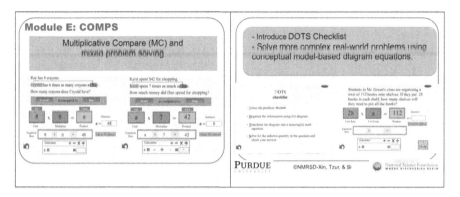

Fig. 14.5 Tasks/activities and sample screenshots from Module E of the PGBM-COMPS computer program

Specifically, in the left panel we see two multiplicative compare (MC) problems. The first tells the child that Ray has 8 crayons, Crystal has 6 times as many crayons as Ray, and asks: "How many crayons does Crystal have?" Prompted by a set of *word problem [WP] story grammar* questions (Xin, 2012; see Figure 14.6b), the child will first figure out "who" (i.e., the compared) is compared to "whom" (i.e., the referent "unit") and name the *referent* Unit box in the model equation as well as the "compared" (product) box. Then the child will represent the corresponding information in the model equation and solve for the unknown quantity for solution.

Notice that at this point the COMPS model equation uses generic terms for the three units involved in a multiplicative situation, similar to how they are taught in schools (Unit, Multiplier, Product). The second problem tells the child that Kerry spent $42 for shopping, and she spent 7 times as much as Dan, then asking: "How much money did Dan spend for shopping?" We note that this problem reverses the order, but with proper scaffolds (e.g., the *WP story grammar*, highlighting, and labeling) as needed, children with whom we worked could at this point assimilate it into a scheme in which they anticipate the need to place each given where it belongs in the model equation and only then solve for the unknown. It is important to note that after problem representation, it is the mathematical model equation that determines what operation to use for solution (depending on the position of the unknown quantity). The "equation" box, as shown in Fig. 14.5, offers the child opportunities to "peel off" the boxes and write it as a true math equation. The "calculator" box allows the child to convert the math equation, as needed, so that the unknown quantity is isolated on one side of the equation for solution. Similarly, with the support of self-regulated WP story grammar questions (Fig. 14.6a), students will use a similar model equation to represent and solve a range of equal-group problems (Fig. 14.5, right panel).

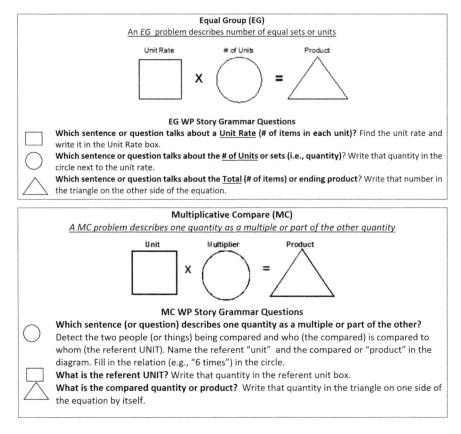

Fig. 14.6 (a) Mathematical model of equal groups problems (Xin, 2012, p. 105). (b) Mathematical model of multiplicative compare problems (Xin, 2012, p. 123)

14.3 Bridging Elementary Arithmetic and Geometry

According to Ma (2020), there is often an absence of connection between mathematical topics in the intended as well as implemented mathematics curriculum, particularly in the US. Connections, as one of the five *Process Standards* stipulated by the National Council of Teachers in Mathematics (NCTM, 2000), emphasize students' ability to understand the connections among mathematical ideas. Multiplicative reasoning (MR) is fundamental to all advanced mathematical thinking (Mulligan, 2002). Bridging the learning of elementary geometry and arithmetic promote deeper conceptual understanding of high-level mathematics through building on big mathematical ideas.

As MR encompasses both arithmetic and geometry, the multiplicative conceptual model (i.e., unit rate × number of units = product, Xin, 2012) can serve as an overarching big idea that bridges the learning of geometry and arithmetic. For instance, Simon and Blume (1994) related the concept of *dimensions* to MR while explaining how *area of a rectangle* problems can be used as a tool for promoting MR. Specifically, Simon and Blume demonstrated how the area of a rectangular region can be evaluated "as a multiplicative relationship between the lengths of the sides" (p. 472). Geometry serves as an ideal context for the extension of students' MR from how it is first introduced for learning the concepts of multiplication and division to more advanced mathematical concepts. Below we describe how teachers can enable and facilitate this connection.

During the beginning stage of learning about the area of a rectangular figure, concrete modeling can be used to facilitate conceptual understanding of the area of rectangles. For instance, to help students understand the area model (area = length × width), unit squares can be used for concrete modeling. First, students could be encouraged to use unit squares to cover the length of one side of the rectangle (for instance, 4 unit-squares). Students will count the number of unit squares (in one row) used to cover the *length* of a rectangular figure. Second, with the first row as the composite unit (i.e., a strip of 4 unit-squares), students could then move on to find out how many such "strips" or how many rows of 4-unit-squares will cover the entire area of the rectangle; that number (e.g., 3) would be the "number of [composite] units" or the *width of the rectangle*. Students will then count the total number of unit squares (12) and that "12" would be the area of the rectangle.

After using unitary counting (count unit squares one by one) or skip counting (count by 4: 4, 8 and 12) to find out the area of a rectangle, students were guided to think about a more efficient way to find the total of the unit squares to cover the area of a rectangle. That is, rather than using unitary counting, students were guided to use multiplicative reasoning (they learned from multiplication and division problem-solving) to figure out the total number of unit squares. That is, the total number of unit squares can be solved by multiplying the number of unit squares in each strip/row (i.e., *unit rate*: 4) and the number of rows (i.e., *number of [composite] units*: 3) to cover the area (i.e., $4 \times 3 = 12$, or unit rate × # of *composite* units = total/product). Finally, students were guided to compare the answer they obtained from unitary counting with the answer they acquired from multiplying the unit rate and the # of composite unit, to realize that the answer they got from the two different approaches would be the same. As such, the idea of multiplicative reasoning they learned from arithmetic was reinforced and *connected* to their geometry learning.

Following the concrete modeling stage (as described above), a semi-concrete level of instruction with graph paper could be used for students to create the unit squares for covering the area based on the given scale on the two dimensions (i.e., length and width) (see Fig. 14.7 left panel for reference). Building on their understanding from the concrete/semi-concrete modeling, students transition to the mathematical model equation for area problem-solving (see Fig. 14.7, right panel). The teacher engages students in a discussion to elicit that the model equation tells the story about the multiplicative relationship between the length and the width. That is, the area is expressed as a multiple of the composite unit (the base strip), or as a mathematical expression A (length) $\times B$ (the width) = Area.

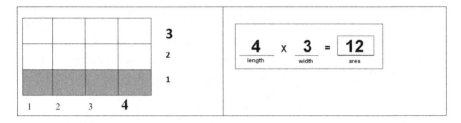

Fig. 14.7 Semi-concrete [left panel] and abstract [right panel] models for the area of rectangle

At this point, students should be ready to advance to the use of an abstract model for area problem-solving. For instance, to solve "How much carpet do you need to cover the floor of a room that is 18 feet long and 9 feet wide," students would first identify from the problem the number of units (of 1s) for the length (i.e., the *unit rate* = 18) and the number of *composite units* for the width (9). Next, students would map the information of the two dimensions onto the model equation (see Fig. 14.7 right panel) and use the equation to solve for the area ($18 \times 9 = ?$ [162]). To strengthen conceptual understanding, the teacher could guide the students to verify the answer they obtained from the abstract model (the formula) through building a mini-sized concrete model and counting out the actual unit squares that cover the area. Through this "two-way" connection (from concrete to abstract, and then from abstract back to concrete), the teacher help students realize that the symbolic model equation indeed represents the concrete model, and therefore can be used directly for solving area problems. Through these activities, the abstract model will no longer be "abstract" or "alien" to the students as it is attached to a concrete model, and therefore, students can make sense of the abstract model.

Following solving problems with the area as the unknown, the teacher will then provide students with opportunities to solve problems with a missing factor (e.g., "what is the *length* of a rectangle that has an area of 54 square feet and a width of 18 feet?"). Students will engage in representing the problem in the area model equation with a letter x to represent the unknown quantity. Then, they will solve the problem through finding the unknown quantity in the equation (Length [?] × Width = Area; or Length [?] = Area / Width).

After students have learned to use the model equation (Length × Width = Area) for solving relevant area problems, students can be introduced to the model equation for solving for the volume of rectangular prisms (see Fig. 14.8). The same instructional sequence applies to the learning of solving problems involving the volume of rectangular prisms. Again, the teacher will start the instruction with concrete manipulatives for modeling the concept. Students will engage in covering the base of the prism with the unit cubes, and they will then count the number of unit cubes that make up the base (that number should be the same as the answer they would obtain from the area model equation, Area = Length × Width). Next, students will count the number of layers of the base to fill the rectangular prism. The teacher will guide students to discover that the number of iterations of the base-unit would be the

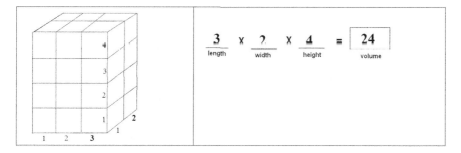

Fig. 14.8 Semi-concrete [left] and abstract [right] models for the volume of rectangular prisms

height (or thickness) of the prism. Finally, students could find out the volume by counting all the unit cubes that make up the prism (unitary counting or skip counting), or multiplying the base (the size of the *composite unit* or the "*unit rate*") by the number of layers of the base (i.e., the height). Again, multiplicative reasoning was reinforced by solving for the volume through multiplying the area of the base (obtained by multiplying the length and width) by the number of layers (or the thickness or the height of the rectangular prism) rather than executing the unitary counting of the cubes. Comparing the results from both approaches (unitary counting or skip counting and multiplication) will help students to understand the meaning of the formula for solving for the volume of a rectangular prism (i.e., Volume = length × width × height). Similar to the learning of the area model, semi-concrete modeling serves to ease the transition from the concrete to the abstract model.

Once the connection is established, students can engage in representing the information from a word (or pictograph) problem directly to the model equation (i.e., Volume = Length × Width × Height) with a letter x to represent the unknown quantity. Then, they will complete the problem-solving through finding for the unknown quantity in the equation. Following solving problems with the volume as the unknown, it is important to provide students with opportunities to solve problems with a missing factor (i.e., the length, width, or height) using the model equation Volume = Length × Width × Height. Providing students opportunities to solve a range of problems, not only those with the product as the unknown but also problems with one of the factors as the unknown, facilitates students' construction of the mathematical model through making connections among seemingly different problems.

14.4 Empirical Evidence That Support PGBM-COMPS Computer Tutor

We culminate this chapter with a few research studies that have investigated the effectiveness of the PGBM-COMPS computer tutor in enhancing problem-solving of students with LDM. For instance, Xin et al. (2020) studied the impact of the

PGBM-COMPS tutoring program on multiplicative reasoning and problem-solving of students with learning disabilities (LDs). The study employed a mixed methods design involving three elementary students (8–11 years old) with LDs. We found that the PGBM-COMPS program significantly improved participating students' problem-solving performance. This was seen not only on researcher developed criterion tests (percentage of nonoverlapping data PND = 100%; Scruggs & Mastropieri, 1998) but also on a norm-referenced standardized test (the Stanford Achievement Test, SAT-10, Pearson Inc., 2004). Qualitative and quantitative data analyses supported our claim that the integration of conceptual-focused instruction (the PGBM) and explicit model-based problem-solving (COMPS) allowed students with LDs to construct multiplicative reasoning, and contributed to their performance on solving multiplicative word problems. In particular, all participating children's improvement on the SAT-10 indicated that they were able to transfer learned skills to solve dissimilar problems in the standardized test.

We note that, data from the study indicated that two participating students had difficulty in solving *Mixed Unit Coordination problems* in Module B of the PGBM-COMPS program. The difficulties the students experienced may be due to their deficiency in working memory, which is one of the common characteristics of students with LD (Swanson & Sachse-Lee, 2001). Specifically, in order to solve a *Mixed Unit Coordination* problem, the child would need to (a) figure out the number of composite units (e.g., bags of 6 candies each) that were needed for twelve 1s (e.g., organize candies into groups of 6; in this case, 2 bags are needed), then (b) take the sub-answer (2 bags) from the first step and add it to the given number of bags in the problem (e.g., 3 bags that a person had to start with). The challenge is for a student to hold the information (and partial answer) from the first step and then apply it in the second step to arrive at the solution. Hord et al. (2016) studied how more scaffolding could promote the success of these children when learning to solve such problems.

In a different study, Xin et al. (2017a) explored the potential effects of the PGBM-COMPS tutor program on enhancing the multiplicative problem-solving skills of seventeen 3rd or 4th grade children with LDM. In that study, the experimental group received instruction through the PGBM-COMPS computer tutor, whereas the control group received a "business as usual" (BAU) instruction provided by two licensed school teachers. Teachers in the BAU condition implemented the instruction consistently as they would do in their regularly scheduled mathematics classes, using the mathematics curriculum (textbook) adopted by the school district. Overall, the findings (see Fig. 14.9 below) supported the potential for using the PGBM-COMPS intelligent tutor to address the knowledge gap of the students with LDM and enhance their problem-solving performance. Students in the PGBM-COMPS (experimental) group outperformed the BAU group on the criterion test. More importantly, following the intervention they showed statistically significant larger improvement on the *Problem Solving* subtest of a norm-referenced standardized assessment, SAT-10. The authors concluded that the unique features of the PGBM-COMPS program, specifically basing mathematical model-equation

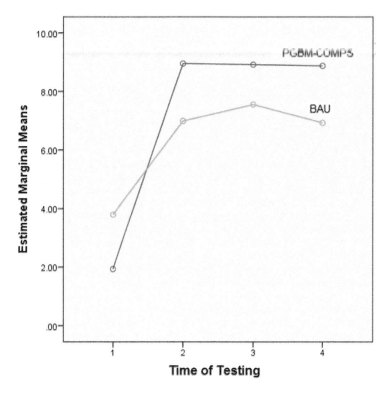

Fig. 14.9 Performance of the two comparison groups (PGBM-COMPS vs. BAU) on the criterion test before (Time = 1) and after the intervention (Time = 2, 3, & 4)

problem-solving on children's understanding of fundamental mathematical ideas (i.e., the six schemes for multiplicative reasoning), have likely contributed to better acquisition and transfer.

14.5 Concluding Remarks

We have presented a progression of six schemes for multiplicative reasoning as well as how they were integrated with the COMPS program in promoting students multiplicative reasoning and problem-solving. We believe our work lends support for teachers who strive to nurture progress in students with LDM. The main point is that using the intervention programs described in this chapter allows engaging the students in constructivist-oriented teaching-learning processes coordinated with abstract, symbolic model equations. Assuming your students have established the fundamental concept of number as composite unit, you can help them conceptualize, and understand, underlying mathematical structures. As our rigorous research studies have shown, this understanding paves the road to mathematical model-based problem representation and generalized problem-solving skills.

Acknowledgement ©All screenshots from the PGBM-COMPS Tutor presented in this chapter are copyrighted by the NMRSD project ([1]Xin et al., 2008). All rights reserved. Therefore, reproduction, modification, and storage, in any form or by any means, is strictly prohibited without prior written permission from the project director (yxin@purdue.edu) and the authors.

References

Baroody, A. J. (2006). Why children have difficulties mastering the basic number combinations and how to help them. *Teaching Children Mathematics, 13*(1), 22–31. www.jstor.org/stable/41198838

Clark, F. B., & Kamii, C. (1996). Identification of multiplicative thinking in children in grades 1-5. *Journal for Research in Mathematics Education, 27*(1), 41–51.

Davydov, V. V. (1992). The psychological analysis of multiplication procedures. *Focus on Learning Problems in Mathematics, 14*(1), 3–67.

Ding, R., Tzur, R., Wei, B., Jin, X., Jin, X., & Davis, A. (2019). Conceptual analysis of an error pattern in Chinese elementary students. In M. Graven, H. Vekat, A. Essien, & P. Vale (Eds.), *Proceedings of the 43rd conference of the International Group for the Psychology of Mathematics Education* (Vol. 4, p. 93). PME.

Fuson, K. C., Smith, S. T., & Lo Cicero, A. M. (1997a). Supporting Latino first graders' ten-structured thinking in urban classrooms. *Journal for Research in Mathematics Education, 28*(6), 738–766.

Fuson, K. C., Wearne, D., Hiebert, J. C., Murray, H. G., Human, P. G., Olivier, A. I., Carpenter, T. P., & Fennema, E. (1997b). Children's conceptual structures for multidigit numbers and methods of multidigit addition and subtraction. *Journal for Research in Mathematics Education, 28*(2), 130–162.

Greer, B. (1992). Multiplication and division as models of situations. In D. Grouws (Ed.), *Handbook of research on mathematics teaching and learning* (pp. 276–295). Macmillan.

Hackenberg, A. J. (2007). Units coordination and the construction of improper fractions: A revision of the splitting hypothesis. *Journal of Mathematical Behavior, 26*, 27–47.

Hackenberg, A. J., & Lee, M. Y. (2015). Relationships between students' fractional knowledge and equation writing. *Journal for Research in Mathematics Education, 46*(2), 196–243.

Hackenberg, A. J., & Tillema, E. S. (2009). Students' whole number multiplicative concepts: A critical constructive resource for fraction composition schemes. *The Journal of Mathematical Behavior, 28*(1), 1–18.

Harel, G., & Confrey, J. (Eds.). (1994). *The development of multiplicative reasoning in the learning of mathematics*. State University of New York.

Hord, C., Tzur, R., Xin, Y. P., Si, L., Kenney, R. H., & Woodward, J. (2016). Overcoming a 4th grader's challenges with working-memory via constructivist-based pedagogy and strategic scaffolds: Tia's solutions to challenging multiplicative tasks. *Journal of Mathematical Behavior, 44*, 13–33.

Kling, G., & Bay-Williams, J. M. (2015). Three steps to mastering multiplication facts. *Teaching Children Mathematics, 21*(9), 548–559. https://doi.org/10.5951/teacchilmath.21.9.0548

Lei, Q., Xin, Y. P., Morita-Mullaney, T., & Tzur, R. (2020). Instructional scaffolds in mathematics instruction for English learners with learning disabilities: An exploratory case study. *Learning Disabilities: A Contemporary Journal, 18*(1), 123–144.

Ma, L. (2020). *Knowing and teaching elementary mathematics: Teachers' understanding of fundamental mathematics in China and the United States*. Routledge.

---

[1]This research was supported by the National Science Foundation, under grant DRL 0822296. The opinions expressed do not necessarily reflect the views of the Foundation.

Mulligan, J. (2002). The role of structure in children's development of multiplicative reasoning. In B. Barton, K. C. Irwin, M. Pfannkuch, & M. O. Thomas (Eds.), *Mathematics education in the South Pacific. Proceedings of the 25th annual conference of the Mathematics Education Research Group of Australasia, Auckland, NZ* (Vol. 2, pp. 497–503). Sydney. MERGA.

National Council of Teachers of Mathematics (NCTM). (2000). *Principles and standards for school mathematics* [Electronic version]. Author. Retrieved December 31, 2007, from http://standards.nctm.org

Schwartz, J. (1988). Intensive quantity and referent transforming arithmetic operations. In J. Hiebert & M. Behr (Eds.), *Number concepts and operations in the middle grades* (pp. 41–52). Lawrence Erlbaum.

Scruggs, T. E., & Mastropieri, M. A. (1998). Summarizing single-subject research: Issues and applications. *Behavior Modification, 22*(3), 221–242. https://doi.org/10.1177/01454455980223001

Simon, M. A., & Blume, G. A. (1994). Building and understanding multiplicative relationships: A study of prospective elementary teachers. *Journal for Research in Mathematics Education, 25*(5), 472–494.

Simon, M. A., Tzur, R., Heinz, K., & Kinzel, M. (2004). Explicating a mechanism for conceptual learning: Elaborating the construct of reflective abstraction. *Journal for Research in Mathematics Education, 35*(3), 305–329.

Simon, M. A., Kara, M., Norton, A., & Placa, N. (2018). Fostering construction of a meaning for multiplication that subsumes whole-number and fraction multiplication: A study of the Learning Through Activity research program. *The Journal of Mathematical Behavior, 52,* 151–173. https://doi.org/10.1016/j.jmathb.2018.03.002

Steffe, L. P. (1992). Schemes of action and operation involving composite units. *Learning and Individual Differences, 4*(3), 259–309. https://doi.org/10.1016/1041-6080(92)90005-Y

Steffe, L. P., & Cobb, P. (1988). *Construction of arithmetical meanings and strategies.* Springer.

Steffe, L. P., & Cobb, P. (1998). Multiplicative and divisional schemes. *Focus on Learning Problems in Mathematics, 20*(1), 45–62.

Steffe, L. P., & Olive, J. (2010). *Children's fractional knowledge.* Springer.

Swanson, H. L., & Sachse-Lee, C. (2001). Mathematical problem solving and working memory in children with learning disabilities: Both executive and phonological processes are important. *Journal of Experimental Child Psychology, 79*(3), 294–321. https://doi.org/10.1006/jecp.2000.2587

Thompson, P. W., & Saldanha, L. (2003). Fractions and multiplicative reasoning. In J. Kilpatrick & G. Martin (Eds.), *Research companion to the NCTM Standards* (pp. 95–113). National Council of Teachers of Mathematics.

Tzur, R. (2019). Hypothetical learning trajectory (HLT): A lens on conceptual transition between mathematical "markers". In D. Siemon, T. Barkatsas, & R. Seah (Eds.), *Researching and using progressions (trajectories) in mathematics education* (pp. 56–74). Sense. https://doi.org/10.1163/9789004396449_003

Tzur, R., Xin, Y. P., Si, L., Woodward, J., & Jin, X. (2009). Promoting transition from participatory to anticipatory stage: Chad's case of multiplicative mixed-unit coordination (MMUC). In M. Tzekaki, M. Kaldrimidou, & H. Sakonidis (Eds.), *Proceedings of the 33rd conference of the International Group for the Psychology of Mathematics Education* (Vol. 5, pp. 249–256). Thessaloniki, Greece: PME.

Tzur, R., Johnson, H. L., McClintock, E., Kenney, R. H., Xin, Y. P., Si, L., Woodward, J., Hord, C., & Jin, X. (2013). Distinguishing schemes and tasks in children's development of multiplicative reasoning. *PNA, 7*(3), 85–101.

Tzur, R., Johnson, H. L., Hodkowski, N. M., Nathenson-Mejía, S., Davis, A., & Gardner, A. (2020). Beyond getting answers: Promoting conceptual understanding of multiplication. *Australian Primary Mathematics Classroom, 25*(4), 35–40.

Tzur, R., Johnson, H. L., Norton, A., Davis, A., Wang, X., Ferrara, M., Harrington, C., & Hodkowski, N. M. (2021). Children's spontaneous additive strategy relates to multiplicative reasoning. *Cognition and Instruction* (online first), 1–26. https://doi.org/10.1080/07370008.2021.1896521

Vergnaud, G. (1983). Multiplicative structures. In R. Lesh & M. Landau (Eds.), *Acquisition of mathematical concepts and processes* (pp. 127–174). Academic Press.

von Glasersfeld, E. (1995). *Radical constructivism: A way of knowing and learning* Falmer

Whitacre, I., & Nickerson, S. D. (2016). Prospective elementary teachers making sense of multidigit multiplication: Leveraging resources. *Journal for Research in Mathematics Education, 47*(3), 270–307. https://doi.org/10.5951/jresematheduc.47.3.0270

Xin, Y. P. (2012). *Conceptual model-based problem solving: Teach students with learning difficulties to solve math problems*. Sense. https://doi.org/10.1007/978-94-6209-104-7_1

Xin, Y. P. (2015). Research related to modeling and problem solving: Conceptual model-based problem solving: Emphasizing pre algebraic conceptualization of mathematical relations. In E. A. Silver & P. A. Kenney (Eds.), *More lessons learned from research: Useful and useable research related to core mathematical practices* (pp. 235–246). National Council of Teachers of Mathematics (NCTM).

Xin Y. P., Tzur, R., & Si, L. (2008a). *Nurturing multiplicative reasoning in students with learning disabilities in a computerized conceptual-modeling environment*. National Science Foundation (NSF) Funded Project

Xin, Y. P., Wiles, B., & Lin, Y. (2008b). Teaching conceptual model-based word-problem story grammar to enhance mathematics problem solving. *The Journal of Special Education, 42, 163–178.* https://doi.org/10.1177/0022466907312895

Xin, Y. P., Liu, J., Jones, S., Tzur, R., & SI, L. (2016). A preliminary discourse analysis of constructivist-oriented math instruction for a student with learning disabilities. *The Journal of Educational Research, 109*(4), 436–447. https://doi.org/10.1080/00220671.2014.979910

Xin, Y. P., Tzur, R., Hord, C., Liu, J., Park, J. Y., & Si, L. (2017a). An intelligent tutor-assisted math problem-solving intervention program for students with learning difficulties. *Learning Disability Quarterly, 40*(1), 4–16.

Xin, Y. P., Tzur, R., & Si, L. (2017b). *PGBM-COMPS Intelligent Tutor©*. Purdue Research Foundation, Purdue University. (Registration # 1-510-4735781).

Xin, Y. P., Park, J., Tzur, R., & Si, L. (2020). The impact of a conceptual model-based mathematics computer tutor on multiplicative reasoning and problem-solving of students with learning disabilities. *The Journal of Mathematical Behavior, 58*, 100762. Available online Feb 27, 2020. https://doi.org/10.1016/j.jmathb.2020.100762

Xin, Y. P., Zhang, D., Park, J. Y., Tom, K., Whipple, A., & Si, L. (2011). A Comparison of Two Mathematics Problem-Solving Strategies: Facilitate Algebra-Readiness. *The Journal of Educational Research, 104, 381–395.* https://doi.org/10.1080/00220671.2010.487080

Chapter 15
Nurturing Fractional Reasoning

Ron Tzur and Jessica H. Hunt

Abstract While central to a person's life and study of mathematics, learning to reason and operate with fractions is among the most relentless sources of difficulty for students and teachers. To support teachers' design and implementation of instruction that nurtures fractional reasoning in students with learning disabilities (LD) or difficulties, in this chapter we present a learning progression (trajectory) comprised of eight concepts, the early four based on the mental activity of iteration and the last four on recursive partitioning. This progression, unlike practices that focus on fractions as part-of-a-whole, focuses on and nurtures conceptualizing fractions as multiplicative relations. We organize the chapter to address two essential questions teachers frequently ask: "What does it mean for a student to know fractions well?" and, "What are some pedagogical moves, and the rationale for those moves, that teachers can use to foster intended conceptual advancements?" To inspire and assist teachers for using this progression to guide their work on fractions, we interweave descriptions of each concept and its importance with instructional tasks that help foster reflection on activities and abstraction of those concepts.

Keywords Fractional reasoning · Learning difficulties · Conceptual progression · Unit iteration · Recursive partitioning · Fractional tasks

R. Tzur (✉)
University of Colorado Denver, Denver, CO, USA
e-mail: RON.TZUR@UCDENVER.EDU

J. H. Hunt
North Carolina State University, Raleigh, NC, USA

© The Author(s), under exclusive license to Springer Nature Switzerland AG 2022   315
Y. P. Xin et al. (eds.), *Enabling Mathematics Learning of Struggling Students*,
Research in Mathematics Education,
https://doi.org/10.1007/978-3-030-95216-7_15

15.1   Nurturing Fractional Reasoning

Conceptual foundations in fractions, which underlie rational number sense, are a prerequisite for students' understanding of more advanced mathematical topics like ratio, rates, proportions, and algebraic reasoning (Hackenberg & Lee, 2015; Thompson & Saldanha, 2003). The importance of fractions is evident in the USA's *Common Core State Standards for Mathematics* (CCSSM, 2010), which emphasized understanding fractions throughout elementary school and tie fractions to rational numbers and proportional reasoning in middle school. Yet, Hunting et al.'s (1996) decades-old assertion still holds: "The teaching and learning of fractions is not only very hard, it is, in the broader scheme of things, a dismal failure" (p. 63). In Chap. 3, we highlighted two issues with commonly used resources. First, they focus the teaching of fractions on the limiting part-of-whole concept. Second, they reinforce instructional practices that diminish opportunities for students to engage in actions and reflective processes that bring about fractional concepts, stunting access to and advancement of conceptual understanding (Hunt et al., 2019; Woodward & Tzur, 2017).

Examples abound of conceptual and procedural issues that many students (LD and non-LD) manifest, which can help illustrate what knowing fractions well might mean. Two difficulties tend to stand out among nearly all students. First, when comparing fractions, students may say that 1/10 is greater than 1/4 because 10 is larger than 4. Similarly, students often misuse whole number reasoning and procedures for fractions (Mack, 1990; Van Hoof et al., 2015). For example, students may erroneously reason that 1/2 + 1/4 = 2/6 due to adding the numerators and denominators separately. Results of the USA's National Assessment of Educational Progress (NAEP, 2012, 2017, 2019) have repeatedly shown erroneous solutions of this kind, signaling students are not reasoning about fractions as "numbers in their own right" (Hackenberg, 2007, p. 28). These issues demonstrate a need to nurture reasoning about fractions as multiplicative relations, that is, as measures (Simon et al., 2018a; Tzur, 2000, 2019a).

A chief source of these difficulties is instruction (Tzur, 2013; Woodward & Tzur, 2017). Part-whole teaching approaches provide students with pre-partitioned shapes and ask them to respond by shading a requested number of parts. Modeled strategies and procedures for finding equivalent fractions are often presented next, followed by abstract procedures used to add, subtract, multiply, and divide these new kinds of numbers. The problem with this approach is the lack of link between students' available concepts, knowledge of how they come to build up understandings of fractions as quantities, and teachers' plans for instruction.

Standards and textbooks are useful in that they provide teachers with desired endpoints of learning. Yet, they fall short of guiding teachers how students might come to understand fractions in these desired ways. To address this issue, we present an 8-concept progression depicting how student reasoning about fractions may originate and develop. We interweave the conceptual discussion of each concept with a brief description of research-based activities/tasks that, by drawing on students' available concepts, proved effective in promoting students' construction of the (next) intended concept. We share teaching methods that we have been using in our own work with LD and non-LD students alike, as well as with their teachers. By such interweaving of the conceptual progression with ways to nurture it we attempt to address two essential questions teachers frequently ask: (a) What does it mean for a student to know fractions well? And (b) What are some pedagogical moves and the rationale for them that teachers can use to foster intended conceptual advancements?

15.2   Nurturing Students' Fractional Reasoning: An 8-Concept Progression

In Chap. 3 (this volume), we presented a theoretical explanation of how a student may come from not having to having a particular mathematical concept by bringing forth and transforming their available concepts. Using an example of the transformation of whole number concepts into the first fractional concept in our progression below, we pointed out three core tenets that underlie effectively designing, selecting, and implementing interactions with students that promote intended conceptual advances. In this chapter, we present an 8-concept progression, summarized by Tzur (2019a, b) based on the work of constructivist researchers (Hackenberg, 2013; Hackenberg et al., 2016; Hackenberg & Tillema, 2009; Norton & Boyce, 2013; Norton & Wilkins, 2009; Norton et al., 2018; Steffe & Olive, 2010). This progression commences with unit fractions as the first of a 4-concept cluster rooted in the mental activity of *unit iteration*. It advances to a latter, 4-concept cluster rooted in activities of partitioning the results of previous partitions, aka, recursive partitioning (e.g., 1/5 of 1/7, 2/5 of 3/7, etc.). Unlike a "part-of-whole" focus, nurturing those eight concepts is geared for yielding thinking about fractions as multiplicative relations (Simon et al., 2018a).

In Table 15.1 we first provide the key terms of the eight concepts and corresponding topics that each concept underlies (examples in parentheses). This concise list of eight concepts underlies numerous terms/procedures typically taught disjointedly. Immediately following the table, we describe each concept. Due to space, we do not articulate the participatory (prompt-dependent) and anticipatory (spontaneous/independent) stages through which each of those concepts evolves (see Chap. 3).

The Equi-partitioning (EP) concept   Student fractional reasoning begins with the *Equi-partitioning (EP) concept*, which supports understanding of and operations on unit fractions ($1/n$). As we explained in Chap. 3, students construct the EP concept as a transformation in their available concept of number as composite unit (Steffe & von Glasersfeld, 1985). The EP concept comprises two, interrelated sub-concepts. First, it involves conceiving of *each unit fraction as a multiplicative relation* between two units, one of which is designated as the unit of 1 ("whole"). Specifically, a unit fraction is quantified as a measure of the whole, that is, it fits n times within

Table 15.1 Eight concepts for fractional reasoning underlie various mathematical concepts

Concept	Familiar math supported
1. Equi-partitioning (EP) *Unit fractions (1/n)*	Understand unit fractions (four 1/4s make a whole) Order unit fractions (1/5 > 1/7)
2. Partitive Fractions (PF) *Within whole (m/n, $m \leq n$)*	Understand proper fractions (2/3 is two times 1/3) Add/subtract/order like-denominator fractions (2/7 + 3/7 = 5/7) Multiply fractions by small whole numbers (3 * 2/7 = 6/7) Compare/order like-denominator fractions (2/7 < 3/7 < 5/7)
3. Iterative Fractions (PF) *Exceed whole (m/n)*	Understand improper fractions (11/6 is 11 times 1/6) Understand/link to mixed numbers (17/6 is also 2 + 5/6) Multiply fractions by any whole number (10 * 2/7 = 20/7) Divide fractions by small whole number (6/7÷3 = 2/7)
4. Reversible Fractions (RF) *First within whole, then any m/n*	"Undo" non-unit fractions (divide 3/5 by 3 to make 1/5, then multiply by 5 to recreate whole or by 6 for 6/5) Solve price-change problems (if 70% sale price is $24– what was the original price)
5. Recursive Partitioning (RP) *First unit fractions only*	Multiply unit fractions (1/3 * 1/7 = 1/21) Understand decimals/percentages (1/10 * 1/10 = 1/100; thus 0.1 * 0.1 = 0.01 and 1/10 of 10% = 1%)
6. Unit Fraction Composition (UFC)	Multiply unit by non-unit fractions (5/7 * 1/3 = 5/21) *Then* multiply the other way (1/3 * 5/7 = 5/21) Understand equivalent fractions/decimals (1/3 * 7/7 = 7/21; 1/10 * 10/10 = 10/100 thus 0.1 = 0.10 and = 10%) Expand (or simplify to) unit fractions Add/subtract unlike denominator fractions (1/3 + 5/7 = 7/21 + 15/21 = 22/21 = 1+ 1/21)
7. Distributive Partitioning (DP)	Understand the result of division as a fraction Divide whole numbers (3 pizzas shared equally among 4 people = 3/4 pizza per person)
8. Any Fraction Composition (aFC)	Multiply any fraction by any fraction (5/7 * 2/3 = 10/21) Understand equivalency among any fraction type Translate between all fractional types (130/100 = 13/10 = 1.3 = 130% = 39/30, etc.) Add/subtract any type of fractions (2/3 + 8/7 = 14/21 + 24/21 = 14/21 + 1 + 3/21 = 1 + 17/21) Divide fractions and understand the algorithm (3/5÷2/7 = 3/5 * 7/2 = 21/10 = 2.1)

the whole – or the whole is *n* times as much of 1/*n*. For example, the fraction 1/7 assumes its meaning based on the whole being seven times as much of another piece. That is, neither the part alone nor the whole alone (even if equally partitioned) form a unit fraction, but rather the relationship between them. Second, the EP concept involves conceiving of the *inverse relationships between the number and size of unit fractions* (e.g., 1/5 > 1/6 because 6 > 5). When generalized and further symbolized (see Tzur, 2000), it provides a conceptual foundation for students' understanding of the inverse relationship and thus for ordering any collection of unit fractions (1/*n* < 1/*m* if *n* > *m*). We presented instructional tasks to promote the EP

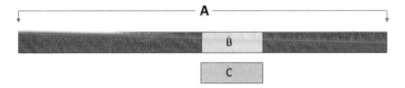

Fig. 15.1 Whole "A" with five unequal parts; pieces "B" (green) and "C" (blue) are equal

concept in Chap. 3, with a reference to a paper (Tzur & Hunt, 2015) that further details those.

Accordingly, when students who have constructed the EP concept are shown a "confusing" diagram (Fig. 15.1), they correctly solve two interrelated problems: What fraction of the whole (A) is the green piece (B) and what fraction of the whole is the blue piece (C)?

Using EP, their goal is to figure out how many times the embedded piece (B), or the disembedded piece (C), fit within whole A. Simply put, they intend to figure out how many times is whole A as much as piece B or Piece C. This goal triggers for them the activity of iterating either piece (or a copy of its length). The effect of their activity would be recognizing each piece as 1/6 of the whole, because it fits six times within the whole and thus the whole is six times as much of it. One could readily imagine erroneous solutions that people conceiving of fractions as part-of-whole might have, such as "B is 1/5 of A, because it is one-of-five parts," or "C is not a fraction of A as it is not part of it," or "I cannot tell what fraction is B of A because the parts are not equal" (see Tzur, 2019a).

Similarly, when shown another "confusing" diagram (Fig. 15.2), people who have established the EP concept would correctly solve the problem: "Pat ate 1/6 of a large chocolate bar and Sam ate 1/4 of a small chocolate bar. Which *fraction* is larger, 1/4 or 1/6?" Using EP, their goal would be to figure out, and compare, how many times each piece fits within *its respective whole* (bar). Their iteration-related activity implies 1/4 fits four times within the smaller bar and 1/6 fits six times within the larger bar (and the respective pieces of chocolate). The effect of their activity includes comparison of the fractions based on the number of times each unit fraction fits within its respective whole. Thus, overcoming the "temptation" to incorrectly compare the amounts of chocolate each person ate, a person's EP would yield a correct response that the fraction 1/4 is (always) larger than the fraction 1/6.

As Patrick Thompson (personal communication) suggested, such a "confusing" problem is a good litmus test for assessing a person's understanding of unit fractions.

The partitive (proper) fractions (PF) concept Students who have established the EP concept and a multiplicative concept for whole numbers can advance to constructing the second, *partitive fractions* (PF) concept (Olive & Steffe, 2010; Tzur, 2000). The PF concept gives rise to understanding composite fractions (*m/n*) as multiplicative relations and meaningfully operating on those numbers (e.g., 5/7 = 5 * 1/7). The PF concept is also based on the mental operation of iterating a unit. In

Fig. 15.2 A large
6-equal-parts chocolate bar
(left) and a small
4 equal-parts bar (right)

abstract symbols, the PF concept indicates an anticipation that iterating a given unit fraction ($1/n$) a few (m) times, not exceeding the n/n whole (i.e., $m \leq n$), yields a *composite fraction* that is m *times as much as* $1/n$. As Tzur (2000) asserted, only after a student establishes this anticipation do we present and help them link it to the symbol (m/n). The PF concept provides a conceptual foundation for adding and subtracting same-denominator fractions ($2/7 + 3/7 = 5/7$), as well as multiplying unit and non-unit fractions by a small whole number (e.g., 3 times $2/7 = 6/7$).

Tasks that support construction of the PF concept might include a scenario where two friends each eat $1/5$ of a pizza and think about the combined quantity as a fraction of the whole pizza. These students may initially consider the result as $1/5 + 1/5$, and the teacher will also emphasize that it is twice as much as $1/5$, hence two-fifths, $2 * 1/5$, and simply $2/5$. Critically, for students who have constructed the PF concept, adding $3/6 + 2/6$ preserves the underlying unit fraction (here, $1/6$, as opposed to adding $6 + 6$ and creating the unit fraction of $1/12$) (Tzur, 2000, 2007). Likewise, ordering the magnitudes of same-denominator fractions is accessible and meaningful to students with the FP concept. They anticipate that if the size of the unit fraction relative to the whole remains the same (e.g., $1/7$), iterating it more times yields, necessarily, a larger composite fraction (hence, for example, $2/11 < 3/11 < 4/11 \ldots < 11/11$).

Iterative/improper fractions (IF) Students who have established the PF concept can advance to constructing the third, *iterative fractions* (IF) concept. The same goal-directed activities of iterating a unit fraction ($1/n$) or a composite fraction (m/n) yields novel effects for students, namely, composite fractions larger than the n/n whole (Olive & Steffe, 2010; Tzur, 1999). A task to promote the IF concept could include the following activities in the JavaBars (Biddlecomb et al., 2013) software: "Create a long 'stick' (whole), then create $1/11$ of it. Now, repeat the $1/11$ twice and write what fraction is the new one of the whole ($2/11$). Next, repeat the last one three times and again write what fraction is it is of the whole ($6/11$). Finally, repeat the last fraction you made twice and tell what fraction is the resulting stick of the whole" (see Fig. 15.3). Reflecting on iteration and abstracting the meaning of novel composite fractions that exceed the "boundaries" of the given whole supports the iterative fraction (IF) concept. The IF concept supports not only the meaning for improper fractions but also their equivalence with mixed numbers (hence, meaningful back-and-forth conversions). It also opens the way to multiplication of unit and composite fractions by any whole number (e.g., $10 * 2/7 = 20/7$).

We cannot stress enough a crucial point. An adult who knows both the partitive (proper) fraction and the iterative (improper) fraction concepts might consider advancing to the IF concept as a clear-cut extension of the partitive fraction (PF)

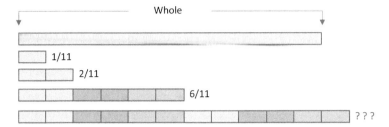

Fig. 15.3 A multistep task using iteration of units to create an improper fraction

concept. After all, if one can iterate 1/7, say, six times to produce 6/7 it would be an "easy" extension to iterate it eight times and produce 8/7. As we have experienced with students, and teachers reported to us similarly, this is not the case. Conceptualizing the iteration of a unit fraction beyond the whole is quite a conceptual leap. It requires anticipating that a unit fraction, once conceived of through iteration within a given whole – can be "freed" from that whole (Hackenberg, 2007; Tzur, 1999). That is, with an IF concept, the student considers a unit fraction ($1/n$) in relation to (but apart from) the potentially n-part whole of which it is a fraction.

Students who have constructed the iterative fraction (IF) concept can correctly answer questions such as, "What if 4 friends ate 2/7 of a pizza each?" (here, 8/7). The reason is they anticipate the effect of such iteration to be a quantity that is four times as much as the composite fraction 2/7. For these students, 8/7 of a pizza is a sensible composite fraction of a 7/7-pizza. Understanding this entails there is more than one whole pizza. Accordingly, such students can also learn to anticipate the resulting quantity (8/7) as 1 whole (e.g., 7/7-pizza) and one more 1/7 of that whole ($1\frac{1}{7}$) – and recognize the equivalence of the two quantities. Thus, the IF concept also supports equivalence of and conversions between improper fractions and mixed numbers. Conversely, a student who is yet to abstract the IF concept is likely to view the result of four iterations of two-sevenths as impossible: "Not sure; maybe 7/8, because we have 8 and 7, but the numerator cannot be larger than the denominator." The IF concept thus demarcates between people who conceive of fractions as part-of-whole and those who conceive of fractions as multiplicative relations. For the former, the literal meaning of something being a part of something else entails the part must be smaller than the whole. This meaning contradicts the possibility of "improper" fractions. For the latter, a fraction larger than the whole simply entails a multiplicative relation resulting from iterating a unit fraction ($1/n$) more times than needed to produce the given, n/n whole.

Reversible (de-composing) fractions Students who have established the IF concept can advance to constructing the fourth, *reversible fractions* (RF), concept. Tasks to promote the RF concept involve a given composite fraction (e.g., 3/5) and a question – also the goal – to use it for reproducing the whole of which it is such a fraction (Boyce & Norton, 2017; Simon et al., 2018b; Steffe, 2003; Tzur, 2004). This concept is termed reversible because, essentially, a student who has abstracted the RF concept can reverse ("undo") the sequence of activities leading from taking

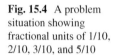

Fig. 15.4 A problem situation showing fractional units of 1/10, 2/10, 3/10, and 5/10

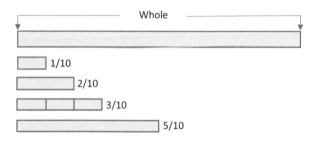

an unpartitioned whole and producing the given composite fraction. That is, the student can reverse the iteration supposedly used to create a composite fraction m/n by partitioning it into m parts to create $1/n$ and then "undo" the initial partitioning of the whole, which created $1/n$, by iterating n times to make the n/n whole.

An effective start for teaching the RF concept is shown in Fig. 15.4. The task is to use any of the given fractions to reproduce the whole. Students often begin by iterating the 5/10 twice, then the 1/10 ten times, then the 2/10 five times. At this point, possibly getting stuck, the teacher would ask if they can find a way to also use the 3/10 to make the whole. They eventually realize the possibility of iterating it three or four times and decomposing one unit of 3/10 to produce a unit of 1/10 that allows them to accomplish their goal. It is this decomposition, using a marked composite fraction, that opens the way to also decomposing unmarked composite fractions (see Fig. 15.5).

An example of an established reversible fraction concept "at work" is to consider an unmarked bar, which is 7/10 of another bar (i.e., the whole). Figure 15.5 shows Sue's chocolate bar, which we are told is 7/10 of Ana's chocolate bar; the task is to produce (e.g., draw) Ana's bar. In considering what the whole bar must be, a student reasoning with the RF concept would partition the 7/10 into 7 equal units, anticipating that one of the equal units is 1/10 of the candy bar. To reproduce the entire bar, the student would then iterate 1/10 ten times. As the studies cited in this paragraph repeatedly show, until constructing the RF concept students would likely incorrectly partition the given 7/10 into ten parts. To them, such a partition seems sensible (and we should expect it) because, to this point in their conceptual development, the denominator has entailed the partition of a given whole bar.

Why is the RF concept relevant? It underlies solutions to challenging problems from other contexts, like price change. For example, a shirt was purchased on sale for $24, which is 4/5 of the original cost. To find the original cost, a student with the RF concept would readily partition 24 into *four* parts to "undo" the iteration of 1/5, finding this unit fraction of the original price is $6. The student would then iterate 1/5 ($6) five times to undo the partition of the whole that created it, arriving at $30 for the original price of the shirt.

Progressing: From iterating to partitioning fractions As a transition from the first cluster of four iteration-based concepts for fractional reasoning to the next cluster of more advanced concepts, we point out two insights. First, iteration under-

Sue's bar is 7/10
of Ana's bar

Draw Ana's
bar here

Fig. 15.5 A problem situation showing Sue's bar, which is 7/10 of Ana's unknown bar

lies and is afforded by a student's concept of number as composite unit (Steffe & von Glasersfeld, 1985). An indicator of this concept can be gleaned from the strategy a student uses *spontaneously* (non-prompted) to add two whole numbers (e.g., $8 + 7$). Counting-on $(8; 9–10–\ldots–15)$ indicates an initial, and likely insufficiently developed concept of number. A much stronger, "safer" indicator is the break-apart-make-ten (BAMT) strategy (Tzur et al., 2021). To assess students' conceptual readiness for all four iteration-based fractional concepts, a teacher is encouraged to determine their spontaneous additive strategy.

Second, and leading to the second cluster of four concepts rooted in recursive partitioning, we stress a key aspect of iteration activities: treating unit fractions as intact entities. The leap to partitioning unit fractions, themselves the result of partitioning, is thus not unlike children's leap from operating on the unit of 1 as an intact entity. As Piaget et al. (1960) had shown, children initially cannot fathom decomposing the unit of 1 into smaller units. Decomposing unit fractions $(1/n)$ is even more challenging, because a child needs not only to accept and make sense of such an activity but also to conceive of its effect as a *fractional unit of the original whole.* For example, two siblings might have equally shared a whole pizza. One of them now decides to create three equal slices, to share the half with two other friends (Fig. 15.6). A child with a part-of-whole concept would have difficulties in overcoming the "confusing" appearance of four unequal parts and likely respond incorrectly (1/4). Instead, a child would need to understand both that one friend's (single) slice is 1/3 of 1/2 – hence a half can itself be decomposed equally – and link that slice back to the whole pizza (1/6 of it). For the former, the child needs to conceive of a particular unit fraction as a "whole in its own right." For the latter – the child would have to conceive of the slice as a unit fraction (1/3), embedded within another unit fraction (1/2), which is embedded within a third unit (the whole, 1). Indeed, such a concept entails coordinating three levels of units (e.g., 1, 1/2, and 1/3 of 1/2), a conceptual leap for students (Hackenberg, 2013; Hackenberg & Lee, 2015; Norton et al., 2015; Ulrich, 2016).

Recursive Partitioning of Unit Fractions Students assessed to have established at least the EP concept (unit fractions), and preferably also the other three, iteration-based concepts (PF, IF, and RF), can advance to constructing the fifth, *recursive partitioning* (RP) concept. Problem situations like the pizza sharing in the previous paragraph can be used to promote the RP concept. A student's goal in such situations would be to figure out how many times a unit fraction of a unit fraction (e.g., 1/3 of 1/2) fits within the given whole, that is, identify the effect of partitioning a

Fig. 15.6 A pizza shared equally between two, with one half equally sliced into three

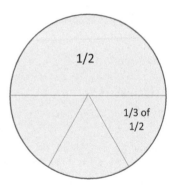

unit fraction into a unit fraction. To this end, they can bring forth the activity of iteration of embedded units. For example, in the situation above they could initially partition the other (top) half into three equal parts and realize the situation now involves two units (halves), each comprised of three subunits, and thus conclude that one of those subunits fits exactly six times within the whole (hence, 1/6 of it).

Next, we can foster further reflection across instances of such recursive partitioning to promote students' conceptualization of the invariant operation involved, namely, multiplication of the denominators. To this end, we may engage them in similar tasks in which we do not change the first partition while changing the second (e.g., 1/4 of 1/2, 1/5 of 1/2, 1/9 of 1/2). After each task, we would orient their reflection onto the relationship between the resulting number of parts (e.g., 8, 10, or 18, respectively) that fit within the entire whole and the two numbers used in the two-step partitioning process (e.g., "You said that 1/4 of 1/2 is 1/8, because there are 4 parts in this half and 4 parts in the other half; Is this surprising? Can you explain why '8 parts' using a different operation?"). Later on, we would promote their ability to anticipate the effect of partitioning both halves without/before actually completing such an activity. For example, we could ask, "You got half of a chocolate bar, and your half shows 6 cubes. Could you tell me, without changing the picture, what fraction is one cube of the entire chocolate bar?" (Fig. 15.7).

Eventually, we would use further variation of the tasks to include also different unit fractions as the first step (e.g., 1/3 of 1/4, 1/5 of 1/4, ... 1/2 of 1/5, 1/3 of 1/5, etc.). Our goal would be to nurture students' abstraction of the intended anticipation: the effect of partitioning a unit fraction ($1/n$) into k subunits is a unit fraction of particular size – a size determined by the multiplication of the two denominators (i.e., $1/n$ of $1/m = 1/(n * m)$). Such an anticipation would serve as a conceptual safeguard against confusions found in the reasoning of students who conceptualize fractions as parts-of-whole (e.g., thinking that one chocolate cube in Fig. 15.7 is 1/7 of the whole bar because they see a whole with seven parts). Here, we typically ask students to come up with *their own rule* of how one solves similar problems, asking the students to explain why it must always work. Their understanding of the rule would allow the teacher to present a symbolic form for it: $1/n * 1/m = (1 * 1)/(n * m)$.

Fig. 15.7 Two halves of a chocolate bar, one further partitioned into 6 cubes

Regarding the EP concept, we asserted above that, just like the unit of 1 is the foundation for all whole numbers, the unit fraction $1/n$ is the foundation for all fractions. Here, similarly, we extend this assertion to unit fraction of a unit fraction $(1/(m * n))$ being the foundation for all fractions that could be produced through recursive partitioning (e.g., $1\% = 1/10$ of $1/10$). As we shall demonstrate in the next three concepts, being able to determine the resulting unit fraction after recursive partitioning would constantly serve as a starting point. Before turning to the next concept, however, we point out two implications of the RP concept for fraction learning.

First, recursive partitioning is a necessary basis for understanding decimals and percentages, in the special case in which all numbers serving as its input are powers of 10. The right side of Fig. 15.8 shows this for $1/100$ being $1/10$ of $1/10$ (also 0.01 and 1%); the reader can imagine extension to show $1/1000$ ($0.001 = 0.1\%$), etc. This understanding leads to understanding why "moving to the right" in a decimal number is equivalent to dividing the previous magnitude by 10. To teach the relationships among those magnitudes, teachers often use "base ten blocks" while reversing the order of the blocks, telling students to consider that what was "1000" is now "1." At issue is that such an approach already requires a student to somehow fathom the process leading from 1 to $1/10$ (a "flat") and then to $1/100$ ("long"), etc., that is, the very concept they need to construct. Instead, we recommend using a software application that allows students to experience the recursive partitioning activity and reflect on its effects (we created Fig. 15.8 this way). Our own preference, and recommendation, is to turn the focus onto $1/10$ of $1/10$ as a special case of recursive partitioning (RP) only after students have established this concept for any denominator.

The second implication of recursive partitioning is pretty straightforward – providing a solid conceptual basis for equivalent fractions. To produce a unit fraction $(1/m)$ of a unit fraction $(1/n)$, the student partitions the first unit into so many (m/m) units, hence creating a composite fraction of $m/(m * n)$. It is easy to illustrate this with our pizza example ($1/3$ of $1/2$, Fig. 15.6). There, students who have constructed the PF and IF concepts would have readily also conceived of the three units that they introduced into $1/2$ (to produce $1/6$) as $3/6$ of the whole pizza. We prefer, and recommend, to postpone orienting students' attention onto such equivalencies until after they have established an anticipatory stage of determining the denominator of the resulting unit fraction ($1/(n * m)$, e.g., $1/(2 * 3) = 1/6$).

We contend that teaching equivalent fractions through recursive partitioning is more sensible than just "showing" that two fractions, such as $1/2$ and $3/6$, have the same size. Our reason is that in recursive partitioning there is self-identity between a unit fraction (e.g., $1/2$) and a composite fraction resulting from a student's

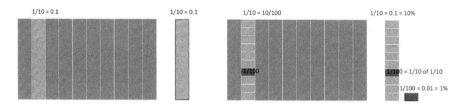

Fig. 15.8 Decimals and percentages understood through recursive partitioning, with units both embedded within and disembedded from their respective, upper-level units

partitioning of that very unit fraction (e.g., 3/6). This way, two equivalent fractions are different representations of, literally, the same quantity. Using task variation similar to the one we presented above would readily also lead students to formulate a rule, linked with a meaningful, abstract symbol to "capture" the process and product: $1/n * m/m = (1 * m)/(n * m) = m/(n * m)$. In turn, a teacher could then foster reversing of the recursive partitioning process as a means to making sense of simplifying fractions.

Unit Fraction Composition Students assessed to have established the RP concept can advance to constructing the sixth, *Unit Fraction Composition* (UFC), concept. They would start by working for quite some time on less demanding problems in which they learn to relate composite fractions of a unit fraction back to the whole (k/n of $1/n$). For example, letting students show the two-step partitioning in the computer (Fig. 15.9), a task may ask: "I ate 4/5 of one slice (1/3) of a whole pizza. What fraction of the whole pizza did I eat?" Before pulling out just one piece, students (like many teachers) have a hard time figuring out the answer. The teacher fosters progress by orienting students' attention onto just one of the 1/5 pieces embedded within one-third: "What fraction of the whole pizza is just this pink piece?" Students may need additional support, for example: "Pretend you pulled-out the pink piece and measured it – what would the number be?" At this point, students bring forth their RP concept, and explain the pink piece is 1/15 of the whole pizza because it is 1/5 of 1/3 (e.g., "if you ate only 1/5 of 1/3 you'd eaten 1/15 of the whole pizza"). To a teacher's follow-up: "So what about all four pieces shown as 4/5 of 1/3?" students are likely to bring forth their PF concept and realize it is four times as much as 1/15 – hence 4/15 (e.g., "But I didn't eat only 1/5 of 1/3 – I ate four-fifths so altogether it's four times 1/15, or 4/15").

Using a variation of such tasks with different numbers and contexts would help students realize two critical effects of their activity. First, they realize that, invariably, the denominator of the answer is determined by multiplying the two denominators of the fractions for the very same reasons it was so determined in recursive partitioning of unit fractions. Second, they realize that, invariably, the effect of the entire composition of fractions yields a composite fraction that is k-times (e.g., 4) as much as the newly produced unit fraction (e.g., $4 * 1/15$). The teacher can then engage them in formulating a rule for multiplication similarly to how they did for recursive partitioning of unit fractions. Initially, they can work the rule for particular

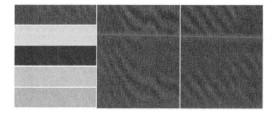

Fig. 15.9 A student's production of two-step composition of 4/5 of 1/3 of a pizza

Fig. 15.10 A whole pizza partitioned into 4/5, then only that fraction partitioned into thirds

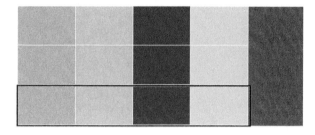

examples (e.g., to find 4/5 of 1/3, I could multiply 4/5 * 1/3 = (4 * 1)/(5 * 3) = 4/15). Eventually, they will suggest a rule for any such composition of fractions (e.g., "When taking a fraction of a unit fraction, we always multiply the numerators and the denominators because …"). At this point, the teacher may add the symbolic rule: $k/m * 1/n = (k * 1)/(m * n) = k/mn$.

The second, more challenging aspect of unit fraction composition is when the order of partitioning changes to first partitioning a whole into k/m and then take $1/n$ of that fraction (e.g., creating 4/5, then focusing on 1/3 of that 4/5; see Fig. 15.10). Before reading forward, we encourage the reader to look at Fig. 15.10 and, *not* using any procedure, algorithm, or property you know, solve what fraction is 1/3 (black-outline at the bottom) of 4/5 of the whole – and why? Based on our work with students and teachers, we expect you would immediately recognize that, in spite of the "same answer," this is a much more demanding task than 4/5 of 1/3 because 1/3 unit is distributed over four units of 1/5 each. To solve such a task, a student would have to set the same subgoal, namely, focusing on 1/3 of just one 1/5 and realize (using their RP concept) it is, again, 1/15 of the whole. With that in mind, the student could then again bring forth the PF concept, realizing that they have 1/15 distributed into each of the four units (of 1/5), hence 4 * 1/15 = 4/15.

To foster this line of reasoning, similarly to the sequence of tasks presented earlier for this unit fraction composition concept, a teacher could ask the students to pull out just one of the pieces (e.g., the pink one, or a blue one, or a gray one). Before the students use the computer to measure it – the teacher would ask: "What fraction do you think the computer would show and why?" Then, the teachers let the students measure and find that one piece is 1/15, before reflecting on their anticipated answer prior to measuring. Similarly, the teacher would now follow with a task: "So you found 1/3 of just 1/5, right? What about 1/3 of all 4/5? Can you tell before using the computer to measure it?" Whatever the response – students will then measure, find out itis 4/15, and again explain why 4/15 would be a reasonable answer – and why it should be expected. Figure 15.11 shows a screen shot after the students pulled out the pink piece and measured it, then also repeated it four times and measured again. The final sequence of tasks would emulate those in the initial part of the unit fraction composition concept – leading students to first formulate a rule and then generalize it into a symbolic form.

At this point, it would be helpful to further students' application of unit fraction composition to decimals and percentages. Consider, for example, a task: "A tank holds 0.4 cubic-meter of water. It takes 10 hours to fill it with water. How much water does it hold after 1 hour (express your answer as a proper fraction, decimal, and percentage of cubic meter)? Figure 15.12 shows a JavaBars diagram students could create to first depict the situation. They would first determine, possibly with a prompt to focus on this issue, that 1 hour of water being filled is 1/10 (0.1, 10%) of the entire time and thus of the amount of water in a full tank. Using their newly constructed UFC concept, they would likely reason, "1/10 (0.1, 10%) of just 1/10 of a cubic-meter is 1/100 (0.01, 1%) of the entire tank; however, we have 4/10 of a cubic-meter, so we need four times as much water. So, in 1 hour 4 * 1/100 = 4/100 (or 0.04, or 4%) of a cubic meter of water is being filled.

When students seem to have established an anticipation of the composition of unit and composite fractions, coupled with their recursive partitioning anticipation, creating common denominators for fractions as a first step in adding/subtracting them is supported conceptually. For example, consider a task: "Water tank A holds 1/3 of cubic-meter and water Tank B holds 5/7 of a cubic-meter. How much water do both tanks hold when full?" Figure 15.13 below shows how a student might use JavaBars to solve such a problem. The left side consists of a gray rectangle representing 1 cubic meter (top), Tank A (1/3 of a cubic meter) below it, and Tank B (5/7 of a cubic meter) below Tank A. The students then realized that adding thirds and sevenths would not work, but each of those could be recursively partitioned to create a unit fraction common to (can measure) both – a commensurate unit (Steffe, 2010). They purposely partitioned the 1/3 unit into 7 subparts, anticipating each would be 1/21 of the whole cubic meter and the entire tank being 7/21 of the whole. They then purposely partitioned, each of the five 1/7th, into three subparts – with the very same anticipation the 5/7 is equivalent to 15/21 cubic meters. Finally, they joined the images of both tanks and placed it under the whole to show the result is

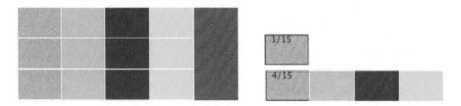

Fig. 15.11 A unit fraction (1/15) pulled out, measured, repeated four times, and measured again

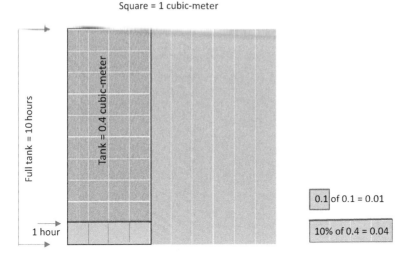

Fig. 15.12 A water tank (4/10 of a cubic meter) fills in 10 hours. How much is filled in 1 hour?

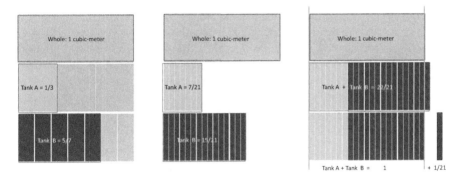

Fig. 15.13 Depicting addition of water in two tanks, holding 1/3 and 5/7 of a cubic meter

1/21 more than one cubic meter. We emphasize that, for teaching, it would make more sense to begin with small, "easy" numbers (e.g., 1/2 + 2/5).

Distributive partitioning Students assessed to have established the UFC concept can advance to constructing the seventh, *Distributive Partitioning* (DP), concept. This concept would underlie their goal-directed activities (and anticipation of effects) in situations involving the sharing of a given number of wholes. For example, when sharing three pizzas equally among four people, a student who has abstracted the DP concept uses the number of sharers in an anticipatory manner and can quantify a share in terms of one whole as a quantifying unit and also in terms of all of the wholes as a quantifying unit. This student might construct three objects ("pizzas") and partition each into fourths so they could distribute to each person 1/4 from each of those wholes (Fig. 15.14 – upper/larger circles). Bringing forth their

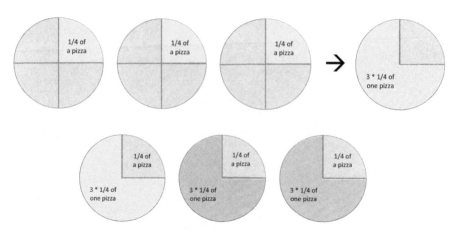

Fig. 15.14 Sharing 3 pizzas among 4 people by distributing 1/4 of each pizza = 3/4 of one pizza to each person

UFC will enable understanding that 1/4 taken from each of the three wholes is equivalent to three times 1/4 (hence 3/4) of one pizza. Furthermore, it would be important to foster an equivalent way of distributing the three pizzas, namely, giving each person 3/4 of a pizza and realizing there are three, 1/4-pizza slices each, given to the fourth person (Fig. 15.14 – lower circles). To further link this way of reasoning to decimals and percentages, a teacher could use variation of the numbers and contexts (e.g., ten children equally share the $7 profit from a lemonade sale; how much money does each child receive?). A teacher can then present the abstract, symbolic linkage between the operation of sharing objects and dividing the corresponding numbers, hence the particular meaning of division indicated by a "fraction-bar."

Any fraction composition Students assessed to have established all (prior) seven concepts can advance to constructing the eighth concept, *Any Fraction Composition* (aFC). These students can conceive of a unit or non-unit fraction of another non-unit fraction as a three-level structure (units of units of units). For example, consider the task: "Each lap in a stadium is 1/4 of a mile. As a warm-up, a runner intends to complete three full laps (that is, 3/4 of a mile in all), while going slow during the first 2/5 of the entire warm-up. What distance (in miles) would the runner go during this first (slow) part?" In Fig. 15.15 we use JavaBars to illustrate the activities a student with the aFC concept would purposely take. A teacher could use similar, yet simpler and perhaps partial problems, to teach this concept. As shown, the top, gray bar represents 1 mile, which in this problem is also "the 4/4 whole." Below it, the student would show 3/4 of a mile, that is, three laps of 1/4 mile each. Using the established unit fraction composition concept, the student then introduced 2/5 of a lap into *each of the three (1/4-mile) laps* by recursively partitioning each 1/4 into five equal parts. Accordingly, the student noted that one such segment (single, gray piece) is 1/5-lap * 1/4-mile = 1/20 mile. Finally, the student reorganized the three

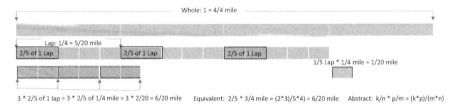

Fig. 15.15 The runner's distance in the slow part – 2/5 of 3/4 of a mile

units of 2/5 (of 1/4-mile; pink) to show that, combined, they are equal to 6/20 of a mile. The teacher would engage them in discussions that lead to reasoning how these results link with symbolized actions (2/5 * 3/4 = (2 * 3)/(5 * 4) = 6/20.

As we noted in Table 15.1, the aFC concept "opens the door" to all operations on fractions, decimals, and percentages – including making sense of division of fractions. Such a sense and how it may be promoted goes beyond the scope of this paper. It may be above what teachers would consider as sensible goal for some students. However, we believe that, at least for the teacher, it is important to make sense of how division of fractions actually operates on quantities. To provide a glimpse into it, we culminate the 8-concept progression with a diagram showing a possible solution with a clear explanation for the task: "A baker has 3/4-cup of flour left. It takes 2/5-cup to bake one pie. How many pies (including parts of pies) can be baked before using all the flour?" We approach this task using a quotitive meaning of division (i.e., how many times the 2/5-cup of flour "goes into" 3/4-cup of flour). Before moving forward to our explanation of the diagram, we encourage the reader to figure out what is going on here on your own.

Supposing the reader has taken the time to interpret the picture, we shall now add our own explanation. The upper section of Fig. 15.16 shows three equal-size wholes, indicating (a) the size of a full cup of flour (gray, middle), with (b) a 3/4-cup above it and (c) a 2/5-cup below it. The lower section shows that the 2/5-cup "goes into" the 3/4-cup more than once (orange copy of 2/5) but less than twice (the additional dark-gray copy of it). At issue is figuring out the precise number of times. To this end, the student faced a subgoal: creating a commensurate unit fraction, that is, a unit that allows measuring both 3/4 and 2/5 (aka 'common denominator'). To accomplish this subgoal, the student brought forth and applied "cross recursive partitioning": partitioning each 1/4-cup into five and each 1/5-cup into four equal parts, which their prior concepts and the aFC concept support. These recursive partitioning actions yielded 1/20 of a full cup as a commensurate unit. Symbolically, these actions entail we multiplied each of three 1/4th by 5 and each of two 1/5th by 4; numerically, to divide 3/4 by 2/5 we so far multiplied 3 * 5 and 2 * 4 (which underlies the invert-and-multiply algorithm).

The "tricky" part now is to realize that the question and the referent whole have changed. Instead of asking about one whole cup, the 3/4-cup became a new whole to be measured by 2/5-cup. Because the student created a commensurate unit, this goal is equivalent to measuring the 15/20-cup by the 8/20-cup. As the 1/20 unit is

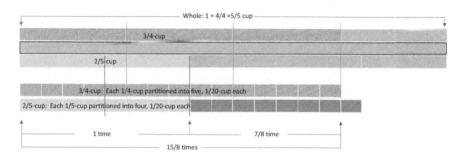

Fig. 15.16 Illustrating quotitive division of 3/4 by 2/5: How many times does 2/5 go into 3/4?

common to both composite fractions, figuring out how many times 2/5 "goes into" 3/4 is equivalent to asking how many times 8 "goes into" 15 (for detailed analysis of teaching leading to such understanding see Simon et al., 2010). The answers to the given division problem (3/4-cup divided by 2/5-cup) are written below the diagrams, as 15/8, or 1 + 7/8, *of the 3/4-cup.*

15.3 Concluding Remarks on Nurturing Students' Development of Fractional Reasoning

In this chapter, we have provided an 8-concept progression in fractional reasoning that, by and large, can be nurtured in students with learning disabilities (LD) and difficulties. Sharply distinguished from approaching fractions as "parts-of-wholes," this progression consists of two clusters, the first rooted in iteration activities and the second (more advanced) in recursive partitioning activities (see Table 15.1):

$$\text{Cluster A} : (1)\text{EP} \rightarrow (2)\text{PF} \rightarrow (3)\text{IF} \rightarrow (4)\text{RF}$$
$$\text{Cluster B} : (5)\text{RP} \rightarrow (6)\text{UFC} \rightarrow (7)\text{DP} \rightarrow (8)\text{AFC}$$

For each of these concepts, we have provided sample tasks that teachers can adopt/adapt to fit with their students' evolving (available) reasoning and pointed out links to curricular topics supported by those concepts.

The progression we presented seems consistent with the literature on fractional reasoning of students with LD, which is complex and diverse. In this wider body of work, two considerations seem to stand out. First, for students with LD, the equipartitioning concept is a necessary and foundational access point to fractions. For instance, Grobecker (2000) researched partitioning activities that were presented within the context of dividing number line wholes into equal sized parts. Seven students with LD were given a line and eight were given blocks. When set adjacently, the measure of the eight blocks equaled the measure of the line. Students were given various problems about the blocks and the line relationship (e.g., modeling and solving 1/4 + 1/8). Twelve-year-old students with LD did understand the relationship between the unit block and the line (e.g., 1/8 iterated eight times is 1), but did not yet meaningfully associate the part and the whole to generate equivalent relationships. This research is one example of both the importance and the accessibility of the equi-partitioning concept for students, especially students who may think differently. It also underscores the importance of multiplicative reasoning and its interconnected nature with fraction concepts (Hackenberg & Tillema, 2009; Tzur, 2019a). In fact, all seven concepts in the progression beyond the first (EP) depend on the child's conceptualization of unit fractions as a multiplicative relation (measure), as well as on their construction of multiplicative concepts for whole numbers (Tzur et al., 2013; Tzur & Xin, this volume).

As we stated across three chapters in this volume, teachers can support their students by viewing "intervention" as a platform to nurture students' access to mathematics by using and reflecting upon their available ways of reasoning. We have discussed how teachers can employ tools within intervention that help students bring about their conceptual change. We have argued that key design and pedagogical moves available to teachers include (a) *bridging, varying, and reinstating* platform tasks and *(b) interactive prompting and gesturing* that support students' noticing and reflection upon their actions toward logical necessity and mathematical connections (Hunt & Tzur, this volume). In this chapter, we further specified how those general ideas about teaching and learning can be applied for advancing students' fractional knowledge. Specifically, we paired a rather succinct conceptual progression (learning trajectory) with the intended endpoints of learning.

The tasks and prompting create a powerful instructional tool that teachers can use to promote their students' fractional thinking. Our instructional suggestions are grounded in the mathematical ideas and development of thinking by students as they grapple with problematic situations in their experience (Clements & Sarama, 2009). Knowing how students come to understand mathematical concepts makes task selection and planning of interactions with them based on learning theory more doable. By explicating the 8-concept progression, we intend to empower educators such that they empower students with LD to develop a conceptual foundation upon which to develop deeper understanding of fractions and beyond.

References

Biddlecomb, B., Olive, J., & Sutherland, P. (2013). *JavaBars 5.3*. http://math.coe.uga.edu/olive/welcome.html#Software%20developed%20through%20the%20Fractions%20Project

Boyce, S., & Norton, A. (2017). Dylan's units coordinating across contexts. *Journal of Mathematical Behavior, 45*, 121–136. https://doi.org/10.1016/j.jmathb.2016.12.009

Clements, D. H., & Sarama, J. (2009). *Learning and teaching early math: The learning trajectories approach*. Routledge.

Grobecker, B. (2000). Imagery and fractions in students classified as learning disabled. *Learning Disability Quarterly, 23*(2), 157–168. https://doi.org/10.2307/1511143

Hackenberg, A. J. (2007). Units coordination and the construction of improper fractions: A revision of the splitting hypothesis. *Journal of Mathematical Behavior, 26*, 27–47.

Hackenberg, A. J. (2013). The fractional knowledge and algebraic reasoning of students with the first multiplicative concept. *The Journal of Mathematical Behavior, 32*(3), 538–563. https://doi.org/10.1016/j.jmathb.2013.06.007

Hackenberg, A. J., & Lee, M. Y. (2015). Relationships between students' fractional knowledge and equation writing. *Journal for Research in Mathematics Education, 46*(2), 196–243.

Hackenberg, A. J., & Tillema, E. S. (2009). Students' whole number multiplicative concepts: A critical constructive resource for fraction composition schemes. *The Journal of Mathematical Behavior, 28*(1), 1–18. https://doi.org/10.1016/j.jmathb.2009.04.004

Hackenberg, A. J., Norton, A., & Wright, R. J. (2016). *Developing fractions knowledge*. Sage.

Hunt, J. H., Silva, J., & Lambert, R. (2019). Empowering students with specific learning disabilities: Jim's concept of unit fraction. *The Journal of Mathematical Behavior, 56*, 100738. https://doi.org/10.1016/j.jmathb.2019.100738

Hunting, R. P., Davis, G., & Pearn, C. A. (1996). Engaging whole-number knowledge for rational-number learning using a computer-based tool. *Journal for Research in Mathematics Education, 27*(3), 354–379.

Mack, N. K. (1990). Learning fractions with understanding: Building on informal knowledge. *Journal for Research in Mathematics Education, 21*(1), 16–32.

National Assessment of Educational Progress (NAEP). (2012). http://nationsreportcard.gov/math_2011/gr4_national.asp?subtab_id=Tab_5&tab_id=tab2#chart

National Assessment of Educational Progress (NAEP). (2017). https://www.nationsreportcard.gov/math_2017/nation/achievement?grade=4

National Assessment of Educational Progress (NAEP). (2019). https://www.nationsreportcard.gov/math_2019/nation/achievement?grade=4

National Governors Association Center for Best Practices. (2010). *Common Core State Standards Initiative*. National Governors Association Center for Best Practices, Council of Chief State School Officers. http://www.corestandards.org/the-standards

Norton, A., & Boyce, S. (2013). A cognitive core for common state standards. *The Journal of Mathematical Behavior, 32*(2), 266–279. https://doi.org/10.1016/j.jmathb.2013.01.001

Norton, A., & Wilkins, J. L. M. (2009). A quantitative analysis of children's splitting operations and fraction schemes. *The Journal of Mathematical Behavior, 28*(2–3), 150–161. https://doi.org/10.1016/j.jmathb.2009.06.002

Norton, A., Boyce, S., Ulrich, C., & Phillips, N. (2015). Students' units coordination activity: A cross-sectional analysis. *Journal of Mathematical Behavior, 39*, 51–66.

Norton, A., Wilkins, J. L. M., & Xu, C. Z. (2018). A progression of fraction schemes common to Chinese and U.S. students. *Journal for Research in Mathematics Education, 49*(2), 210–226.

Olive, J., & Steffe, L. P. (2010). The partitive, the iterative, and the unit composition schemes. In L. P. Steffe & J. Olive (Eds.), *Children's fractional knowledge* (pp. 171–223). Springer.

Piaget, J., Inhelder, B., & Szeminska, A. (1960). *The child's conception of geometry* (E. A. Lunzer, Trans.). W. W. Norton.

Simon, M. A., Saldanha, L., McClintock, E., Akar, G. K., Watanabe, T., & Zembat, I. O. (2010). A developing approach to studying students' learning through their mathematical activity. *Cognition and Instruction, 28*(1), 70–112.

Simon, M. A., Placa, N., Avitzur, A., & Kara, M. (2018a). Promoting a concept of fraction-as-measure: A study of the Learning Through Activity research program. *Journal of Mathematical Behavior, 52*, 122–133. https://doi.org/10.1016/j.jmathb.2018.03.004

Simon, M. A., Placa, N., Kara, M., & Avitzur, A. (2018b). Empirically-based hypothetical learning trajectories for fraction concepts: Products of the Learning Through Activity research program. *Journal of Mathematical Behavior, 52*, 188–200. https://doi.org/10.1016/j.jmathb.2018.03.003

Steffe, L. P. (2003). Fractional commensurate, composition, and adding schemes: Learning trajectories of Jason and Laura: Grade 5. *Journal of Mathematical Behavior, 22*, 237–295.

Steffe, L. P. (2010). The unit composition and the commensurate schemes. In L. P. Steffe & J. Olive (Eds.), *Children's fractional knowledge* (pp. 123–169). Springer.

Steffe, L. P., & Olive, J. (2010). *Children's fractional knowledge*. Springer.

Steffe, L. P., & von Glasersfeld, E. (1985). Helping children to conceive of number. *Recherches en Didactique des Mathematiques, 6*(2/3), 269–303.

Thompson, P. W., & Saldanha, L. (2003). Fractions and multiplicative reasoning. In J. Kilpatrick & G. Martin (Eds.), *Research companion to the NCTM Standards* (pp. 95–113). National Council of Teachers of Mathematics.

Tzur, R. (1999). An integrated study of children's construction of improper fractions and the teacher's role in promoting that learning. *Journal for Research in Mathematics Education, 30*(4), 390–416.

Tzur, R. (2000). An integrated research on children's construction of meaningful, symbolic, partitioning-related conceptions, and the teacher's role in fostering that learning. *Journal of Mathematical Behavior, 18*(2), 123–147.

Tzur, R. (2004). Teacher and students' joint production of a reversible fraction conception. *Journal of Mathematical Behavior, 23*, 93–114.

Tzur, R. (2007). Fine grain assessment of students' mathematical understanding: Participatory and anticipatory stages in learning a new mathematical conception. *Educational Studies in Mathematics, 66*(3), 273–291. https://doi.org/10.1007/s10649-007-9082-4

Tzur, R. (2013). Too often, these children are teaching-disabled, not learning-disabled. In *Proceedings of the 11th annual Hawaii International Conference on Education*. Author (DVD).

Tzur, R. (2019a). Developing fractions as multiplicative relations: A model of cognitive reorganization. In A. Norton & M. W. Alibali (Eds.), *Constructing number: Merging perspectives from psychology and mathematics education* (pp. 163–191). Springer Nature. https://doi.org/10.1007/978-3-030-00491-0_8

Tzur, R. (2019b). Elementary conceptual progressions: Reality check + implications. In J. Novotná & H. Moraová (Eds.), *Proceedings of the international Symposium on Elementary Mathematics Teaching* (Vol. 1, pp. 29–40). Charles University.

Tzur, R., & Hunt, J. H. (2015). Iteration: unit fraction knowledge and the French fry task. *Teaching Children Mathematics, 22*(3), 148–157.

Tzur, R., Johnson, H. L., McClintock, E., Kenney, R. H., Xin, Y. P., Si, L., Woodward, J., Hord, C., & Jin, X. (2013). Distinguishing schemes and tasks in children's development of multiplicative reasoning. *PNA, 7*(3), 85–101.

Tzur, R., Johnson, H. L., Norton, A., Davis, A., Wang, X., Ferrara, M., Harrington, C., & Hodkowski, N. M. (2021). Children's spontaneous additive strategy relates to multiplicative reasoning. *Cognition and Instruction* (online first), 1–26. https://doi.org/10.1080/0737000 8.2021.1896521

Ulrich, C. (2016). Stages in constructing and coordinating units additively and multiplicatively (part 2). *For the Learning of Mathematics, 36*(1), 34–39.

Van Hoof, J., Verschaffel, L., & Van Dooren, W. (2015). Inappropriately applying natural number properties in rational number tasks: Characterizing the development of the natural number bias through primary and secondary education. *Educational Studies in Mathematics, 90*(1), 39–56. https://doi.org/10.1007/s10649-015-9613-3

Woodward, J., & Tzur, R. (2017). Final commentary to the cross-disciplinary thematic special series: special education and mathematics education. *Learning Disability Quarterly, 40*(30), 146–151. https://doi.org/10.1177/0731948717690117

Chapter 16
Proportional Reasoning to Move Toward Linearity

Signe E. Kastberg, Casey Hord, and Hanan Alyami

Abstract Proportional reasoning is central to the development of linear relationships. Existing research suggests a series of interrelated benchmarks all students experience in conceptual development. Yet such development and related benchmarks need support from teachers who understand the powerful cognition and learning needs of students with learning disabilities. In this chapter we outline three benchmarks in the development of proportional reasoning, illustrate problem situations that allow teachers to gain insight into student progress toward the benchmarks, and share strategies to meet the learning needs of students with learning disabilities. These benchmarks set the foundation for students to move toward linearity.

Keywords Proportional reasoning · Ratio · Gestures · Conceptual benchmarks · Learning disabilities · Tasks · Visuals · Linear relationships · Problem situations

The utility of proportional reasoning in human exploration makes such reasoning central to mathematics learning. For example, humans have used proportional reasoning to estimate that the Moon is one-third the size of Earth (Hirshfeld, 2004) and to develop measuring tools that represent systematic proportional conventions (Dilke, 1987). In particular, proportional reasoning is a key mathematical concept students use in constructing understanding of linear relationships among quantities or linearity. In this chapter, we share ways teachers can build on the intuition of students with learning disabilities (LD) using problem situations that involve multiplicative comparisons to construct ratios and set a foundation for linearity.

S. E. Kastberg (✉)
Purdue University, College of Education, Department of Curriculum and Instruction, West Lafayette, IN, USA
e-mail: skastber@purdue.edu

C. Hord
University of Cincinnati, Cincinnati, OH, USA

H. Alyami
Purdue University, West Lafayette, IN, USA

Meeting the mathematical needs of all students has been identified as a critical demand (Tan & Kastberg, 2017). Students with learning disabilities (34% ages 3–17 NCES data 2016–2017) often go on to college and careers that require proportional reasoning for success. To meet the mathematical needs of students with LD, we draw on research that describes student reasoning and ways teachers can support such reasoning. Students' construction of and shifts among additive, multiplicative, and proportional reasoning (Howe et al., 2011; Jones et al., 2009; Staples & Truxaw, 2012) require adequate time for development (e.g., Hilton et al., 2016; Hunter et al., 2014). One source of students' constructions is their mathematical intuition. Student intuition regarding proportional situations has been underutilized (van den Heuvel-Panhuizen, 1996) in schools. Students' intuition, coupled with appropriate problem situations and instructional approaches, such as using visuals and gesturing, can develop into more formal ways of representing one's thinking (Hunt et al., 2016; Tzur et al., 2009; Woodward, 2015).

Building on research that suggests how teachers can support proportional reasoning (e.g., Lobato & Ellis, 2010), we emphasize three benchmarks in conceptual development students exhibit in identifying, understanding, and operating with proportional situations. Benchmark one is the development of ratio as a composed unit. Such ratios involve developing multiplicative relationships between two quantities (Lobato & Ellis, 2010). Situations that provide opportunities to identify an attribute and a multiplicative relationship may support such development. One such situation illustrates two orange drinks made with different quantities of water and orange mix (2 cups water: 2 cups mix or 1:1). Students are to compare the taste of the drinks. Benchmark two is the development of ways of operating with ratios including the generation of ratio tables as students create new ratios from those given. Situations that provide opportunities for such development encourage students to create multiple equivalent ratios. One such example involves pancake batter (Hunt & Vasquez, 2014). If two cups of batter are needed for three pancakes, find (1) how many pancakes can be made from 4 cups of batter (2) how much batter is needed to make six pancakes. Benchmark three is the development of unit ratios (Maloney et al., 2014). One example of a problem situation that draws on unit ratios request the number of food bars needed to feed one alien if 2.5 food bars are needed to feed 3 aliens.

The benchmarks belong to a system; they do not develop discreetly or necessarily in sequence. Thompson et al. (2014) described such a system as a "developmental cloud" (p. 14), where a learner's way of thinking about proportional reasoning includes a multitude of concepts that can be characterized individually (e.g., ratio, fraction, multiplication, division, scaling, conversion). The three benchmarks are not unique for students with LD. Research indicates that all students develop such concepts (Hunt & Tzur, 2017; Sinicrope & Mick, 1983). The factors differentiating students with LD could be time needed to achieve the benchmarks and instructional techniques needed to properly support students while leaving room for critical thinking.

In this chapter we focus on proportional reasoning as the development of equivalent ratios and comparing ratios in preparation for using such reasoning in constructing linear relationships. Awareness of benchmarks involved in ratio equivalence and ratio comparison problems can be valuable for teachers as they support students' conceptual development of proportional reasoning, while honoring the interconnectedness of these benchmarks. In the following sections we discuss strategies for working with students with learning disabilities in general. In the remaining sections we elaborate on each benchmark, suggest characteristics of problem situations that encourage conceptual development, and describe instructional supports teachers can provide.

16.1 Working with Students with Learning Disabilities

Teachers working with students with LD need to balance demands regarding the need to support students while still challenging them to meet high expectations (Hord et al., 2016b; Smith, 2000). In many cases, the dilemma is that teachers need to attend to learning differences students with LD experience, while leaving the time and space to encourage their critical thinking (Hord et al., 2016b). Such differences may include memory and processing capacities. Collectively, working memory refers to the processing, storing, and integration of information (Baddeley, 2003; Swanson & Siegel, 2001). For example, students with LD may have trouble remembering all components of a multistep problem while trying to make connections between steps to find a solution (Swanson & Beebe-Frankenberger, 2004).

When students are presented with challenging problem situations involving multiple quantities, teachers may need to help students offload information (i.e., store on paper or with their hands) so they can devote more attentional resources to thinking critically about the problem situation (Hord et al., 2016a; Risko & Dunn, 2015; Xin, 2008). Said differently, if students have to devote too much attention to remembering information in the problem, they may not have the cognitive capacity left to do the critical thinking needed to succeed with the problem (Hord et al., 2016b). For example, as a teacher reads a mathematics problem aloud to a student with reading difficulties, the student can benefit from writing down information as the problem is read to offload some of the information. Written information is then stored in drawings of proportional relationships or even representations of mathematics relationships (e.g., equations). Teachers can support students to offload information in visuals, such as ratio tables (Middleton & van den Heuvel-Panhuizen, 1995), to build insights into ratio equivalences and linearity. Students with LD can offload information in the problem into the tables (as the problem is being read aloud) and then concentrate on the information stored on paper (after the problem is read aloud) to think critically about the mathematics.

In other situations, it may be more useful for the teacher to gesture to key parts of the problem, on a projected image of the problem (Hord et al., 2016a). As text is read or discussed, teachers point to key information in the problem when it is mentioned (for more on gestures and student learning, see Alibali & Nathan, 2007).

For example, when information is presented in visuals, such as in a drawing, a ratio table, or an equation, teachers may need to motion with a pointing finger from one quantity to another to highlight an existing relationship between those quantities. The teacher can then carefully reduce support to provide students time and space to think critically about the problem (Hord & Marita, 2014). Once teachers use offloading and gesturing to help students with LD make sense of the expectations of the problem, students can often be provided with opportunities to work independently on the problem and have a high likelihood of success. However, without initial support to orient students with LD to key information in the problem and the expectations of the problem, students may struggle with working memory as they try to process, store, and integrate the multiple pieces of information in the problem (Hord et al., 2019).

In the following sections each of the benchmarks is discussed and descriptions of how teachers might use offloading and gestures to help students process, store, and integrate information is provided. While teachers' approaches will be described in the context of particular problem situations and benchmarks, the instructional strategies described can be used across mathematical concepts.

16.2   Conceptions and Supports: Benchmarks

The benchmarks described in the following sections were derived from research literature on students' conceptions (e.g., Hunt, 2015; Hunt & Vasquez, 2014; Lobato & Ellis, 2010; Thompson, 1994) of ratios and proportional reasoning. We focus on ratios in the development of proportional reasoning because such development has been identified as an essential understanding for proportional reasoning (Lobato & Ellis, 2010). In particular, we provide examples of problem situations (drawn from research) and discuss characteristics of such situations that encourage students to represent their thinking so teachers can gather evidence of that thinking. These examples were selected to meet two goals. First, the problem situations support the students' conceptual development. Second, the problem situations support the teacher in gathering evidence of student progress toward the benchmark. While the problem situations presented in each section can be modified or used to support thinking beyond the benchmark discussed, we also focus on ways the problem situations can be used by teachers to gain insight into student reasoning and to provoke new reasoning. We describe design principles for such problems so that teachers can create or select additional problem situations. Lastly, we describe and illustrate research informed pedagogy that teachers can use to support students with LD to engage and to build conceptions from their intuitive strategies.

16.3 Benchmark One: From One Quantity to Two: Ratio as Composed Unit

Teachers use problem situations such as missing products to create insights about multiplicative relationships between quantities. For example, a mathematics student might be asked to find "the number of books five students have, when each had four books in his or her backpack" (Hord et al., 2021). In this situation, a mathematics student must coordinate (multiplicatively) the number of students and books for each student to solve the problem (Clark & Kamii, 1996). The idea of a ratio requires that students "account for two quantities at the same time" (Lobato & Ellis, 2010, p. 61) and use those quantities in concert to describe an attribute. For example, a learner might be asked to describe charges from a cell phone plan. "A cell phone service provider offers 200 minutes for $10. How much do they charge per minute?" (Hunter et al., 2014, p. 365). Students must use both numbers in the problem situation to create a solution in the form of a ratio, sometimes called a rate or unit ratio, (for more on the idea of rate and unit ratio see Thompson (1994) and Maloney et al. (2014)). Problem situations such as this one invite "students to attend to two quantities in the process of isolating an attribute" (Lobato & Ellis, 2010, p. 61)—in this case the attribute of cost per minute. It is the coordination of two quantities to create a new unit (that includes a multiplicative relationship between the two quantities) that generates a ratio.

Supporting students in shifting from one quantity to multiplicative relationships between two quantities involves creating opportunities for students to make sense of visual information presented and gain experience with the situation posed. For example, van den Heuvel-Panhuizen (1996) created realistic situations, including visual representations (Moreno et al., 2011), to provide opportunities for students with LD to construct meaning for ratio. Figure 16.1 presents a problem situation where a strand of beads is made using two black beads followed by two white beads. With only part of the strand visible (i.e., the ends), students are told that there were 20 white beads and asked to find the number of black beads (Fig. 16.1). Students need to identify and use the relationship between the black and white beads to

Fig. 16.1 Adapted from van den Heuvel-Panhuizen (1996, p. 249)

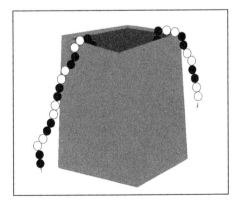

answer the problem. The bead pattern within the strand is the attribute students can isolate by coordinating the quantity of white beads and the quantity of black beads.

van den Heuvel-Panhuizen (1996) explored the bead problem and other problem situations with fifth and sixth graders with mathematics learning difficulties. She reasoned that the informal roots of ratio are "grounded in visual perception" (van den Heuvel-Panhuizen, 1996, p. 239). Problem situations she designed thus used contexts familiar to students that could be "expressed in pictures" (p. 240). Such problem situations can provide students with opportunities to reason and communicate about their reasoning instead of resorting to procedures. In her study, students modeled situations like those in Fig. 16.1 using their view of a pattern. For example, students drew images of 2 white beads alongside 2 black beads demonstrating their understanding that both quantities were needed to create the strand of beads. Students then used representations of the strand to identify the number of black beads given the number of white ones. Observing how students approach the problem using drawings or counting may provide insight regarding student thinking about the relationship between the black and white beads.

Teachers working with students with LD on the problem situation in Fig. 16.1 may need to gesture to parts of the problem to support students to attend to relevant quantities so that they can be productive in thinking critically about the problem. For example, teachers may need to point with one or two fingers to a pair of white beads while describing how pairs of white beads are followed by pairs of black beads. Teachers can then point to a pair of black beads that follows a pair of white beads they just mentioned. As the teacher says that there are 20 white beads overall, they can point to the left of the picture and use a sweeping motion (for more on such gestures see Hord et al., 2019) across all visible and nonvisible beads from left-to-right. This gesture helps the student focus on consideration of the number of beads overall, including how many black beads there could be in the strand. Then, teachers can remove their hands and give the student time and space needed to think critically about the problem. Once the student understands the components of the problem situation and the expectations of the problem, the student is prepared to reason.

Research has shown that student reasoning about numbers can be derived from their reasoning about quantities (e.g., Thompson, 1994). In addition, including visuals without numbers encourages the use and development of powerful "visual intuition" (Cox, 2013, p. 4) about relationships between space represented in the visual. These findings suggest the importance of presenting problem situations without numbers to support students' development and to identify their reasoning. For example, creating lemonade with the same sweetness as a given glass of lemonade (see Fig. 16.2 derived from van den Heuvel-Panhuizen, 1996) provides students with opportunities to visually consider relationships between quantities of lemon (white) to sugar (black) to decide which mixture is sweeter, where the sweetness is the attribute. Additionally, students' descriptions of their reasoning can allow teachers to gain insight into ways students are using quantities involved in the problem situation. For example, a student may (incorrectly) reason that the lemonade on the right is sweeter as it shows more sugar—apparently showing to the teacher a need for further work on the coordination of both quantities.

Fig. 16.2 Non-numerical ratio comparison adapted from van den Heuvel-Panhuizen (1996, p. 241)

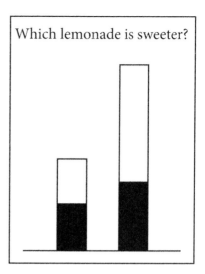

Fig. 16.3 Orange mixture problem situations adapted from Noelting (1980, p. 219)

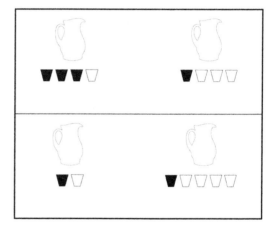

One problem situation commonly used to support student reasoning about quantities in relation to an attribute is the Orange Juice Mixture (Fig. 16.3). Such a problem situation asks students to reason about the relationship between the quantity of orange juice and the quantity of water. Students compare concentrations of orange juice and water mixtures, as represented in the two rows of Fig. 16.3, and decide which mixture would be tastier (Noelting, 1980). Students then explain ways they make sense of the orange flavor (concentration) of the mixture as the attribute described by coordinating the two quantities (water and juice).

To support students to make comparisons in problems situations like the orange mixture, teachers could point to the black cups when mentioning the orange concentrate and slide their pointer finger sideways across the black and white cups to help students focus on the relevant quantities in the visual and the comparison they need to make (see Fig. 16.4). As students with LD gain more experience with such

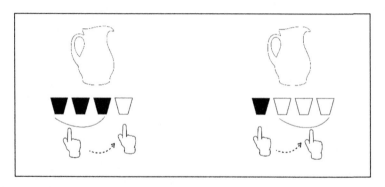

Fig. 16.4 Gestures to help students focus on comparing relevant quantities in the visual

problem situations, they may not need these cues and can benefit from being able to think more independently about the problem. Teachers using a series of orange mixture comparisons can thus gradually fade their use of gestures while checking for the development of independent thinking. Teachers judge the degree to which such fading could proceed by attending to student reasoning (Hord et al., 2016a; Hord et al., 2019).

Sinicrope and Mick (1983) administered the juice mixture problems to students with and without LD. They reported that students with LD were able to determine which concentration was orangier but had difficulty expressing the strategy they used. This is an additional challenge for students with LD and thus for their teachers. Gestures support students to attend to appropriate visual space in the problem situation, but students also need encouragement to represent how they are thinking. Research illustrates that learners who are asked to represent their thinking may have difficulty, but if given consistent opportunities over time they develop structures of mathematical communication (Hackenberg et al., 2020; Xin et al., 2020).

While the lemonade (Fig. 16.2) and juice (Fig. 16.3) situations ask for ratio comparison, teachers might also ask students to construct equivalent ratios using visual intuition (Cox, 2013). Non-numerical continuous contexts such as the lemonade contexts (see Fig. 16.2), as opposed to discrete contexts such as the strand context (see Fig. 16.1), encourage students to use their visual intuition. Boyer and Levine (2012) reported that young children successfully constructed proportional models of a given mixture. For example, given a drink mixture as a non-numeric continuous bar (see Fig. 16.5), students successfully generated a proportional mixture by creating a line between the two ingredients. However, given a picture of unifix cubes representing the ratio of sugar to water (e.g. 1 unit of sugar to 2 units of water), students incorrectly reasoned using the difference between the units of sugar and water (e.g. 2 units of sugar: 3 units of water).

Findings from the work of Cox (2013) and Cox and Lo (2012) suggested that providing units (numbers) in the visual may elicit additive approaches by introducing additional complexity regarding distances between lines in the visual. These findings suggest students with LD may benefit from opportunities to construct

Fig. 16.5 Non-numeric task adapted from Dwyer and Levine (2012)

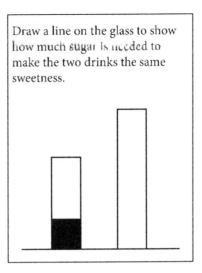

Draw a line on the glass to show how much sugar is needed to make the two drinks the same sweetness.

equivalent ratios using visuals and reflect on these visuals (Hunt & Tzur, 2017) prior to an introduction to ratios with numbers.

Comparison of problem situations with visuals, such as the lemonade and orange juice problems, provides offloaded information for students with LD to consider and thus minimizing attentional resources devoted to maintaining information in short-term memory. Teachers can leverage these visuals as they discuss the problem with their students. By pointing to parts of the visual, such as pointing to black sections of the bars to show the sweeter portions of the mixtures and then to the whole mixture using a sweeping motion up the bars, teachers can help students attend to and compare the sweeter part of the mixture and the overall composition of the mixture. As students develop attention to the quantities in the visual, teachers may fade such gestures, to provide the students with more time and space to think critically. In other cases, when the student may be stuck, gestures like these overlaid on top of the visuals can be crucial for students with LD in making progress and maintaining a productive disposition toward mathematics (Hord et al., 2019). Key in use of gestures is giving students opportunities to illustrate their own relationships and thinking using gestures and communication. Teacher's observations of student gesturing provide information about student thinking foundational to instructional decisions.

When teachers have evidence that students are reasoning about an attribute with two quantities, such as attending to both the orange concentrate and the water in Fig. 16.3, numerical problem situations that build on students' visual intuition can be given. One such problem is the "connected gear problem" (Ellis, 2009) shown in Fig. 16.6:

You have two gears on your table. Gear A has 8 teeth and gear B has 12 teeth. Answer the following questions.

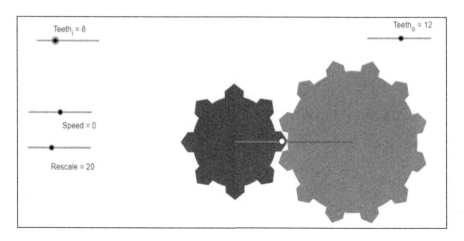

Fig. 16.6 An image adapted from the GeoGebra app, developed by O'Connor (2017)

1. If you turn gear A a certain number of times, does gear B turn more revolutions, fewer revolutions, or the same number? How can you tell?
2. Devise a way to keep track of how many revolutions gear A makes. Devise a way to keep track of gear B's revolutions. How can you keep track of both at the same time? (p. 485).

The connected gear problem presents a situation where two gears exist as parts of one mechanism. While the number of teeth can draw the students' attention, the two teacher-posed questions above provide opportunities for students to coordinate the number of revolutions one gear would make in relation to a given number of turns of the other gear (see Ellis, 2007, 2009). The attribute involved is the gear ratio that represents the relationship between the number of turns of the two gears in relation to each other. This problem situation may be less familiar to some students and will require more support prior to exploring the attribute. Teachers and students can initially use interlocking fingers to explore ways the gears operate (e.g., Alibali, 2005). Then, teachers and students can use moving gestures to track the movements of the teeth and the gears, if the gear mechanism is available as a manipulative. Students could put a finger from each hand on a tooth of each gear and turn the finger one step at a time as the gear turns to keep track of how far the gears moved. Eventually, gestures used by the teacher provide students with LD with a model of a coping skill that students can use when they are stuck on a problem. Students often benefit from their use of gestures and reflection on such gestures as well as those of their teachers (Goldin-Meadow & Alibali, 2013).

If models of gears are not available, teachers can introduce dynamic representations of the connected gear problem that help students become familiar with the problem situations and relationships between the rotations. For example, using an online application (https://www.geogebra.org/m/j6ajgbra) students can explore how the gears turn, rotating one of the gears and anticipating the effect of the rotation on the related gear.

In these situations, gestures are useful for keeping track of movement in the problem as well as offloading information such as the position of teeth when the rotation of one gear stops (see Hunt & Tzur, Chap. 5, this volume). Gestures can also be overlaid on the visuals (e.g., pointing to the teeth of the gears in Fig. 16.6) where tracking the movement of a tooth in each gear would enable students to offload (rather than imagine in their mind) how far each tooth in the gears moved to make comparisons between the two. For example, students make motions over the visual in the app to track where rotations ended. Students can then focus on what they can see and decide based on the dynamic, then static, visuals they create (rather than trying to rotate the gears mentally and store the effects of such rotation in short-term memory). If students try to do this process mentally, without the gestures overlaid on the visual, there is an increased chance they will struggle with working memory (Barrouillet et al., 2007; Hord et al., 2019). Teachers can also observe and use student actions to information instruction.

16.4   Benchmark Two: Operating with a Composed Unit

Students construct ratio by multiplicatively comparing two quantities. However, since students have much more experience comparing two quantities additively (Misailidou & Williams, 2003), middle school students with LD often use additive reasoning in problem situations that require proportional reasoning (Dougherty et al., 2016; Im & Jitendra, 2020). Thompson (1994) distinguished between reasoning about quantities and the mental process of operating. Reasoning about quantities is the basis for applying a mathematical operation (e.g., addition, multiplication) to numbers given in a problem situation. In the development of ratios this involves shifting between reasoning with quantities that are used to describe an attribute and operating with the numbers associated with the quantities.

In this section we emphasize that teachers use gestures as a tool when formulating and asking questions to provide balance between support and challenge needed for teaching mathematics to students with LD. In these situations, teachers can support with gestures, use questions to elicit thinking, and do so in ways that position students with LD to succeed, but still be challenged to think critically.

For example, in the gear problem above (Fig. 16.6), identifying that one gear rotates twice while the other rotates thrice can result in students' reasoning that the rotations are always one apart (additive reasoning). However, continued use of the gears to find other rotation pairs can result in equivalent ratios students can reflect on. The following question could help elicit such reflective thinking:

> You found out that when the small gear turns 3 times, the big gear turns 2 times. What are some other rotation pairs for the gears? (Lobato & Ellis, 2010, p. 63).

Such reflection allows for construction of relationships between ratios. Looking across rotation pairs can support the student to identify relationships between the pairs and develop a "composed unit" (Lobato & Ellis, 2010, p. 19). Students link two units such as 3 rotations to 2 rotations and operate with the ratio as a single entity. Evidence of a composed unit emerges when students are asked to find an unknown quantity. For example, if the small gear turns 12 times, how many times would the large gear turn? To solve this problem students may add ratios by finding that 3:2; 6:4; 9:6; 12:8 all correspond to this situation. Sometimes called the build-up strategy (Hart, 1983; Hunt & Vasquez, 2014), the approach involves students using a ratio additively to build up to the given quantity, in this case 12 rotations for the small gear, to determine how many rotations are needed for the large gear. The explicit use of the composed unit is called iterating.

Iterating with a Composed Unit After students are aware that an attribute is described with two quantities such as beads on a strand, taste (e.g., oranginess) of a drink, or gear ratio, the ratio as a composed unit is used to generate additional equivalent ratios as in the gear problem. For example, in an adaptation of the strand problem situation (in Fig. 16.1), if the bead ratio is 2 white beads to 3 black (Fig. 16.7), then finding strands with different numbers of beads would allow students to use the build-up strategy by iterating 2:3 and developing 4:6, 8:12, …. Here, the student iterates the initial ratio to create other strand ratios. Similarly, in the oranginess situation, if 3 cups of an orange drink mix are used with 4 cups of water, then asking students, say, how much drink mix is needed for 20 cups of water, could foster a building up way of reasoning.

Many problem situations can provide opportunities for students to build up and to reflect on results of iterating a composed unit. For example, Hunt and Vasquez (2014) implemented an instructional sequence of 15 lessons that began with the build-up strategy to assist students with LD in the coordination of two quantities. The sequence focused on ratios using whole number multiples of the composed unit. One such problem situation, with an associated visual (Fig. 16.8), involved 4 cups of batter to make 6 pancakes.

Beginning with visuals like those in Fig. 16.8, students generate equivalent ratios by considering the number of pancakes that could be made with a given number of cups of batter (e.g., 8 cups of batter). Some students create drawings to achieve this goal. Teachers who observe students generating these drawings may introduce a

Fig. 16.7 Strand problem situation with a ratio of 2 white beads to 3 black beads

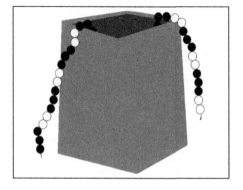

Fig. 16.8 Visual of 4 cups of butter to make 6 pancakes adapted from Hunt & Vasquez (2014, p. 181)

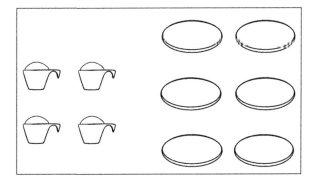

Fig. 16.9 Ratio table constructed from build-up strategy adapted from Hunt & Vasquez (2014, p. 182)

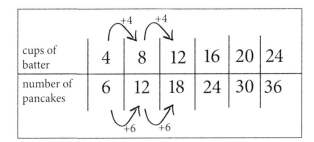

useful tool: the ratio table. Typically, students do not generate ratio tables on their own. Instead, once students have generated iterations of composed units presented as visuals, teachers know students are likely ready to consider the ratio table as a useful, organizing tool (Hunt & Vasquez, 2014; Middleton & van den Heuvel-Panhuizen, 1995; Sozen-Ozdogan et al., 2019). This order of introduction differs from the typical order of providing a ratio table as a structure and asking students to use it, in that the student's approach to the problem situation has demonstrated readiness to move to the table. Figure 16.9 illustrates how students' use of iterating the composed unit can be represented in a ratio table. The table can then be used to consider relationships and patterns in the table (Ellis, 2007).

The ratio table provides a high-quality visual that can be overlaid with gestures. By pointing and arching over the arrows to encourage attention to relationships between problem elements, teachers can help students relate increasing number of cups of batter and a corresponding, increasing number of pancakes. Hunt described using her middle finger arching over the cups of batter and her thumb curving under the number of pancakes "in a sweeping gesture to explain the quantity (numerals) changing together" (personal communication June 30, 2020). Teachers could also point to the number of cups of batter and ask for the corresponding number of pancakes to encourage generalization across the equivalent ratios creating the possibility for students to attend to the multiplicative relationship between 4 and 8 cups of batter and 6 and 12 pancakes. In the pancake situation, the offloaded information in the visual can be leveraged by the teacher using overlaid gestures (Hord et al., 2019). As with other situations in this chapter, the teacher can decide and gradually change the balance between support with gestures and challenging the student by avoiding those gestures. Providing less information may help the student mentally sort through multiple pieces of information and provide support for making connections between problem elements.

Partitioning with a Composed Unit Iterating a composed unit involves whole number multiples of each quantity of a composed unit, but equivalent ratios also involve preserving the ratio for non-integer multiples of each quantity. Im and Jitendra (2020), building on the substantial work of Jitendra and colleagues, identified students with mathematics learning difficulties as having more difficulty with problems involving non-integer multiples.

To promote reasoning that preserves ratio for non-integer multiples, problem situations such as Mr. Short and Mr. Tall (Riehl & Steinthorsdottir, 2014), referred to as Sam and Tal in this chapter, can be useful in exploring student partitioning of a composed unit. Students are told Sam can be thought of as 4 buttons tall or as 6 paper-clips tall (Fig. 16.10) and asked to find Tal's height in paper-clips if he is 6 buttons tall.

This problem situation can be used to elicit student reasoning (see Riehl & Steinthorsdottir, 2014), particularly to distinguish those who reason additively from those who reason multiplicatively. One example of additive reasoning is that Tal is 2 more buttons tall than Sam, and thus Tal is 2 more paper-clips tall than Sam (i.e., 8 paper-clips tall). This approach is based on (incorrect) additive relationships between the quantities. Students may also demonstrate their attention to a composed unit by circling 2 buttons and 3 paper-clips or 1 button and 1 ½ paper-clips. Students who use this approach illustrate their partitioning of the given ratio 4:6 while (correctly) reasoning multiplicatively. The visual enables the use of visual intuition (Cox, 2013; Cox & Lo, 2012). Students use this intuition to create a model of 8 buttons, then iterate their new ratio of 2:3 or 1:1½ to draw 9 paper clips.

To support students in reasoning multiplicatively about this problem situation, we encourage teachers to read this situation aloud and consider pointing to elements in the visual as they are read. For example, we recommend pointing to the buttons

Fig. 16.10 Sam's height in buttons and paper-clips (Riehl & Steinthorsdottir, 2014, p. 222)

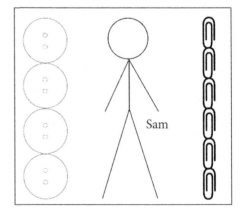

as they are mentioned and doing the same with the paper clips. After the students are familiar with the situation, they are asked to find the height of Tal in paper clips, given that Tal is 6 buttons tall. Once students are familiar with the situation, teachers may encourage students to offload some of the information in the problem by creating an associated drawing of 6 buttons. As with iterating, teachers may also want to introduce the ratio table and use the table to encourage reflection on the relationships between ratios in the table. This table differs from the one in Fig. 16.9, because the pattern in numbers in the row of buttons does not involve repeatedly adding the same composed unit.

Buttons	1	2	4	6			
Paper clips	1 ½	3	6	?			

Using this table, students may reason that 6 buttons would require 6 groups of 1 ½ paper clips beginning to move from one composed unit to another using multiplication by a unit. Other students may persist in iterating 2:3, while possibly comparing that to a strategy such as multiplying 2 buttons by 3 and 3 paper clips by 3 (Riehl & Steinthorsdottir, 2014). If the teacher decides to offer the student support toward multiplying 6 by 1½, the teacher may gesture to the 1 on the top row and then the 1½ in the bottom row while saying, "for the height of every 1 button we have, we have a height of 1½ paper clips." Then, the teacher could gesture to the 6 on the top row while saying, "if we have 6 buttons, how many paper clips would we have?" This use of gestures to draw attention to key parts of the problem and describing the relationships between these parts may be appropriate to support a student who is stuck, without solving the problem for them. In this case, the student may simply multiply 6 by 1½ to find the answer or iterate the composed unit 1:1½. In either case, the student will hopefully be able to progress. And, once students succeed with this support, it may be appropriate to withdraw the overlaid gestures and see if the student can progress with the concept independently.

To encourage partitioning, teachers may return to earlier problem situations such as the orange mix problem. One such problem would involve making orange drink by putting 3 scoops of mix into 4 cups of water. Students could then be asked how much orange mix would be needed to create the same orangiment for a drink if 10 cups of water are provided. The selection of 10 cups of water is strategic since iterating the given composed unit will not produce the number 10. The students need a new equivalent ratio. Perhaps beginning with 3:4 the student might iterate the composed unit creating 6:8. Noticing that 2 more cups of water are needed, the student might then partition 3 and 4 to arrive at 1½ scoops are needed for 2 cups of water. The student could then add the equivalent ratios to create a new ratio of 7½ cups of orange mix for 10 cups of water.

Partitioning composed units may develop more slowly than iterating. Even recognition of equivalence between, for example, 3:4 and 1½:2 can be challenging for students (Im & Jitendra, 2020). Creating problem situations where partitioning is an efficient way of creating solutions can support students' use of this strategy (Lobato

& Ellis, 2010). Problem situations that involve partitioning are created from continuous quantities such as distances and times (Thompson & Thompson, 1996), or flavored drinks that involve ounces of water and syrup (Kastberg et al., 2014). An example of the latter could be a problem situation: "If 10 ounces of cherry syrup is added to 32 ounces of water to make a cherry drink, how many ounces of cherry syrup might be needed to make a drink of the same taste by using 16 ounces of water?" This problem situation sets up the opportunity to partition the original quantities in a way that can be repeated to find the syrup for 8 and then 4 ounces of water. Such a problem enables students to use the same operation multiple times. Other problem situations may encourage different partitions, such as thirds. For example, "If 9 ounces of chocolate are added to 24 ounces of milk to make a chocolate drink of particular taste, how many ounces of chocolate are needed to create a chocolate drink with the same taste while using 8 ounces of milk?" Results from problem situations such as these generate equivalent ratios that can then be documented in ratio tables for further discussion as teachers build toward creating linear relationships (Ellis, 2007; Lobato & Ellis, 2010). Teachers should work to link visual representations and ratio tables using "generalization questions" as they move toward constructing graphs from ratio tables (Dougherty et al., 2016, p. 103). Such questions support students to identify patterns within a problem and across problems (see Tzur & Hunt Chap. 3, this volume). After completing problems where partitioning is used, teachers could ask students to compare the problems and share what they have noticed about processes they used to solve the problems. Key in such interactions is to have visuals and make gestures across artifacts from the problem-solving process, such as pointing and arching over and curving under entries in ratio tables.

Often, teachers can utilize gestures and well-crafted questions in tandem, using gestures to support students with LD to notice and connect key information and questions to push students to engage in challenging mathematics (DeJarnette & Hord, 2020). Such moves can be interconnected for teachers, sometimes quite intuitively, in that talking with students along with gestures can relieve the burden on cognition. During conversations, gestures encourage attention to information in accessible ways, making it easier for the person communicating to think clearly and more easily about (and eventually express) the information they are trying to convey (Cook et al., 2013; Goldin-Meadow et al., 2001).

16.5 Benchmark Three: Developing Unit Ratios

A significant development in proportional reasoning is moving from using composed units, through iterating and partitioning such units, to creating and using a unit ratio. A unit ratio is a ratio that can be written "in the form 1/a or a/1" (Maloney et al., 2014, p. 152). Such ratios are important since they allow students to solve problems that move beyond situations where iterating and partitioning are straight forward. For example, unit ratios can help in finding equivalent ratios when the quantities are not integers, such as in determining the value of 2.1 ounces of gold, if 0.5 ounces of gold is valued at 867.5 dollars. If the value of 1 ounce of gold is known, computing the value of 2.1 ounces of gold can be done by iterating and partitioning, but it is less efficient than multiplying both the ounce of gold and the value of the gold by 2.1. This second approach to problem situations often seems easy to adults, but moving from iterating and partitioning composed units to this computational approach with sense making can take significant time for students.

Students, with or without LD, can solve many problems using a composed unit, so creating situations that build opportunities to related composed units and unit ratios is a significant challenge in problem situation design. One approach to achieve this is to continue from problem situations already used and ratio tables constructed from a composed unit as in the pancake problem in Fig. 16.8. Such situations can be carefully constructed by identifying ways in which iterating and partitioning produce a unit of one for a quantity involved in the ratio. For example, in the pancake situation, asking how many cups of batter would be needed to make 3 pancakes would make use of a partitioning strategy. After finding that two cups of batter make 3 pancakes, the student could then be asked how many pancakes can be made if using only 1 cup of batter. The previously used halving strategy may result in one of the two possible unit ratios. Another question could focus on the other unit ratio: "How many cups of batter are needed to make one pancake?" Whereas unit ratios can be inferred using a ratio table, Confrey et al. (2012) contended that when given a ratio between a pair of numbers such as a and b, curricular treatment tends to focus on generating one unit ratio (i.e., a/b:1). Having the flexibility of reasoning with another unit ratio (1: b/a) to find a missing value and an equivalent ratio is important (Maloney et al., 2014).

Some useful problem situations are presented with a unit ratio given. For example, a problem presented in Sozen-Ozdogan et al. (2019) indicates that 1 food bar feeds 3 aliens (Fig. 16.11). Students are then asked if 2 food bars would be enough to feed 6 aliens. Students are then asked to decide if 12 food bars would be enough to feed 36 aliens. To show their reasoning students created representations using drawings or lists (Fig. 16.12).

Teachers can capitalize on students' representations of the food bar problem to build a ratio table (Fig. 16.13) (Stephan et al., n.d.). To generate the other, missing unit ratio, teachers can ask what would happen when one more alien is added. This question encourages students to develop the other unit ratio of 1/3 of a bar per alien. Including both unit ratios in the table interrupts the pattern in the table, providing an opportunity for students reflecting on the table to reason in new ways about the equivalent ratios.

Fig. 16.11 Adapted from Sozen-Ozdogan et al. (2019, p. 17)

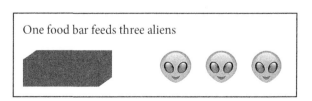

One food bar feeds three aliens

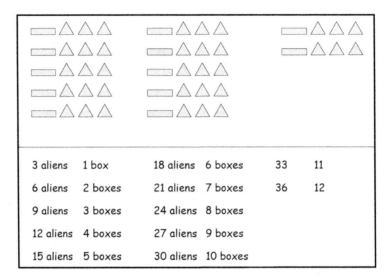

Fig. 16.12 Student representations adapted from Sozen-Ozdogan et al. (2019, p. 17)

Food bars	1	2	3	4	5	6	7	8	9	10	11	12
Aliens	3	6	9	12	15	18	21	24	27	30	33	36

Fig. 16.13 Ratio table documenting student representations adapted from Stephan et al. (2017, p. 92)

Depending on the given problem situation, the use of ratio tables in the process of iterating and partitioning a composed unit can involve the generation of a unit ratio. For example, given an onion soup recipe that requires 2 pints of water to make soup for 8 people, students are asked how much water is needed to make onion soup of the same taste to feed 6 people (Streefland, 1985). This problem situation involves partitioning. Students given a ratio table may approach the problem by partitioning to decide the amount of water to make the soup for one person (finding a unit ratio), then derive the amount of water needed for 6 persons (Fig. 16.14). Alternatively, students may also continue to use their composed unit with iterating and partitioning strategies that do not involve the use of a unit ratio to find the missing value (Fig. 16.15).

Teachers can draw students' approaches together and use them as a basis for a discussion. It is important to note that the students' work, building on the problem situation as in the case of Stephan et al. (2017), informs the focus of the discussion. Critical in this discussion is connecting different approaches to solving these problems. The unit ratio approach in Fig. 16.14 and the iterating and partitioning approach in Fig. 16.15 provide opportunities for students to reason about how the two processes result in the appropriate amount of water.

Fig. 16.14 A student approach involving a mult ratio adapted from Streefland (1985, p. 90)

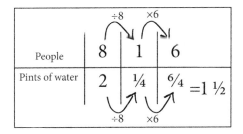

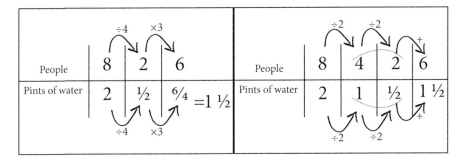

Fig. 16.15 Student approaches using partitioning adapted from Streefland (1985, p. 90)

As with other similar situations, these visuals can be leveraged by teachers as a tool for offloading information for students with LD as well as for overlaying gestures to support students with LD as they think about the problem. As students become more advanced in their thinking about proportional reasoning, it is possible that they will gradually become more independent and need less support for their working memory. Struggle with processing, storing, and integrating information is reduced as we become more familiar with a concept (Barrouillet et al., 2007). However, students with LD may still need this kind of support, even when they are demonstrating success with a concept due to the tendency of many of these students to struggle with working memory (Swanson & Siegel, 2001). Despite this need, however, we encourage teachers to progressively look for more opportunities for students with LD to demonstrate independence in their thinking especially as they become more successful and more confident with a concept.

16.6   Conclusions: Linearity at the Horizon

In this chapter we have focused on problem situations and teacher actions that support three developmental benchmarks: ratios as composed units, operating with ratios (including the generation of ratio tables), and linking such generation to unit ratios. When students develop these approaches to such problem situations, they are creating a foundation for linearity. Research has shown that developing the benchmarks discussed in this chapter sets the stage for reasoning about linear relationships (Dougherty et al., 2016; Ellis, 2009; Lobato & Ellis, 2010). Students engaged in creating equivalent ratios are well positioned to use ratio tables in the construction of graphs and to reason about the constant of proportionality or slope, m, in $y = mx$ relationships (Lobato & Ellis, 2010, p. 49). While the three benchmarks discussed in this chapter are only a small part of the "developmental cloud" (Thompson et al., 2014, p. 14) for proportional reasoning, they are significant in linking students' whole number reasoning (multiplicative and additive) to more advanced mathematics.

The use of student's approaches to problems and the thinking behind those approaches stands in contrast to the often-taught procedure for solving proportions based on finding the missing value in a proportion. Teachers can incorporate students' creation of many equivalent ratios in proportional situations and encourage reflections on these situations to create notions of unit ratio and rate (Thompson, 1994; Tzur & Hunt Chap. 3, this volume). The overused approach of cross-multiplication to find solutions to equivalent ratio problems (proportions) can result in some students temporarily improving performance. However instructional approaches that focus on cross-multiplication, limit students' mathematical flexibility and do little to support student with LD to continue growth as mathematical thinkers (Hunt & Tzur, 2017; Tan & Kastberg, 2017).

As we have repeatedly emphasized in this chapter, teachers play a vital role in promoting students' development of the benchmarks discussed in this chapter. Designing problem situations, creating visuals, using gestures, and encouraging reflection across problems support students with LD to create networks of connections between concepts (Hunt & Tzur, 2017). Such networks stored in long-term memory and recalled to working-memory allow students to work flexibly with concepts that are foundational for understanding linearity.

References

Alibali, M. W. (2005). Gesture in spatial cognition: Expressing, communicating, and thinking about spatial information. *Spatial Cognition & Computation, 5*(4), 307–331. https://doi.org/10.1207/s15427633scc0504_2

Alibali, M. W., & Nathan, M. J. (2007). Teachers' gestures as a means of scaffolding students' understanding: Evidence from an early algebra lesson. In R. Goldman, R. P. B. Barron, & S. J. Denny (Eds.), *Video research in the learning sciences* (pp. 349–365). Routledge.

Baddeley, A. (2003). Working memory and language: An overview. *Journal of Communication Disorders, 36*(3), 189–208. https://doi.org/10.1016/S0021-9924(03)00019-4

Barrouillet, P., Bernardin, S., Portrat, S., Vergauwe, E., & Camos, V. (2007). Time and cognitive load in working memory. *Journal of Experimental Psychology: Learning, Memory, and Cognition, 33*(3), 570–585. https://doi.org/10.1037/0278-7393.33.3.570

Boyer, T. W., & Levine, S. C. (2012). Child proportional scaling: Is $1/3 = 2/6 = 3/9 = 4/12$? *Journal of Experimental Child Psychology, 111*(3), 516–533. https://doi.org/10.1016/j.jecp.2011.11.001

Clark, F. B., & Kamii, C. (1996). Identification of multiplicative thinking in children in grades 1-5. *Journal for Research in Mathematics Education, 27*(1), 41–51.

Confrey, J., Nguyen, K. H., Lee, K., Panorkou, N., Corley, A. K., & Maloney, A. P. (2012). *Turn-on common core math: learning trajectories for the common core state standards for mathematics*. Retrieved from http://www.turnonccmath.net

Cox, D. C. (2013). Similarity in middle school mathematics: At the crossroads of geometry and number. *Mathematical Thinking and Learning, 15*(1), 3–23. https://doi.org/10.1080/1098606 5.2013.738377

Cox, D. C., & Lo, J.-J. (2012). Discuss similarity using visual intuition. *Mathematics Teaching in the Middle School, 18*(1), 30–37. https://doi.org/10.5951/mathteacmiddscho.18.1.0030

Cook, S. W., Duffy, R. G., & Fenn, K. M. (2013). Consolidation and transfer of learning after observing hand gesture. *Child Development, 84*, 1863–1871.

DeJarnette, A. F., & Hord, C. (2020, October). *Pre-service teachers' patterns of questioning while tutoring students with learning disabilities in Algebra 1*. Paper prepared for the annual meeting of the North American Chapter of the International Group for the Psychology of Mathematics Education, Mazatlán, Mexico.

Dilke, O. A. W. (1987). *Reading the past: Mathematics and measurement*. British Museum Publications.

Dougherty, B., Bryant, D. P., Bryant, B. R., & Shin, M. (2016). Helping students with mathematics difficulties understand ratios and proportions. *Teaching Exceptional Children, 49*(2), 96–105. https://doi.org/10.1177/0040059916674897

Ellis, A. B. (2007). Connections between generalizing and justifying: Students' reasoning with linear relationships. *Journal for Research in Mathematics Education, 38*(3), 194–229. https://doi.org/10.2307/30034866

Ellis, A. B. (2009). Patterns, quantities, and linear functions. *Mathematics Teaching in the Middle School, 14*(8), 482–491.

Goldin-Meadow, S., & Alibali, M. W. (2013). Gesture's role in speaking, learning, and creating language. *Annual Review of Psychology, 64*(1), 257–283. https://doi.org/10.1146/annurev-psych-113011-143802

Goldin-Meadow, S., Nusbaum, H., Kelly, S. D., & Wagner, S. (2001). Explaining math: Gesturing lightens the load. *Psychological Science, 12*(6), 516–522. https://doi.org/10.1111/1467-9280.00395

Hackenberg, A. J., Creager, M., & Eker, A. (2020). Teaching practices for differentiating mathematics instruction for middle school students. *Mathematical Thinking and Learning, 23*(2), 95–124. https://doi.org/10.1080/10986065.2020.1731656

Hart, K. M. (1983). I know what I believe; do I believe what I know? *Journal for Research in Mathematics Education, 14*(2), 119–125. https://doi.org/10.2307/748580

Hilton, A., Hilton, G., Dole, S., & Goos, M. (2016). Promoting middle school students' proportional reasoning skills through an ongoing professional development programme for teachers. *Educational Studies in Mathematics, 92*(2), 193–219. https://doi.org/10.1007/s10649-016-9694-7

Hirshfeld, A. W. (2004). The triangles of Aristarchus. *The Mathematics Teacher, 97*(4), 228–231.

Hord, C., Kastberg, S., & Draeger, A. L. (2021). Teaching and learning in multiplicative situations: A case study of tutoring an elementary school student with a learning disability. *School Science and Mathematics, 121*(4), 223–233.

Hord, C., Kastberg, S., & Marita, S. (2019). Layering visuals to empower struggling students. *Australian Mathematics Education Journal, 1*(2), 6–10.

Hord, C., & Marita, S. (2014). Students with learning disabilities tackling multi-step problems: Better opportunities through math conversations supported by tables. *Mathematics Teaching in the Middle School, 19*(9), 548–555. https://doi.org/10.5951/mathteacmiddscho.19.9.0548

Hord, C., Marita, S., Walsh, J. B., Tomaro, T.-M., Gordon, K., & Saldanha, R. L. (2016a). Teacher and student use of gesture and access to secondary mathematics for students with learning disabilities: An exploratory study. *Learning Disabilities: A Contemporary Journal, 14*(2), 189–206.

Hord, C., Tzur, R., Xin, Y. P., Si, L., Kenney, R. H., & Woodward, J. (2016b). Overcoming a 4th grader's challenges with working-memory via constructivist-based pedagogy and strategic scaffolds: Tia's solutions to challenging multiplicative tasks. *The Journal of Mathematical Behavior, 44*, 13–33. https://doi.org/10.1016/j.jmathb.2016.09.002

Howe, C., Nunes, T., & Bryant, P. (2011). Rational number and proportional reasoning: Using intensive quantities to promote achievement in mathematics and science. *International Journal of Science and Mathematics Education, 9*(2), 391–417. https://doi.org/10.1007/s10763-010-9249-9

Hunt, J. (2015). Notions of equivalence through ratios: Students with and without learning disabilities. *Journal of Mathematical Behavior, 37*, 94–105. https://doi.org/10.1016/j.jmathb.2014.12.002

Hunt, J., & Tzur, R. (2017). Where is difference? Processes of mathematical remediation through a constructivist lens. *Journal of Mathematical Behavior, 48*, 62–76. https://doi.org/10.1016/j.jmathb.2017.06.007

Hunt, J., Tzur, R., & Westenskow, A. (2016). Evolution of unit fraction conceptions in two fifth-graders with a learning disability: An exploratory study. *Mathematical Thinking and Learning, 18*(3), 182–208. https://doi.org/10.1080/10986065.2016.1183089

Hunt, J., & Vasquez, E. (2014). Effects of ratio strategies intervention on knowledge of ratio equivalence for students with learning disability. *The Journal of Special Education, 48*(3), 180–190. https://doi.org/10.1177/0022466912474102

Hunter, A. E., Bush, S. B., & Karp, K. (2014). Systematic interventions for teaching ratios. *Mathematics Teaching in the Middle School, 19*(6), 360–367. https://doi.org/10.5951/mathteacmiddscho.19.6.0360

Im, S.-H., & Jitendra, A. K. (2020). Analysis of proportional reasoning and misconceptions among students with mathematical learning disabilities. *Journal of Mathematical Behavior, 57*, 100753. https://doi.org/10.1016/j.jmathb.2019.100753

Jones, G., Taylor, A., & Broadwell, B. (2009). Estimating linear size and scale: Body rulers. *International Journal of Science Education, 31*(11), 1495–1509. https://doi.org/10.1080/09500690802101976

Kastberg, S. E., D'Ambrosio, B. S., Lynch-Davis, K., Mintos, A., & Krawczyk, K. (2014). Ccssm challenge: Graphing ratio and proportion. *Mathematics Teaching in the Middle School, 19*(5), 294–300. https://doi.org/10.5951/mathteacmiddscho.19.5.0294

Lobato, J., & Ellis, A. (2010). *Developing essential understanding of ratios, proportions, and proportional reasoning for teaching mathematics: Grades 6–8*. National Council of Teachers of Mathematics.

Maloney, A. P., Confrey, J., & Nguyen, K. H. (2014). *Learning over time: Learning trajectories in mathematics education*. Information Age Publishing, Inc.

Middleton, J. A., & van den Heuvel-Panhuizen, M. (1995). The ratio table. *Mathematics Teaching in the Middle School, 1*(4), 282–288.

Misailidou, C., & Williams, J. (2003). Children's proportional reasoning and tendency for an additive strategy: The role of models. *Research in Mathematics Education, 5*(1), 215–247. https://doi.org/10.1080/14794800008520123

Moreno, R., Ozogul, G., & Reisslein, M. (2011). Teaching with concrete and abstract visual representations: Effects on students' problem solving, problem representations, and learning perceptions. *Journal of Educational Psychology, 103*(1), 32–47. https://doi.org/10.1037/a0021995

Noelting, G. (1980). The development of proportional reasoning and the ratio concept part I—differentiation of stages. *Educational Studies in Mathematics, 11*(2), 217–253. https://doi.org/10.1007/BF00304357

O'Connor, M. (2017). *Gears*. https://www.geogebra.org/m/j6ajgbra

Riehl, S. M., & Steinthorsdottir, O. B. (2014). Revisiting Mr. Tall and Mr. Short. *Mathematics Teaching in the Middle School, 20*(4), 220–228. https://doi.org/10.5951/mathteacmiddscho.20.4.0220

Risko, E. F., & Dunn, T. L. (2015). Storing information in-the-world: Metacognition and cognitive offloading in a short term memory task. *Consciousness and Cognition, 36*, 61–74. https://doi.org/10.1016/j.concog.2015.05.014

Sinicrope, R., & Mick, L. B. (1983). A comparison of the proportional reasoning abilities of learning disabled and non-learning disabled children and adolescents. *Learning Disability Quarterly, 6*(3), 313–320. https://doi.org/10.2307/1510442

Smith, M. S. (2000). Balancing old and new: An experienced middle school teacher's learning in the context of mathematics instructional reform. *The Elementary School Journal, 100*(4), 351–375. https://doi.org/10.1086/499646

Sozen-Ozdogan, S., Akyuz, D., & Stephan, M. (2019). Developing ratio tables to explore ratios. *Australian Mathematics Education Journal, 1*(2), 16–21.

Staples, M., & Truxaw, M. (2012). An initial framework for the language of higher-order thinking mathematics practices. *Mathematics Education Research Journal, 24*(3), 257–281. https://doi.org/10.1007/s13394-012-0038-3

Stephan, M., McManus, G., Smith, J., & Dickey, A. (n.d.). *Ratio and rates*. https://cstem.uncc.edu/sites/cstem.uncc.edu/files/media/Ratio%20T%20Manual.pdf

Stephan, M., Pugalee, D., Cline, J., & Cline, C. (2017). *Lesson imaging in math and science: Anticipating student ideas and questions for deeper STEM learning*. Association for Supervision and Curriculum Development (ASCD).

Streefland, L. (1985). Search for the roots of ratio: Some thoughts on the long term learning process (towards...a theory). *Educational Studies in Mathematics, 16*(1), 75–94. https://doi.org/10.1007/BF00354884

Swanson, H. L., & Beebe-Frankenberger, M. (2004). The relationship between working memory and mathematical problem solving in children at risk and not at risk for serious math difficulties. *Journal of Educational Psychology, 96*(3), 471–491. https://doi.org/10.1037/0022-0663.96.3.471

Swanson, H. L., & Siegel, L. (2001). Learning disabilities as a working memory deficit. *Issues in Education, 7*(1), 1–48.

Tan, P., & Kastberg, S. (2017). Calling for research collaborations and the use of dis/ability studies in mathematics education. *Journal of Urban Mathematics Education, 10*(2), 25–38.

Thompson, A. G., & Thompson, P. W. (1996). Talking about rates conceptually, part II: Mathematical knowledge for teaching. *Journal for Research in Mathematics Education, 27*(1), 2–24. https://doi.org/10.2307/749194

Thompson, P. W. (1994). The development of the concept of speed and its relationship to concepts of rate. In G. Harel & J. Confrey (Eds.), *The development of multiplicative reasoning in the learning of mathematics* (pp. 179–234). SUNY Press.

Thompson, P. W., Carlson, M. P., Byerley, C., & Hatfield, N. (2014). Schemes for thinking with magnitudes: A hypothesis about foundational reasoning abilities in algebra. In K. C. Moore, L. P. Steffe, & L. L. Hatfield (Eds.), *Epistemic algebra students: Emerging models of students' algebraic knowing. WISDOMᵉ monographs* (Vol. 4, pp. 1–24). University of Wyoming.

Tzur, R., Xin, Y. P., Si, L., Woodward, J., & Jin, X. (2009). Promoting transition from participatory to anticipatory stage: Chad's case of multiplicative mixed-unit coordination. In M. Tzekaki, M. Kaldrimidou, & H. Sakonidis (Eds.), *Proceedings of the 33rd conference of the International Group for the Psychology of Mathematics Education* (Vol. 5, pp. 249–256). PME.

van den Heuvel-Panhuizen, M. (1996). *Assessment and realistic mathematics education* (Vol. 19). CD-β Press.

Woodward, J. (2015). *Algebraic reasoning in and through quantitative conceptions* (Unpublished doctoral dissertation). Purdue University.

Xin, Y. P. (2008). The effect of schema-based instruction in solving mathematics word problems: An emphasis on prealgebraic conceptualization of multiplicative relations. *Journal for Research in Mathematics Education, 39*(5), 526–551.

Xin, Y. P., Chiu, M. M., Tzur, R., Ma, X., Park, J. Y., & Yang, X. (2020). Linking teacher–learner discourse with mathematical reasoning of students with learning disabilities: An exploratory study. *Learning Disability Quarterly, 43*(1), 43–56. https://doi.org/10.1177/0731948719858707

Chapter 17
Commentary on Part IV

Lieven Verschaffel and Wim Van Dooren

Abstract This commentary addresses the five contributions to Part IV, dealing with counting, additive reasoning, multiplicative reasoning, fractional reasoning, and proportional reasoning. The aim of the commentary is twofold. First, we point to elements and aspects of the authors' work that we think are the most important contributions for pre-service and in-service teachers, as well as to elements of their ideas and materials that were somewhat less clear or deserve further research. Second, we raise some issues that the authors left out or could only address partially, but that, based on our research expertise and background, also deserve attention from practitioners who want to build their instructional practice on a sound research base.

Keywords Counting · Executive functions · Spontaneous numerical focusing tendencies · Additive reasoning · Problem-solving · Multiplicative reasoning · Word problems · Part-whole relation · Fractional reasoning · Graphical representations · Proportional reasoning · Mathematical modeling

17.1 Introduction

The five contributions to Part IV of the book provide prospective and practicing teachers with valuable research-based insights into the mathematical difficulties of students with and without LD, as well as into classroom practices that aim to effectively address these difficulties, in major curriculum domains ranging from counting to proportional reasoning.

L. Verschaffel (✉) · W. Van Dooren
KU Leuven, Leuven, Belgium
e-mail: lieven.verschaffel@kuleuven.be

Besides being research-based, this set of five chapters makes a very coherent and integrative impression, in the sense that these chapters share a number of basic assumptions. These assumptions include the commonly shared constructivist view on children's mathematics thinking, learning and teaching, the great attention to the role of external representations and working memory, the common aversion for superficial or narrow learning outcomes, and a striving for a deep conceptual understanding of core mathematical ideas for all children, including the younger and mathematically weaker ones. This coherence should not come as surprise, given that the authors of these chapters have been collaborating extensively over the past years on various common research projects, working groups, and symposia at national and international scientific conferences. As a result, the reader will find many similarities and connections between the various chapters, for instance, between the chapters on additive and multiplicative reasoning, or between the chapters dealing with fractions and proportions.

Without doubt, the ideas and recommendations presented in the various chapters are very ambitious and challenging, notably with respect to the authors' common strive for deep and abstract learning, even for the younger and weaker learners. In general, we strongly support and share this ambitious attitude. In fact, our own ongoing research project on the analysis and stimulation of early core mathematical competencies (see https://ppw.kuleuven.be/o_en_o/CIPenT/WisCo/development-and-stimulation-of-childrens-core-mathematical-competencies) has the same ambition. Yet, at times when reading these chapters we had a feeling that the authors risk being somewhat overambitious in their educational approaches, in relation to children's level of cognitive development and prior knowledge.

Furthermore, the restricted available space forced the authors to focus on some topics and to leave out others within the broad and complex curricular subdomains to be covered in the different chapters. Our commentary is thus twofold. We will point to elements and aspects of the authors' work that we think are the most important contributions for pre-service and in-service teachers, as well as to elements of their ideas and materials that were somewhat less clear or deserve further research. We will also raise some issues that the authors left out or could only address partially, but that, based on our research expertise and background, also deserve attention from practitioners who want to build their instructional practice on a sound research base.

17.2   Chapter 12: Counting

Chapter 12 starts by defining what it means to be a counter and then explains what children need to be able to know and do to be successful at counting, as well as the role of working memory for counting. Additionally, the authors define subitizing, explain its importance, and how it is different from, although related to, counting. They complete their chapter by discussing the importance of counting as a prerequisite to other aspects of arithmetic, as well as learning difficulties related to counting and ways to optimally assess and stimulate counting.

Especially from the perspective of practitioners, the chapter provides a very clear and comprehensive overview of the theme. It includes interesting observations and findings, and recommendations that may help to open (future) teachers' eyes to this fascinating topic and how to address it in their diagnostic and instructional practices. Examples relate to the role of finger gnosis in counting, the facilitating or hindering impact of number words in different languages, and the relation between the development of counting and additive skills.

Evidently, the restricted available space forced the authors to focus on some counting-related topics and to leave out others. Hereafter, we raise two issues that the authors left out or could only discuss partially.

First, the authors point to the great importance of working memory in counting and early numerical competence more broadly and, therefore, recommend to support children's working memory. Importantly, working memory is only a part of what cognitive psychologists call "executive functions," which have been identified as an important domain-general factor affecting mathematical performance and development. The term, executive functions, refers to a group of closely related cognitive processes that are required when engaging in deliberate, goal-directed thought and action (Bellon et al., 2019; Miyake et al., 2000). Typically, three executive functions are distinguished: updating, inhibition, and shifting (Bellon et al., 2019; Miyake et al., 2000). The first, *updating*, refers to the central executive component of working memory, allowing individuals to temporarily hold and manipulate pieces of information in memory. For example, when a child solves a simple addition such as $3 + 5$ by means of a counting-on strategy (i.e., "3…, 4, 5, 6, 7, 8"), the child has to keep track of how many times (s)he added 1. The second, *inhibition*, refers to individuals' ability to control their own attention, action, or thoughts, allowing them to override strong external cues or internal tendencies to focus attention on irrelevant information or to initiate inappropriate actions or thoughts. For example, when a child is asked which of two rows of blocks has the largest number of blocks—a row of seven blocks placed close to each other or a row of five blocks that are spread further apart—the child has to inhibit the misleading information about the greater length of the second row. The third, *shifting*, refers to the ability to flexibly switch between different mindsets, concepts, or strategies, rather than rigidly operating within one single mindset or continuously applying a given concept or following a given strategy. For example, children may achieve a better performance on a series of one-digit additions if they are able to switch flexibly between counting on from the first given number and counting on from the largest given number, depending on the placement of the larger number. Developmental (neuro) psychological research has shown that these three executive functions emerge during the first few years of life and continue to develop in childhood and even adolescence, at slightly different rates, as a result of brain maturation and educational experiences. Various empirical studies have started to document the theoretically claimed importance of all three executive functions for arithmetic (Bellon et al., 2019; Hecht, 2002; Lemaire & Lecacheur, 2011). Based on the above-mentioned theoretical and empirical evidence, we would like to broaden the authors' plea for paying more attention to the role of working memory in early numerical development and education towards all three executive functions.

Second, as nicely reviewed by the authors, the past decades have witnessed the emergence of a remarkably productive and highly influential line of research on the development of young children's early numerical knowledge and skills, including counting. This line of research extended to other themes such as: the association with school mathematics, the stimulation of early numerical knowledge and skills in the home, and to their enhancement in preschool and beginning elementary school environments (Andrews & Sayers, 2015; Torbeyns et al., 2015; Verschaffel et al., 2017). In this line of research, children are explicitly prompted to use their mathematical knowledge and skills, and assessed in terms of their ability to deal with the task at hand. In other words, all these studies take an "ability" perspective on children's early mathematical development and stimulation. While this perspective is evidently a very important one, we want to draw (future) teachers' attention to another relevant aspect of young children's early numerical competence, namely, their inclination or tendency to attend to and focus on number. We emphasize that these "numerical focusing tendencies," as we tend to call them (Verschaffel et al., 2020a), are not about what children think and do when they are instructed or guided to nonsymbolic and symbolic numbers in the situation. Rather, they are about what children spontaneously think and do when there is no explicit instruction or guidance to focus on numbers. For instance, when looking at a picture of a real-world situation (e.g., a family trip) some children will primarily notice that all the family members are dressed the same or that a thunderstorm is coming; whereas other children's attention goes to the fact that the family has three children or that the boat they want to take has too few seats to accommodate all family members. The basic claim underlying this latter line of research is that, besides having different abilities with respect to the above-mentioned distinct elements or aspects of numerical and counting ability, young children also demonstrate different numerical focusing tendencies when exploring, describing, and organizing their everyday world. Furthermore, we argue that this tendency to focus on the numerical aspects of a situation can trigger self-initiated practice of the corresponding numerical knowledge skills. Thus, if some children are more prone to focus on numerical aspects of situations, it will provide them further support to develop their number and counting skills compared to children who are not (Hannula-Sormunen, 2015). During the past 10–15 years, researchers have started to empirically investigate these various numerical focusing tendencies in young children, their development, their concurrent or predictive relation to children's mathematical achievement, and their stimulation by means of particular interventions (Hannula-Sormunen, 2015; Verschaffel et al., 2020a). So far, this research has largely attended to children's spontaneous focusing on nonsymbolic numerosities (SFON, Hannula & Lehtinen, 2005). More recently, similar efforts have been made for their spontaneous focusing on Arabic number symbols (SFONS, Rathé et al., 2019).

17.3 Chapter 13: Additive Reasoning and Problem-Solving

This chapter first describes current common practices in the United States in teaching mathematics word problem-solving. We believe that this description essentially also applies to current practices in many other countries. Then the authors introduce two paradigms in solving word problems: an inappropriate and an appropriate one, which they call, respectively, operational and relational. Next, they review recent mathematics intervention research in additive reasoning and problem-solving involving students who are struggling in mathematics, followed by a detailed description of a research-based intervention program, called *Conceptual Model-based Problem Solving* (COMPS). Finally, a web-based mathematics problem-solving computer tutor is presented that has incorporated the curriculum design features of the COMPS program.

We consider this chapter as a nice successor of the previous one, as it takes the step from counting to additive reasoning and problem-solving. There is, again, much in the chapter that we appreciate and endorse: (1) the authors' sharp attack on the "key-word strategy" as a way to approach one-step word problems around the four arithmetic operations; (2) their plea for a more meaningful conceptual approach wherein the choice of operation is rooted in a deep mathematical understanding of the problem situation rather than a superficial screening of the text of the word problem; (3) their suggestion to make instructional use of carefully chosen graphic representations to help learners with the construction of a proper understanding of problem situation; (4) their use of the powerful facilities of information and communication technology (ICT) to support this instructional process; and (5) the idea that programs that are aimed at teaching and practicing word problem-solving in the elementary school should attempt to bridge arithmetic with algebra learning. At the same time, we do have some questions and queries we would like to share with the authors and readers. But before doing so we summarize the quintessence of the authors' perspective on this complex and challenging issue.

Essential to the authors' program is the distinction between the operational and relational approaches. In a nutshell, the operational approach is driven by the search for arithmetical operations; whereas, according to the adherents of the relational approach, it is the underlying mathematical relation that should be in the focus of attention. According to the authors, the analysis of word problems at this deeper mathematical relational level will lead to the development of problem-solving concepts and skills that will be generalizable across a range of structurally similar problems and will lay good foundations for learning algebra later on. More specifically, when confronted with an additive word problem situation (i.e., a word problem that can be solved with an addition or a subtraction with the two given numbers), learners are expected to identify this relation as additive (with or without the help of semi-concrete bar models), to interpret it in terms of the abstract part-part-whole relation and to represent it with a symbolic representation of the form $a + b = x$, $x + a = b$ or $a + x = b$ (with "x" being the unknown in the situation), but never in a symbolic representation containing a subtraction sign.

Having done research in elementary arithmetic word problem-solving for a very long time, we fully support the authors' need to distinguish between a superficial, meaningless "keyword" strategy versus a deep, thoughtful approach to elementary word problem-solving, and their plea for the latter one. We also fully agree that, at a certain moment in their mathematical development, children will have to be able to conceive all additive word problems (a) in terms of the abstract part-part-whole schema and (b) as symbolically representable by means of a number sentence of the type $a + b = x$ or $a + x = b$. But we also would like to caution against this being done prematurely in the lower and middle grades of elementary school, without being preceded by an instructional stage wherein children are allowed, and even stimulated, to think about these problems in terms of semantic structures that describe *a change, a combination, or a comparison* of sets. And, shouldn't there be room, in that preceding instructional stage, for modeling such word problems *either* as additions *or* subtractions? Hereafter we will briefly elaborate these two instructional queries.

First, as stated before, we agree with Chapter 13 authors' claim that "ultimately understanding mathematical relations (e.g., additive or multiplicative relations) in word problems is fundamental to accurate problem-solving." Still, we would like to reemphasize that: (1) learners should always start solving a problem by building/comprehending a situational model of the problem statement; (2) the basic semantic schemas underlying addition and subtraction word problems (i.e., change, combine, compare, see De Corte & Verschaffel, 1987; Riley et al., 1983), each having its own semantic peculiarities, will inevitably play a role in that initial comprehension stage of the solution process; and (3) attempts to prematurely jump to the identification of the abstract part-part-whole structure without priming students for this understanding are doomed to failure.

Second, the authors seem to adopt one unified mathematical modeling of word problems, that is, $P + P = W$ for solving elementary additive word problems. We would suggest there is value to flexibly allowing children from the early and middle grades of elementary school to think of a symbolic representation with either a plus or a minus sign, instead of an abstract and generic symbolic representation making exclusive use of the plus sign. Elementary school children have to learn to see the four basic mathematical operations—including subtraction—as "models of situations" (Greer et al., 2007). Therefore, we have some reservations about instructional approaches that would allow young learners to only make use of the plus sign in their mathematical modeling attempts.

A final query we have relates to their choice for a computer-based instructional approach. Surely, using ICT in the learning and teaching of word problem-solving and mathematical modeling may have various advantages, but there may also be a risk: It may lead to a learning environment exclusively built around a restricted set of standard word problems that may fit nicely into predesigned approach but is problematic from a genuine mathematical modeling perspective (e.g., Greer et al., 2007; Savard & Polotskaia, 2017; Verschaffel et al., 2020b). Word problem-solving viewed from such a genuine mathematical modeling viewpoint is more than transforming a textual problem statement into an algebraic equation and giving the numerical outcome of the equation as the solution to the word problem. From such a viewpoint, the modeling process is considered as a cycle proceeding from the complex, fuzzy real-world to the mathematical world and back to the real-world by carrying out various cognitive activities. These activities do not only include the two above-mentioned solution steps, but also trying to understand the real-world situation and structure, simplifying the information in order to create a mental model of the real-world situation, interpreting the mathematical solution and, finally, validating the solution with regards to the original real-world problem situation. Such a genuine mathematical modeling process also involves competencies such as specifying the modeling goal, distinguishing important from unimportant information, identifying missing values, and filling in information gaps by making assumptions based on real-world knowledge. Although it is, in principle, possible to include these genuine modeling competencies in a computerized learning environment, it may not be ideally suited for incorporating such complex and fuzzy mathematical modeling tasks and for teaching these competencies. So, besides the valuable strive for mathematical abstraction and generalization, which rightly gets great attention and is intelligently implemented in the authors' computer-based instructional program, instruction should also help making children aware of these real-world complexities of the modeling process and make them competent in handling them. The latter might require complementary instructional settings, characterized by more interactive instructional approaches, such as small-group work and whole-class discussions around tasks that are constructed and presented from a more genuine mathematical modeling perspective.

17.4   Chapter 14: Multiplicative Relations and Problem-Solving

Multiplicative reasoning is, without any doubt, one of the most important concepts for children to develop throughout their elementary mathematics education years (NCTM, 1989). After a brief introduction to the conceptual field of multiplication, and the important task of differentiating it from additive thinking, the authors demonstrate how the teacher can facilitate students' construction of multiplicative reasoning and modeling of multiplicative problem-solving at the symbolic level. More concretely, they provide a detailed presentation and account of an intelligent computer tutor, called PGBM-COMPS, that draws on a constructivist view of math learning and a conceptual model-based view on (word) problem-solving. They also nicely show how bridging the learning of the arithmetic operation of multiplication with elementary geometry can promote deeper conceptual understanding of high-level mathematics through building on the mathematical idea of "dimensions." Finally, they briefly report some well-designed but small-scale intervention studies that demonstrate the effectiveness of the computer tutor, particularly with children with learning difficulties in mathematics.

Again, this chapter offers many important research-based ideas, recommendations, and materials for teachers who want to advance their learners' multiplicative reasoning and problem-solving, including: (a) the authors' warning that true multiplicative reasoning requires a major conceptual shift (i.e., a coordination of operations on composite units); (b) the authors' justified concerns about an overly simplified and narrow perception and treatment of multiplication as "repeated addition" and division as "repeated subtraction," (c) the recommendation to include in the problem set used for instruction not only equal groups and multiplicative compare problems, which are quite commonly seen in elementary mathematics curriculum, but also rectangular area and Cartesian product problem types; (d) the authors' plea for making more use, particularly in the initial phases of the teaching/learning process, of situations to be represented symbolically rather than of problems the answer has to be found through computation; (e) their conceptual attention to the similarities as well as the differences between additive and multiplicative reasoning; (f) their design of an elaborated and comprehensive web-based computer tutoring program wherein instruction in conceptual structures and heuristic instruction go hand in hand.

While reading this interesting chapter, we also had some questions and queries, which partially overlap with those raised with respect to the previous chapter.

First, the chapter pays substantial conceptual attention to the similarities as well as to the differences between additive and multiplicative reasoning. However, it does not pay a lot of instructional attention to the discovery of and reflection on these similarities and differences, given that their (computer-based) intervention programs for additive and multiplicative thinking seem to be completely separated. This may be addressed in other chapters, but the computer-based intervention programs presented in these chapters do not make this link. Consequently, with these programs children learn to reason and operate within both additive and multiplicative (problem) situations, but it also seems important that they learn to compare and differentiate between the two conceptual fields. Surely, this problem of a dissociation between the two fields – both at the conceptual and instructional level – is a much broader issue, to which we have pointed repeatedly (e.g., Verschaffel et al., 2007). In our own recent work, we have touched upon this issue by empirically investigating elementary school children's preference for additive versus multiplicative thinking when confronted with problems that do not unquestionably belong to either of the two conceptual fields. We have found that such a preference for additive or multiplicative thinking is predictive of students' errors in word problems, in the sense that they erroneously give additive answers to multiplicative problems and vice versa (Degrande et al., 2019). The authors might consider to add a module into their computer tutoring program addressing this issue.

Second, in the last modules of the authors' computer tutoring program, children explore and learn to represent quotitive and partitive division (problem) situations. In contrast to common practice, they start with quotitive division. The authors' reason for this choice to deviate from common practice is that the operation of iterating a unit is available to the children. Given that, thinking about the question "How many times do I have to take the unit rate to arrive at the product" ($a \times . = c$) is indeed (psycho)logically a better next step to take, after having studied multiplication problems, than teaching children to represent and solve (problem) situations built around the question "What is the unit rate if we divide the product into a given number of units?" However, further research might look into how this design choice can be reconciled with research evidence about the development of children's (intuitive) conceptions of division (Greer, 1992). Of course, Greer investigated children who received education (informal as well as formal) wherein partitive division had received priority, which might have significantly affected this result. And importantly, the authors have already obtained some empirical evidence of the efficacy of their own instructional approach starting with the quotitive division, although it should be added that this evidence is still somewhat tentative due to the very small number of children involved in these studies. So, the question how to sequence and integrate the teaching of partitive and quotitive division is still an open issue that deserves further research.

Third, the chapter spends extensive attention to "word problem story grammar": Students need to be assisted in comprehending word problems and to identify and relate the key elements that make up the mathematical model for multiplicative problem structures. Indeed, linguistic scaffolding seems a viable approach to assist learners in making sense of the variety of multiplicative situations that are discussed in the chapter. The role of language abilities has been recognized for many years (e.g., Pimm, 1991; Staples & Truxaw, 2012). Recently, research on the role of language has started to focus on the specific terminology that is involved in specific mathematical areas (Purpura & Reid, 2016; Purpura et al., 2017). Such domain-specific mathematical terminology in the area of multiplicative reasoning has also been shown to predict performance on multiplicative reasoning 1 year later (Vanluydt et al., 2021).

Finally, given the strong overlap in the theoretical background and the instructional elaboration of the computer tutoring program for the additive and conceptual field, our queries about the suitability of the computer-based setting for the development of genuine mathematical modeling, as stated at the end of our discussion of the previous chapter, may also apply to this chapter.

17.5 Chapter 15: Nurturing Fractional Reasoning

The chapter describes a progression in eight concepts across two clusters that shows how learners (with but also without learning difficulties) can develop fraction concepts. The importance of a good fraction understanding cannot be underestimated (e.g., Booth et al., 2014). Fractions are an essential part in the mathematics curriculum, forming the transition between basic mathematics involving mostly natural numbers and higher order mathematics. Fraction understanding has been found to be predictive of general mathematical competence, even after controlling for a wide range of other variables such as IQ, working memory, number knowledge, reading ability, and family income (Siegler et al., 2012).

A thorough understanding of a possibly fruitful learning trajectory that leads to fraction understanding is therefore of utmost importance for educational practice. Ample research has shown that teachers do not always have the required pedagogical content knowledge that allows them to understand the difficulties learners encounter when learning fractions, and/or to choose the adequate actions and representations in order to address or prevent such difficulties (e.g., Depaepe et al., 2015; Newton, 2008; Tirosh, 2000).

The authors explicitly choose to develop the fraction concept *not* commencing from the part/whole idea of a fraction. Rather, they commence from fractions as a multiplicative relation. This is a potentially powerful idea, allowing also the later linking of fractions to other multiplicative concepts, and thus to develop a thorough understanding of the multiplicative conceptual field (Vergnaud, 1994).

The 8-concept progression that is presented in the chapter is explicitly described as based on the work of constructivist researchers. Still, the specific constructivist stance and the empirical underpinning of the progression are not entirely clarified in the chapter. Basically, constructivism assumes that learners construct knowledge through their experiences and interactions with the world. Thereby, learners bring in their prior knowledge, and their interactions with the surrounding world can be facilitated and directed by others (a teacher, other learners, artifacts, …). An instructional sequence can then depart from the prior knowledge that learners already have. By its careful sequencing, pointing out the key insights that need to be built up before one can meaningfully think about the next concept, the 8-concept progression also follows this line. The chapter explains that the prior knowledge of number that learners bring is one of number as a composite unit and iteration as the foundational operation. The chapter also explicitly avoids a focus on part-whole concepts, while this kind of idea of a fraction is probably also available as prior knowledge from a very young age: From playful contexts, children know what taking half of a quantity is, and some even understand taking a third or a quarter. Thus, while such knowledge is brought to the classroom, the 8-concept progression intentionally does *not* build on this knowledge. The part-whole teaching approaches whereby children are offered pre-partitioned shapes and need to shade a requested number is explicitly not given a place. The "measure" concept may indeed be more promising for enabling extensions of the fraction concept, and staying too close to the part-whole idea is indeed bound to elicit misconceptions (see below). The same holds for decimals and percentages: Children will have encountered them in their everyday life (contexts of money, packages, price reductions, …) before the start of this learning sequence. Still, they are only addressed quite late in the learning progression, as equivalent expressions of fractions. While these elements of prior knowledge may not be particularly productive in the initial stages of the proposed learning progression, it is a pronounced choice to ignore them in the beginning of the learning progression, as children bring them to the classroom.

The chapter briefly goes into the conceptual and procedural obstacles that many students encounter. Several of these can be traced back to the inadequate application of whole number knowledge. The chapter mentions examples such as thinking that $1/10$ is greater than $1/4$ because 10 is greater than 4, or that $1/2 + 1/4 = 2/6$. Another case worth mentioning is students thinking that half of 1/8 is 1/4. Indeed, a large range of errors that learners make when reasoning about rational numbers are due to the fact that rational numbers do not necessarily behave in the same way as whole numbers, but learners are inclined to apply their prior knowledge which is (initially) heavily based on whole numbers (Van Dooren et al., 2015). Research suggests this is quite a persistent phenomenon, also occurring in adults (e.g., Vamvakoussi et al., 2013). The chapter indicates that such findings demonstrate a need for nurturing reasoning about fractions as multiplicative relations, i.e. as measures. Indeed, the inadequate application of whole number knowledge may be rooted in the early teaching of fractions as parts-of-wholes, which this 8-concept learning progression approach explicitly doesn't do. It is worthwhile investigating whether the 8-concept progression approach that is proposed in the chapter can really prevent and/or remedy this tendency in learners to apply their previously acquired whole number knowledge. The learning progression starts with unit fractions, as the result of equi-partitioning. Reasoning about such unit fractions happens also in a contextualized way, for instance by dividing a chocolate bar into six equal pieces. Such contexts allow learners to think about the unit fractions as "pieces of chocolate": Sixths become a synonym for "pieces." This might enable learners to continue reasoning in terms of (labeled) whole numbers, i.e. a number of objects. This is quite productive along many steps in the 8-concept learning progression: understanding proper fractions (2 pieces is 2 sixths), adding proper fractions (2 and 3 pieces is 5 sixths), improper fractions (11 pieces is 11 sixths), multiplying and dividing fractions by a whole number (2 times 2 pieces is 4 sixths). If contextualized, none of these steps really necessitate the learner to consider the fractions truly as a multiplicative relation. Considering them as "labeled" whole numbers (2 sixths is 2 pieces of chocolate) is sufficient. The chapter authors do a valuable attempt to offer an alternative approach of teaching fractions for struggling learners, and it is worth investigating whether the proposed learning progression would really be effective in avoiding the widely documented whole number-based errors.

The chapter brings into the picture several representations to illustrate the kind of thinking steps that learners need to take, and to show how a graphical representation of a certain situation can enhance the reasoning about fractions and operations with fractions, which in turn may lead to a more general rule or procedure. While this is undoubtfully very valuable, the question of which representation is most suited along the learning progression seems to remain a bit under the surface. It is a well-known idea in mathematics education that well-chosen external representations and combinations of external representations may facilitate the learning and the problem-solving process (e.g., Acevedo Nistal et al., 2014). However, which representations would be most suited in the 8-concept learning progression for fraction understanding, and at which moments are they crucial? The chapter uses circular diagrams (pizzas), bar diagrams, and rectangular diagrams, and it invites the reader several times to study a given diagram from a given perspective. However, it does not treat explicitly what might be strengths and weaknesses of one of these diagrams in a given context, for a given question, or to make a specific learning progression. This question certainly deserves some reflection: pizza diagrams may remain close to a well-known context, but their circular shape makes it difficult to partition a given quantity in equal parts, especially when prime numbers beyond 3 are involved: It is very difficult to divide a pizza into 7 or 11 equal pieces, for instance. More importantly, the further partitioning of a given unit fraction becomes difficult and imprecise. Bar and rectangle representations seem more suitable for such practices (which is also why these are also the starting point and circular representations come later). The rectangular representation is two-dimensional, and therefore very suitable if two different ways of partitioning need to be combined, for instance, when multiplying fractions or adding two fractions with different denominators. But the cubic meter tank example in the chapter shows it can still be solved using the bar model (i.e. not exploiting the two-dimensional nature of the representation, but only using the width), even though this may not be the most efficient approach, and even though exploiting the two dimensions in the representation may make the underlying procedure more insightful. At the same time, the one-dimensional nature of the bar representation comes closer to the number line representation. Students often have trouble understanding that fractions are numbers in their own right, with a magnitude that can be represented on a number line (on which also natural numbers and rational numbers in a decimal representation are situated) (Vamvakoussi et al., 2011). Taking that into consideration, the bar model may be particularly valuable too, and important learning activities can be built around the choice of representations and the combination of representations along the learning trajectory.

17.6   Chapter 16: Proportional Reasoning to Move Toward Linearity

The chapter on proportional reasoning rightfully indicates the importance of the concept. The National Council of Teachers of Mathematics (1989) argued that "whatever time and effort must be expended to assure its careful development." It is a stepping-stone for various important mathematical ideas throughout elementary, secondary, and even higher education, including rational numbers probability, linear algebra, and linear models in calculus and statistics. Proportional reasoning is one of the central "big ideas" in the math curriculum. It can help students to see mathematics as an integrated body of interrelated concepts that revolve around a unifying idea (Van Dooren et al., 2018).

Still, a thorough understanding of proportionality is known to be a notoriously difficult achievement for many learners, and particularly those with learning difficulties. The traditional Piagetian stance on proportional reasoning would consider it as a rather late achievement, typically not fully reached until the age of 11 (Inhelder & Piaget, 1958). However, many studies suggest that it starts its development in early childhood: Resnick and Singer (1993) saw 5- to 7-year-olds giving proportionally larger amounts of food to larger fish; Boyer and Levine (2012) saw 6- to

9-year-old children matching equal mixtures, and Vanluydt et al. (2020) observed 5- to 9-year-olds successfully solving fair sharing tasks.

How can one reconcile the observation that proportional reasoning seems feasible at a young age with the fact that at the same time learners often struggle with proportional reasoning tasks and commit additive errors (Im & Jitendra, 2020)? The authors of the chapter suggest that one of the reasons may be that tasks in which concrete numbers are given (and/or where a concrete number is asked for an answer), learners may be inclined to conduct additive rather than multiplicative calculations. It is suggested that tasks that invoke intuitions lead to fewer additive answers. The example given is a lemonade task by van den Heuvel-Panhuizen (1996), as shown in Fig. 17.1. The task indeed invokes a visual approach, as no numbers are given. It is argued in the chapter that this encourages the use and development of powerful "visual intuition" (Cox, 2013, p. 4). Indeed, where many cognitive and educational psychologists in the past decades focused on an "approximate number system" (as measured by the ability to compare nonsymbolic numerosities, among others) predictive of later mathematical achievement (e.g., Schneider et al., 2017), researchers now have a renewed interest in learners' (equally innate) ratio processing system (as measured by the processing and comparing of nonsymbolically presented ratios). This ratio processing system has been shown to be highly predictive for higher order mathematics, such as fractions and algebra (Lewis et al., 2015).

While such tasks indeed appeal to the "visual intuition" and may invoke correct answers in learners, one wonders whether the underlying reasoning is indeed multiplicative in nature, or whether it remains at a purely visual level, judging (geometrical) similarity. Additive reasoning would be particularly difficult in the situation sketched in Fig. 17.1, unless concrete numbers are used (e.g., by measuring the bars and their parts). However, when turning the problem into one involving concrete

Fig. 17.1 Nonnumeric task derived from (van den Heuvel-Panhuizen, 1996)

Draw a line on the glass to show how much sugar is needed to make the two drinks the same sweetness.

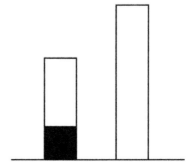

numbers, learners may again be inclined to turn to additive operations. The chapter does not explicitly discuss how the step to concrete numbers in this kind of situations can be taken, how additive reasoning by learners can be avoided, and/or how additive errors can be discussed in the classroom.

Problems that appeal to the visual intuition of learners often involve continuous quantities. Discrete quantities (or using concrete numbers to quantify continuous ones) seem to elicit more additive errors. While much research points in this direction (e.g., Boyer et al., 2008), the opposite is found too (e.g., Resnick & Singer, 1993). Whatever the impact of the discrete or continuous nature of the quantities in proportional problems is, it seems that this issue deserves explicit attention and discussion in classrooms as learners may turn to a correct or incorrect approach for mathematically very similar problems when other quantities are involved.

Mixture problems such as the one in Fig. 17.2 raise yet another important issue: the role of the problem context. Throughout the chapter, various real-life situations are used to introduce problems to learners, and to provoke new insights. However, not all contexts may be equally accessible to students. It is well known that situations in which a concrete one-to-many correspondence can be established between discrete quantities in a certain problem are well-suited stepping stones for multiplicative reasoning (e.g., Kouba, 1989). Still, many of the problems that are used throughout the chapter do not allow for a meaningful one-to-many correspondence. One of the problems in the chapter deals with two interconnected gears, one gear having 8 teeth while the other has 12 teeth. The underlying proportional relation is much more abstract for learners, and importantly, the "unit ratio" (an important step in the development) has no concrete referent: A gear with 1 tooth does not exist. The various mixture problems may be even more difficult for learners. "Taste," "sweetness," or "oranginess" are very abstract notions, and the fact that they imply a multiplicative relation between the amount of water and the amount of sugar or juice may be not at all evident for learners. Learners have been documented to think that a mixture having more sugar is sweeter, disregarding the total amount of the mixture (thereby applying a so-called More A – More B intuitive rule, Van Dooren et al., 2004). Thus, their struggle to give the correct proportional response may be simply due to not understanding the real-life situation and the multiplicative relation underlying it. This holds even more strongly for probabilistic situations: Two urns with black and white balls have the same chance of producing a winning white ball in a random draw if the black/white ratio is the same. However, a great variety of

Fig. 17.2 Visual of four cups of batter to make six pancakes (Hunt & Vasquez, 2014, p. 181)

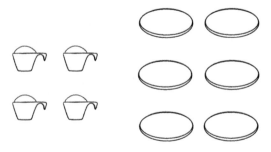

strategies have been reported in children: focusing only on the number of (winning) balls, only on the number of (losing) black balls, on the total number of balls, on the difference, etc. (Falk et al., 2012). While it is of course important that learners are able to reason proportionally in a variety of contexts, and learn to recognize which of such contexts are proportional in nature, it seems that the choice of such contexts is not thoroughly considered in the chapter.

An important observation made in the chapter is that the tendency of learners to use additive reasoning can be grasped using Thompson's (1994) distinction between reasoning about quantities and the process of operating. When given concrete numbers, learners may not always completely reason through the problem context and its involved quantities, but rather focus on conducting operations with the given numbers. This may partly explain the additive errors given to proportional problems. However, we want to stress that the opposite occurs too: Van Dooren et al. (2005) gave a test containing both proportional and various kinds of nonproportional word problems to primary school children. Among the nonproportional problems, there were additive problems including the following:

> Ellen and Kim are running around a track. They run equally fast but Ellen started later. When Ellen has run 4 laps, Kim has run 8 laps. When Ellen has run 12 laps, how many has Kim run?

The percentage of wrong proportional responses ("24 laps") increased from 10% in third grade to more than 50% in sixth grade, whereas correct additive answers ("16 laps") decreased from 60% in third grade to 30% in sixth grade. Along with learning to correctly solve proportional problems, the tendency to use that skill also beyond its applicability (and particularly in additive situations) seems to occur as well. As mentioned in our discussion on Chap. 14 on multiplicative reasoning, this phenomenon and the preference of some learners for additive relations and the preference of others for multiplicative relations deserve further attention in research, as well as in the development of learning environments that aim to develop proportional reasoning abilities.

As in previous chapters, the role of adequate representations in the development of mathematical insight is stressed. This seems of particular importance when learning difficulties are involved, as representations allow children to offload information from the working memory to the external representation. The chapter on proportional reasoning clearly puts forward the ratio table as a valuable representation for that purpose. A lot of importance seems to be given to the steps that learners need to take to turn textual or verbally presented problems into written notes, in which ratio tables indeed may provide the necessary tool to grasp the key characteristics of the problem situation. However, also visual representations, such as in Fig. 17.2, seem useful, and can even serve as a stepping stone for creating a ratio table: Two cups of batter can be grouped together with three pancakes, clarifying the two-cups-for-three-pancakes relation. The cells in the ratio table could then visually depict a number of cups and a number of pancakes before turning to a table involving symbolically represented numbers (Fig. 17.3).

Fig. 17.3 Ratio table
constructed from build-up
strategy (Hunt & Vasquez,
2014, p. 182)

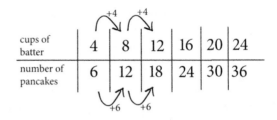

In the chapter, the ratio table with the pancake recipe is then used to show how the relations in the proportional situation at hand can be studied further. The building-up example that is elaborated in the text (adding four cups of batter allows us to make six additional pancakes) is certainly worthwhile. We would, however, suggest to also explore the other properties: not only the constant batter: pancakes ratio of 2:3, but also the fact that for instance by adding a batch of 8 cups and a batch of 12 cups, one can make 12 plus 18 pancakes, or that by tripling the number of cups, the number of pancakes also triples. All these properties are inherent to the underlying linear relation. However, it would need to be stressed that such properties do not necessarily hold in nonlinear situations, as students have been reported to use them beyond their applicability range (e.g., Stacey, 1989). Understanding when the properties hold and when they do not seem an inextricable part of proportional reasoning.

17.7   Conclusion

In sum, the five contributions to Part IV of the book provide prospective and practicing teachers with valuable research-based insights into the mathematical difficulties of students with and without LD, as well as into classroom practices that aim to effectively address these difficulties, in major curriculum domains ranging from counting to proportional reasoning. These insights, and related practical recommendations, are presented by the authors as research-based, which is a very important quality, if not an absolute requirement, for books of this kind. Understandably, the authors have essentially drawn on their own conceptual frameworks and empirical research and, consequently, have paid less attention to alternative perspectives and findings about elementary mathematics education. Indeed, research in mathematics education is a relatively young and growing scientific discipline, characterized by a lot of remaining unresolved questions and conflicting views on major issues of mathematics learning and teaching. For some of these authors' own conclusions and recommendations, the available research evidence is still somewhat scarce. Given that much of this research is only very recent, or even still "work in progress," we would have preferred to see more empirical evidence. Nevertheless, we believe the authors should be applauded for their decision to share their current research-based insights and recommendations to (future) teachers, as a source of inspiration and reflection.

References

Acevedo Nistal, A., Van Dooren, W., & Verschaffel, L. (2014). Improving students' representational flexibility in linear-function problems: An intervention. *Educational Psychology, 34*(6), 763–786. https://doi.org/10.1080/01443410.2013.785064

Andrews, P., & Sayers, J. (2015). Identifying opportunities for grade one children to acquire foundational number sense: Developing a framework for cross cultural classroom analyses. *Early Childhood Education Journal, 43*(4), 257–267. https://doi.org/10.1007/s10643-014-0653-6

Bellon, E., Fias, W., & De Smedt, B. (2019). More than number sense: The additional role of executive functions and metacognition in arithmetic. *Journal of Experimental Child Psychology, 182*, 38–60. https://doi.org/10.1016/j.jecp.2019.01.012

Booth, J. L., Newton, K. J., & Twiss-Garrity, L. K. (2014). The impact of fraction magnitude knowledge on algebra performance and learning. *Journal of Experimental Child Psychology, 118*(1), 110–118.

Boyer, T. W., & Levine, S. C. (2012). Child proportional scaling: Is 1/3 = 2/6 = 3/9 = 4/12? *Journal of Experimental Child Psychology, 111*(3), 516–533. https://doi.org/10.1016/j.jecp.2011.11.001

Boyer, T. W., Levine, S. C., & Huttenlocher, J. (2008). Development of proportional reasoning: Where young children go wrong. *Developmental Psychology, 44*(5), 1478–1490. https://doi.org/10.1037/a0013110

Cox, D. C. (2013). Similarity in middle school mathematics: At the crossroads of geometry and number. *Mathematical Thinking and Learning, 15*(1), 3–23. https://doi.org/10.1080/10986065.2013.738377

De Corte, E., & Verschaffel, L. (1987). The effect of semantic structure on first graders' solution strategies of elementary addition and subtraction word problems. *Journal for Research in Mathematics Education, 18*(5), 363–381. https://doi.org/10.2307/749085

Degrande, T., Verschaffel, L., & Van Dooren, W. (2019). To add or to multiply? An investigation of the role of preference in children's solutions of word problems. *Learning and Instruction, 61*, 60–71. https://doi.org/10.1016/j.learninstruc.2019.01.002

Depaepe, F., Torbeyns, J., Vermeersch, N., Janssens, D., Janssen, R., Kelchtermans, G., Verschaffel, L., & Van Dooren, W. (2015). Teachers' content and pedagogical content knowledge on rational numbers: A comparison of prospective elementary and lower secondary school teachers. *Teaching and Teacher Education, 47*, 82–92. https://doi.org/10.1016/j.tate.2014.12.009

Falk, R., Yudilevich-Assouline, P., & Elstein, A. (2012). Children's concept of probability as inferred from their binary choices—Revisited. *Educational Studies in Mathematics, 81*(2), 207–233. https://doi.org/10.1007/s10649-012-9402-1

Greer, B. (1992). Multiplication and division as models of situations. In D. A. Grouws (Ed.), *Handbook of research on mathematics teaching and learning* (pp. 276–295). McMillan.

Greer, B., Verschaffel, L., & Mukhopadhyay, S. (2007). Modelling for life: Mathematics and children's experience. In W. Blum, P. L. Galbraith, H.-W. Henne, & M. Niss (Eds.), *Modelling and applications in mathematics education (ICMI Study 14)* (pp. 89–98). Springer. https://doi.org/10.1007/978-0-387-29822-1_7

Hannula, M. M., & Lehtinen, E. (2005). Spontaneous focusing on numerosity and mathematical skills of young children. *Learning and Instruction, 15*(3), 237–256. https://doi.org/10.1016/j.learninstruc.2005.04.005

Hannula-Sormunen, M. M. (2015). Spontaneous focusing on numerosity and its relation to counting and arithmetic. In R. Cohen Kadosh & A. Dowker (Eds.), *The oxford handbook of mathematical cognition* (pp. 275–290). University of Oxford.

Hecht, S. A. (2002). Counting on working memory in simple arithmetic when counting is used for problem solving. *Memory and Cognition, 30*(3), 447–455. https://doi.org/10.3758/BF03194945

Hunt, J., & Vasquez, E. (2014). Effects of ratio strategies intervention on knowledge of ratio equivalence for students with learning disability. *The Journal of Special Education, 48*(3), 180–190. https://doi.org/10.1177/0022466912474102

Im, S.-H., & Jitendra, A. K. (2020). Analysis of proportional reasoning and misconceptions among students with mathematical learning disabilities. *Journal of Mathematical Behavior, 57*, 100753. https://doi.org/10.1016/j.jmathb.2019.100753

Inhelder, B., & Piaget, J. (1958). *The growth of logical thinking from childhood to adolescence.* Routledge. https://doi.org/10.1037/10034-000

Kouba, V. (1989). Children's solution strategies for equivalent set multiplication and division word problems. *Journal for Research in Mathematics Education, 20*(2), 147–158. https://doi.org/10.2307/749279

Lemaire, P., & Lecacheur, M. (2011). Age-related changes in children's executive functions and strategy selection: A study in computational estimation. *Cognitive Development, 26*(3), 282–294. https://doi.org/10.1016/j.cogdev.2011.01.002

Lewis, M. R., Matthews, P. G., Hubbard, E. M., & Matthews, P. G. (2015). Neurocognitive architectures and the non-symbolic foundations of fractions understanding. In D. B. Berch, D. C. Geary, & K. M. Koepke (Eds.), *Development of mathematical cognition: Neural substrates and genetic influences* (pp. 141–160). Elsevier. https://doi.org/10.1016/B978-0-12-801871-2.00006-X

Miyake, A., Friedman, N. P., Emerson, M. J., Witzki, A. H., Howerter, A., & Wager, T. D. (2000). The unity and diversity of executive functions and their contributions to complex "frontal lobe" tasks: A latent variable analysis. *Cognitive Psychology, 41*(1), 49–100. https://doi.org/10.1006/cogp.1999.0734

National Council of Teachers of Mathematics. (1989). *Curriculum and evaluation standards for school mathematics.* Author.

Newton, K. J. (2008). An extensive analysis of preservice elementary teachers' knowledge of fractions. *American Educational Research Journal, 45*(4), 1080–1110. https://doi.org/10.3102/0002831208320851

Pimm, D. (1991). Communicating mathematically. In K. Durkin & B. Shire (Eds.), *Language in mathematical education. Research and practice* (pp. 18–23). Open University Press.

Purpura, D. J., Logan, J. A. R., Hassinger-Das, B., & Napoli, A. R. (2017). Why do early mathematics skills predict later reading? The role of mathematical language. *Developmental Psychology, 53*(9), 1633–1642. https://doi.org/10.1037/dev0000375

Purpura, D. J., & Reid, E. E. (2016). Mathematics and language: Individual and group differences in mathematical language skills in young children. *Early Childhood Research Quarterly, 26*, 259–268. https://doi.org/10.1016/j.ecresq.2015.12.020

Rathé, S., Torbeyns, J., De Smedt, B., & Verschaffel, L. (2019). Spontaneous focusing on Arabic number symbols and its association with early mathematical competencies. *Early Childhood Research Quarterly, 48*(3), 111–121. https://doi.org/10.1016/j.ecresq.2019.01.011

Resnick, L. B., & Singer, J. A. (1993). Protoquantitative origins of ratio reasoning. In T. P. Carpenter, E. Fennema, & T. A. Romberg (Eds.), *Rational numbers: An integration of research* (pp. 107–130). Erlbaum.

Riley, M. S., Greeno, J. G., & Heller, J. I. (1983). Development of children's problem-solving ability in arithmetic. In H. P. Ginsburg (Ed.), *The development of mathematical thinking* (pp. 153–196). Academic Press.

Savard, A., & Polotskaia, E. (2017). Who's wrong? Tasks fostering understanding of mathematical relationships in word problems in elementary students. *ZDM Mathematics Education, 49*(6), 823–833. https://doi.org/10.1007/s11858-017-0865-5

Schneider, M., Beeres, K., Coban, L., Merz, S., Schmidt, S. S., Stricker, J., & De Smedt, B. (2017). Associations of non-symbolic and symbolic numerical magnitude processing with mathematical competence: A meta-analysis. *Developmental Science, 20*(3), e12372. https://doi.org/10.1111/desc.12372

Siegler, R. S., Duncan, G. J., Davis-Kean, P. E., Duckworth, K., Claessens, A., Engel, M., Susperreguy, M. I., & Chen, M. (2012). Early predictors of high school mathematics achievement. *Psychological Science, 23*(7), 691–697. https://doi.org/10.1177/0956797612440101

Stacey, K. (1989). Finding and using patterns in linear generalizing problems. *Educational Studies in Mathematics, 20*(2), 147–164. https://doi.org/10.1007/BF00579460

Staples, M. E., & Truxaw, M. P. (2012). An initial framework for the language of higher-order thinking in mathematics practices. *Mathematics Education Research Journal, 24*(3), 257–281. https://doi.org/10.1007/s13394-012-0030-0

Thompson, P. W. (1994). The development of the concept of speed and its relationship to concepts of rate. In G. Harel & J. Confrey (Eds.), *The development of multiplicative reasoning in the learning of mathematics* (pp. 179–234). SUNY Press.

Tirosh, D. (2000). Enhancing prospective teachers' knowledge of children's conceptions: The case of division of fractions. *Journal for Research in Mathematics Education, 31*(1), 5–25. https://doi.org/10.2307/749817

Torbeyns, J., Gilmore, C., & Verschaffel, L. (2015). The acquisition of preschool mathematical abilities: Theoretical, methodological and educational considerations. An introduction. *Mathematical Thinking and Learning, 17*(2), 99–115. https://doi.org/10.1080/10986065.2015.1016810

Vamvakoussi, X., Christou, K. P., Mertens, L., & Van Dooren, W. (2011). What fills the gap between discrete and dense? Greek and Flemish students' understanding of density. *Learning and Instruction, 21*(5), 676–685. https://doi.org/10.1016/j.learninstruc.2011.03.005

Vamvakoussi, X., Van Dooren, W., & Verschaffel, L. (2013). Brief Report. Educated adults are still affected by intuitions about the effect of arithmetical operations: Evidence from a reaction-time study. *Educational Studies in Mathematics, 82*(2), 323–330. https://doi.org/10.1007/s10649-012-9432-8

van den Heuvel-Panhuizen, M. (1996). *Assessment and realistic mathematics education* (Vol. 19). CD-β Press.

Van Dooren, W., De Bock, D., Hessels, A., Janssens, D., & Verschaffel, L. (2005). Not everything is proportional: Effects of age and problem type on propensities for overgeneralization. *Cognition and Instruction, 23*(1), 57–86. https://doi.org/10.1207/s1532690xci2301_3

Van Dooren, W., De Bock, D., Weyers, D., & Verschaffel, L. (2004). The predictive power of intuitive rules: A critical analysis of the impact of "More A-more B" and "Same A-same B". *Educational Studies in Mathematics, 56*(2), 179–207. https://doi.org/10.1023/B:EDUC.0000040379.26033.0d

Van Dooren, W., Lehtinen, E., & Verschaffel, L. (2015). Unravelling the gap between natural and rational numbers. *Learning and Instruction, 37*, 1–4. https://doi.org/10.1016/j.learninstruc.2015.01.001

Van Dooren, W., Vamvakoussi, X., & Verschaffel, L. (2018). *Proportional reasoning*. (Educational Practices Series, 30). International Academy of Education (IAE).

Vanluydt, E., Supply, A.-S., Verschaffel, L., Van Dooren, W., & with Vanluydt, E. (2021). The importance of specific mathematical language for early proportional reasoning. *Early Childhood Research Quarterly, 55*(2), 193–200. https://doi.org/10.1016/j.ecresq.2020.12.003

Vanluydt, E., Degrande, T., Verschaffel, L., & Van Dooren, W. (2020). Early stages of proportional reasoning: A cross-sectional study with 5-to 9-year olds. *European Journal of Psychology of Education, 35*(3), 529–547.

Vergnaud, G. (1994). Multiplicative conceptual field: What and why? In G. Harel & J. Confrey (Eds.), *The development of multiplicative reasoning in the learning of mathematics* (pp. 41–59). State University of New York.

Verschaffel, L., Greer, B., & De Corte, E. (2007). Whole number concepts and operations. In F. K. Lester (Ed.), *Second handbook of research on mathematics teaching and learning* (pp. 557–628). Information Age Publishing.

Verschaffel, L., Rathé, S., Wijns, N., Degrande, T., Van Dooren, W., De Smedt, B., & Torbeyns, J. (2020b). Young children's early mathematical competencies: The role of mathematical focusing tendencies. In M. Carlsen, I. Erfjord, & P. S. Hundeland (Eds.), *Mathematics education in the early years. Results from the POEM4 Conference, 2018* (pp. 23–42). Springer Nature. https://doi.org/10.1007/978-3-030-34776-5_2

Verschaffel, L., Schukajlow, S., Star, J., & Van Dooren, W. (2020a). Word problems in mathematics education. A survey. *ZDM Mathematics Education, 52*(1), 1–16. https://doi.org/10.1007/s11858-020-01130-4

Verschaffel, L., Torbeyns, J., & De Smedt, B. (2017). Young children's early mathematical competencies: Analysis and stimulation. In T. Dooley & G. Gueudet (Eds.), *Proceedings of the Tenth Congress of the European Society for Research in Mathematics Education (CERME10, February 1–5, 2017)* (pp. 31–52). DCU Institute of Education and ERME.

Index